Lecture Notes in Computer Science 14194

The series Lecture Notes in Computer Science (LNCS), including its subseries Lecture Notes in Artificial Intelligence (LNAI) and Lecture Notes in Bioinformatics (LNBI), has established itself as a medium for the publication of new developments in computer science and information technology research, teaching, and education.

LNCS enjoys close cooperation with the computer science R & D community, the series counts many renowned academics among its volume editors and paper authors, and collaborates with prestigious societies. Its mission is to serve this international community by providing an invaluable service, mainly focused on the publication of conference and workshop proceedings and postproceedings. LNCS commenced publication in 1973.

Mickael Coustaty · Alicia Fornés
Editors

Document Analysis and Recognition – ICDAR 2023 Workshops

San José, CA, USA, August 24–26, 2023
Proceedings, Part II

Editors
Mickael Coustaty
University of La Rochelle
La Rochelle, France

Alicia Fornés
Autonomous University of Barcelona
Bellaterra, Spain

ISSN 0302-9743 ISSN 1611-3349 (electronic)
Lecture Notes in Computer Science
ISBN 978-3-031-41500-5 ISBN 978-3-031-41501-2 (eBook)
https://doi.org/10.1007/978-3-031-41501-2

This Springer imprint is published by the registered company Springer Nature Switzerland AG
The registered company address is: Gewerbestrasse 11, 6330 Cham, Switzerland

Foreword from ICDAR 2023 General Chairs

We are delighted to welcome you to the proceedings of ICDAR 2023, the 17th IAPR International Conference on Document Analysis and Recognition, which was held in San Jose, in the heart of Silicon Valley in the United States. With the worst of the pandemic behind us, we hoped that ICDAR 2023 would be a fully in-person event. However, challenges such as difficulties in obtaining visas also necessitated the partial use of hybrid technologies for ICDAR 2023. The oral papers being presented remotely were synchronous to ensure that conference attendees interacted live with the presenters and the limited hybridization still resulted in an enjoyable conference with fruitful interactions.

ICDAR 2023 was the 17th edition of a longstanding conference series sponsored by the International Association of Pattern Recognition (IAPR). It is the premier international event for scientists and practitioners in document analysis and recognition. This field continues to play an important role in transitioning to digital documents. The IAPR-TC 10/11 technical committees endorse the conference. The very first ICDAR was held in St Malo, France in 1991, followed by Tsukuba, Japan (1993), Montreal, Canada (1995), Ulm, Germany (1997), Bangalore, India (1999), Seattle, USA (2001), Edinburgh, UK (2003), Seoul, South Korea (2005), Curitiba, Brazil (2007), Barcelona, Spain (2009), Beijing, China (2011), Washington, DC, USA (2013), Nancy, France (2015), Kyoto, Japan (2017), Sydney, Australia (2019) and Lausanne, Switzerland (2021).

Keeping with its tradition from past years, ICDAR 2023 featured a three-day main conference, including several competitions to challenge the field and a post-conference slate of workshops, tutorials, and a doctoral consortium. The conference was held at the San Jose Marriott on August 21–23, 2023, and the post-conference tracks at the Adobe World Headquarters in San Jose on August 24–26, 2023.

We thank our executive co-chairs, Venu Govindaraju and Tong Sun, for their support and valuable advice in organizing the conference. We are particularly grateful to Tong for her efforts in facilitating the organization of the post-conference in Adobe Headquarters and for Adobe's generous sponsorship.

The highlights of the conference include keynote talks by the recipient of the IAPR/ICDAR Outstanding Achievements Award, and distinguished speakers Marti Hearst, UC Berkeley School of Information; Vlad Morariu, Adobe Research; and Seiichi Uchida, Kyushu University, Japan.

A total of 316 papers were submitted to the main conference (plus 33 papers to the ICDAR-IJDAR journal track), with 53 papers accepted for oral presentation (plus 13 IJDAR track papers) and 101 for poster presentation. We would like to express our deepest gratitude to our Program Committee Chairs, featuring three distinguished researchers from academia, Gernot A. Fink, Koichi Kise, and Richard Zanibbi, and one from industry, Rajiv Jain, who did a phenomenal job in overseeing a comprehensive reviewing process and who worked tirelessly to put together a very thoughtful and interesting technical program for the main conference. We are also very grateful to the members of the Program Committee for their high-quality peer reviews. Thank you to our competition chairs, Kenny Davila, Chris Tensmeyer, and Dimosthenis Karatzas, for overseeing the competitions.

The post-conference featured 8 excellent workshops, four value-filled tutorials, and the doctoral consortium. We would like to thank Mickael Coustaty and Alicia Fornes, the workshop chairs, Elisa Barney-Smith and Laurence Likforman-Sulem, the tutorial chairs, and Jean-Christophe Burie and Andreas Fischer, the doctoral consortium chairs, for their efforts in putting together a wonderful post-conference program.

We would like to thank and acknowledge the hard work put in by our Publication Chairs, Anurag Bhardwaj and Utkarsh Porwal, who worked diligently to compile the camera-ready versions of all the papers and organize the conference proceedings with Springer. Many thanks are also due to our sponsorship, awards, industry, and publicity chairs for their support of the conference.

The organization of this conference was only possible with the tireless behind-the-scenes contributions of our webmaster and tech wizard, Edward Sobczak, and our secretariat, ably managed by Carol Doermann. We convey our heartfelt appreciation for their efforts.

Finally, we would like to thank for their support our many financial sponsors and the conference attendees and authors, for helping make this conference a success. We sincerely hope those who attended had an enjoyable conference, a wonderful stay in San Jose, and fruitful academic exchanges with colleagues.

David Doermann
Srirangaraj (Ranga) Setlur

Foreword from ICDAR 2023 Workshop Chairs

Our heartiest welcome to the proceedings of the ICDAR 2023 Workshops, which were organized as part of the 17th International Conference on Document Analysis and Recognition (ICDAR) held in San José, California, USA on August 21–26, 2023. The workshops were held after the main conference, from August 24–26, 2023, together with the Tutorials and the Doctoral Consortium. All the workshops were in-person events, as was the main conference.

The ICDAR conference included 8 workshops, covering diverse document image analysis and recognition topics, and also complementary topics such as Natural Language Processing, Computational Paleography and Digital Humanities. Overall the workshops received 60 papers and 43 of them were accepted (global acceptance rate 71.6%).

This volume collects the edited papers from 6 of these 8 workshops. We sincerely thank the ICDAR general chairs for trusting us with the responsibility for the workshops, and the publication chairs for assisting us with the publication of this volume. We also thank all the workshop organizers for their involvement in this event of primary importance in our field. Finally, we thank all the workshop presenters and authors.

August 2023

Mickael Coustaty
Alicia Fornés

Organization

General Chairs

David Doermann	University at Buffalo, The State University of New York, USA
Srirangaraj Setlur	University at Buffalo, The State University of New York, USA

Executive Co-chairs

Venu Govindaraju	University at Buffalo, The State University of New York, USA
Tong Sun	Adobe Research, USA

PC Chairs

Gernot A. Fink	Technische Universität Dortmund, Germany (Europe)
Rajiv Jain	Adobe Research, USA (Industry)
Koichi Kise	Osaka Metropolitan University, Japan (Asia)
Richard Zanibbi	Rochester Institute of Technology, USA (Americas)

Workshop Chairs

Mickael Coustaty	La Rochelle University, France
Alicia Fornes	Universitat Autònoma de Barcelona, Spain

Tutorial Chairs

Elisa Barney-Smith	Luleå University of Technology, Sweden
Laurence Likforman-Sulem	Télécom ParisTech, France

Competitions Chairs

Kenny Davila — Universidad Tecnológica Centroamericana, UNITEC, Honduras
Dimosthenis Karatzas — Universitat Autònoma de Barcelona, Spain
Chris Tensmeyer — Adobe Research, USA

Doctoral Consortium Chairs

Andreas Fischer — University of Applied Sciences and Arts Western Switzerland
Veronica Romero — University of Valencia, Spain

Publications Chairs

Anurag Bharadwaj — Northeastern University, USA
Utkarsh Porwal — Walmart, USA

Posters/Demo Chair

Palaiahnakote Shivakumara — University of Malaya, Malaysia

Awards Chair

Santanu Chaudhury — IIT Jodhpur, India

Sponsorship Chairs

Wael Abd-Almageed — Information Sciences Institute USC, USA
Cheng-Lin Liu — Chinese Academy of Sciences, China
Masaki Nakagawa — Tokyo University of Agriculture and Technology, Japan

Industry Chairs

Andreas Dengel — DFKI, Germany
Véronique Eglin — Institut National des Sciences Appliquées (INSA) de Lyon, France
Nandakishore Kambhatla — Adobe Research, India

Publicity Chairs

Sukalpa Chanda	Østfold University College, Norway
Simone Marinai	University of Florence, Italy
Safwan Wshah	University of Vermont, USA

Technical Chair

Edward Sobczak	University at Buffalo, The State University of New York, USA

Conference Secretariat

University at Buffalo, The State University of New York, USA

Program Committee

Senior Program Committee Members

Srirangaraj Setlur
Richard Zanibbi
Koichi Kise
Gernot Fink
David Doermann
Rajiv Jain
Rolf Ingold
Andreas Fischer
Marcus Liwicki
Seiichi Uchida
Daniel Lopresti
Josep Llados
Elisa Barney Smith
Umapada Pal
Alicia Fornes
Jean-Marc Ogier
C. V. Jawahar
Xiang Bai
Liangrui Peng
Jean-Christophe Burie
Andreas Dengel
Robert Sablatnig
Basilis Gatos
Apostolos Antonacopoulos
Lianwen Jin
Nicholas Howe
Marc-Peter Schambach
Marcal Rossinyol
Wataru Ohyama
Nicole Vincent
Faisal Shafait
Simone Marinai
Bertrand Couasnon
Masaki Nakagawa
Anurag Bhardwaj
Dimosthenis Karatzas
Masakazu Iwamura
Tong Sun
Laurence Likforman-Sulem
Michael Blumenstein
Cheng-Lin Liu
Luiz Oliveira
Robert Sabourin
R. Manmatha
Angelo Marcelli
Utkarsh Porwal

Program Committee Members

Harold Mouchere
Foteini Simistira Liwicki
Vernonique Eglin
Aurelie Lemaitre
Qiu-Feng Wang
Jorge Calvo-Zaragoza
Yuchen Zheng
Guangwei Zhang
Xu-Cheng Yin
Kengo Terasawa
Yasuhisa Fujii
Yu Zhou
Irina Rabaev
Anna Zhu
Soo-Hyung Kim
Liangcai Gao
Anders Hast
Minghui Liao
Guoqiang Zhong
Carlos Mello
Thierry Paquet
Mingkun Yang
Laurent Heutte
Antoine Doucet
Jean Hennebert
Cristina Carmona-Duarte
Fei Yin
Yue Lu
Maroua Mehri
Ryohei Tanaka
Adel M. M. Alimi
Heng Zhang
Gurpreet Lehal
Ergina Kavallieratou
Petra Gomez-Kramer
Anh Le Duc
Frederic Rayar
Muhammad Imran Malik
Vincent Christlein
Khurram Khurshid
Bart Lamiroy
Ernest Valveny
Antonio Parziale
Jean-Yves Ramel
Haikal El Abed
Alireza Alaei
Xiaoqing Lu
Sheng He
Abdel Belaid
Joan Puigcerver
Zhouhui Lian
Francesco Fontanella
Daniel Stoekl Ben Ezra
Byron Bezerra
Szilard Vajda
Irfan Ahmad
Imran Siddiqi
Nina S. T. Hirata
Momina Moetesum
Vassilis Katsouros
Fadoua Drira
Ekta Vats
Ruben Tolosana
Steven Simske
Christophe Rigaud
Claudio De Stefano
Henry A. Rowley
Pramod Kompalli
Siyang Qin
Alejandro Toselli
Slim Kanoun
Rafael Lins
Shinichiro Omachi
Kenny Davila
Qiang Huo
Da-Han Wang
Hung Tuan Nguyen
Ujjwal Bhattacharya
Jin Chen
Cuong Tuan Nguyen
Ruben Vera-Rodriguez
Yousri Kessentini
Salvatore Tabbone
Suresh Sundaram
Tonghua Su
Sukalpa Chanda

Mickael Coustaty
Donato Impedovo
Alceu Britto
Bidyut B. Chaudhuri
Swapan Kr. Parui
Eduardo Vellasques
Sounak Dey
Sheraz Ahmed
Julian Fierrez
Ioannis Pratikakis
Mehdi Hamdani
Florence Cloppet
Amina Serir
Mauricio Villegas
Joan Andreu Sanchez
Eric Anquetil
Majid Ziaratban
Baihua Xiao
Christopher Kermorvant
K. C. Santosh
Tomo Miyazaki
Florian Kleber
Carlos David Martinez Hinarejos
Muhammad Muzzamil Luqman
Badarinath T.
Christopher Tensmeyer
Musab Al-Ghadi
Ehtesham Hassan
Journet Nicholas
Romain Giot
Jonathan Fabrizio
Sriganesh Madhvanath
Volkmar Frinken
Akio Fujiyoshi
Srikar Appalaraju
Oriol Ramos-Terrades
Christian Viard-Gaudin
Chawki Djeddi
Nibal Nayef
Nam Ik Cho
Nicolas Sidere
Mohamed Cheriet
Mark Clement
Shivakumara Palaiahnakote
Shangxuan Tian
Ravi Kiran Sarvadevabhatla
Gaurav Harit
Iuliia Tkachenko
Christian Clausner
Vernonica Romero
Mathias Seuret
Vincent Poulain D'Andecy
Joseph Chazalon
Kaspar Riesen
Lambert Schomaker
Mounim El Yacoubi
Berrin Yanikoglu
Lluis Gomez
Brian Kenji Iwana
Ehsanollah Kabir
Najoua Essoukri Ben Amara
Volker Sorge
Clemens Neudecker
Praveen Krishnan
Abhisek Dey
Xiao Tu
Mohammad Tanvir Parvez
Sukhdeep Singh
Munish Kumar
Qi Zeng
Puneet Mathur
Clement Chatelain
Jihad El-Sana
Ayush Kumar Shah
Peter Staar
Stephen Rawls
David Etter
Ying Sheng
Jiuxiang Gu
Thomas Breuel
Antonio Jimeno
Karim Kalti
Enrique Vidal
Kazem Taghva
Evangelos Milios
Kaizhu Huang
Pierre Heroux
Guoxin Wang
Sandeep Tata
Youssouf Chherawala

Reeve Ingle
Aashi Jain
Carlos M. Travieso-Gonzales
Lesly Miculicich
Curtis Wigington
Andrea Gemelli
Martin Schall
Yanming Zhang
Dezhi Peng
Chongyu Liu
Huy Quang Ung
Marco Peer
Nam Tuan Ly
Jobin K. V.
Rina Buoy
Xiao-Hui Li
Maham Jahangir
Muhammad Naseer Bajwa
Oliver Tueselmann
Yang Xue
Kai Brandenbusch
Ajoy Mondal
Daichi Haraguchi
Junaid Younas
Ruddy Theodose
Rohit Saluja
Beat Wolf
Jean-Luc Bloechle
Anna Scius-Bertrand
Claudiu Musat
Linda Studer
Andrii Maksai
Oussama Zayene
Lars Voegtlin
Michael Jungo

Program Committee Sub Reviewers

Li Mingfeng
Houcemeddine Filali
Kai Hu
Yejing Xie
Tushar Karayil
Xu Chen
Benjamin Deguerre
Andrey Guzhov
Estanislau Lima
Hossein Naftchi
Giorgos Sfikas
Chandranath Adak
Yakn Li
Solenn Tual
Kai Labusch
Ahmed Cheikh Rouhou
Lingxiao Fei
Yunxue Shao
Yi Sun
Stephane Bres
Mohamed Mhiri
Zhengmi Tang
Fuxiang Yang
Saifullah Saifullah
Paolo Giglio
Wang Jiawei
Maksym Taranukhin
Menghan Wang
Nancy Girdhar
Xudong Xie
Ray Ding
Mélodie Boillet
Nabeel Khalid
Yan Shu
Moises Diaz
Biyi Fang
Adolfo Santoro
Glen Pouliquen
Ahmed Hamdi
Florian Kordon
Yan Zhang
Gerasimos Matidis
Khadiravana Belagavi
Xingbiao Zhao
Xiaotong Ji
Yan Zheng
M. Balakrishnan
Florian Kowarsch

Mohamed Ali Souibgui
Xuewen Wang
Djedjiga Belhadj
Omar Krichen
Agostino Accardo
Erika Griechisch
Vincenzo Gattulli
Thibault Lelore
Zacarias Curi
Xiaomeng Yang
Mariano Maisonnave
Xiaobo Jin
Corina Masanti
Panagiotis Kaddas
Karl Löwenmark
Jiahao Lv
Narayanan C. Krishnan
Simon Corbillé
Benjamin Fankhauser
Tiziana D'Alessandro
Francisco J. Castellanos
Souhail Bakkali
Caio Dias
Giuseppe De Gregorio
Hugo Romat
Alessandra Scotto di Freca
Christophe Gisler
Nicole Dalia Cilia
Aurélie Joseph
Gangyan Zeng
Elmokhtar Mohamed Moussa
Zhong Zhuoyao
Oluwatosin Adewumi
Sima Rezaei
Anuj Rai
Aristides Milios
Shreeganesh Ramanan
Wenbo Hu
Arthur Flor de Sousa Neto
Rayson Laroca
Sourour Ammar
Wenbo Hu
Gianfranco Semeraro
Andre Hochuli
Saddok Kebairi
Shoma Iwai
Cleber Zanchettin
Ansgar Bernardi
Vivek Venugopal
Abderrhamne Rahiche
Wenwen Yu
Abhishek Baghel
Mathias Fuchs
Yael Iseli
Xiaowei Zhou
Yuan Panli
Minghui Xia
Zening Lin
Konstantinos Palaiologos
Loann Giovannangeli
Yuanyuan Ren
Shreeganesh Ramanan
Shubhang Desai
Yann Soullard
Ling Fu
Juan Antonio Ramirez-Orta
Chixiang Ma
Truong Thanh-Nghia
Nathalie Girard
Kalyan Ram Ayyalasomayajula
Talles Viana
Francesco Castro
Anthony Gillioz
Yunxue Shao
Huawen Shen
Mathias Fuchs
Sanket Biswas
Haisong Ding
Solène Tarride

Contents – Part II

VINALDO

Typefaces and Ligatures in Printed Arabic Text: A Deep Learning-Based OCR Perspective 5
Omar Alhubaiti and Irfan Ahmad

Leveraging Knowledge Graph Embeddings to Enhance Contextual Representations for Relation Extraction 19
Fréjus A. A. Laleye, Loïc Rakotoson, and Sylvain Massip

Extracting Key-Value Pairs in Business Documents 32
Eliott Thomas, Dipendra Sharma Kafle, Ibrahim Souleiman Mahamoud, Aurélie Joseph, Mickael Coustaty, and Vincent Poulain d'Andecy

Long-Range Transformer Architectures for Document Understanding 47
Thibault Douzon, Stefan Duffner, Christophe Garcia, and Jérémy Espinas

KAP: Pre-training Transformers for Corporate Documents Understanding 65
Ibrahim Souleiman Mahamoud, Mickaël Coustaty, Aurélie Joseph, Vincent Poulain d'Andecy, and Jean-Marc Ogier

Transformer-Based Neural Machine Translation for Post-OCR Error Correction in Cursive Text 80
Nehal Yasin, Imran Siddiqi, Momina Moetesum, and Sadaf Abdul Rauf

Arxiv Tables: Document Understanding Challenge Linking Texts and Tables 94
Karolina Konopka, Michał Turski, and Filip Graliński

Subgraph-Induced Extraction Technique for Information (SETI) from Administrative Documents 108
Dipendra Sharma Kafle, Eliott Thomas, Mickael Coustaty, Aurélie Joseph, Antoine Doucet, and Vincent Poulain d'Andecy

Document Layout Annotation: Database and Benchmark in the Domain of Public Affairs 123
Alejandro Peña, Aythami Morales, Julian Fierrez, Javier Ortega-Garcia, Marcos Grande, Íñigo Puente, Jorge Córdova, and Gonzalo Córdova

A Clustering Approach Combining Lines and Text Detection for Table Extraction 139
Karima Boutalbi, Visar Sylejmani, Pierre Dardouillet, Olivier Le Van, Kave Salamatian, Hervé Verjus, Faiza Loukil, and David Telisson

WML

Absformer: Transformer-Based Model for Unsupervised Multi-Document Abstractive Summarization 151
Mohamed Trabelsi and Huseyin Uzunalioglu

A Comparison of Demographic Attributes Detection from Handwriting Based on Traditional and Deep Learning Methods 167
Fahimeh Alaei and Alireza Alaei

A New Optimization Approach to Improve an Ensemble Learning Model: Application to Persian/Arabic Handwritten Character Recognition 180
Omid Motamedisedeh, Faranak Zagia, and Alireza Alaei

BN-DRISHTI: Bangla Document Recognition Through Instance-Level Segmentation of Handwritten Text Images 195
Sheikh Mohammad Jubaer, Nazifa Tabassum, Md Ataur Rahman, and Mohammad Khairul Islam

Text Line Detection and Recognition of Greek Polytonic Documents 213
Panagiotis Kaddas, Basilis Gatos, Konstantinos Palaiologos, Katerina Christopoulou, and Konstantinos Kritsis

A Comprehensive Handwritten Paragraph Text Recognition System: LexiconNet 226
Lalita Kumari, Sukhdeep Singh, Vaibhav Varish Singh Rathore, and Anuj Sharma

Local Style Awareness of Font Images 242
Daichi Haraguchi and Seiichi Uchida

Fourier Feature-based CBAM and Vision Transformer for Text Detection in Drone Images 257
Ayush Roy, Palaiahnakote Shivakumara, Umapada Pal, Hamam Mokayed, and Marcus Liwicki

Document Binarization with Quaternionic Double Discriminator Generative Adversarial Network 272
Giorgos Sfikas, George Retsinas, and Basilis Gatos

Crosslingual Handwritten Text Generation Using GANs 285
Chun Chieh Chang, Leibny Paola Garcia Perera, and Sanjeev Khudanpur

Knowledge Integration Inside Multitask Network for Analysis of Unseen ID Types 302
Timothée Neitthoffer, Aurélie Lemaitre, Bertrand Coüasnon, Yann Soullard, and Ahmad Montaser Awal

Author Index 319

Contents – Part I

ADAPDA

Beyond Human Forgeries: An Investigation into Detecting Diffusion-Generated Handwriting 5
Guillaume Carrière, Konstantina Nikolaidou, Florian Kordon, Martin Mayr, Mathias Seuret, and Vincent Christlein

Leveraging Large Language Models for Topic Classification in the Domain of Public Affairs 20
Alejandro Peña, Aythami Morales, Julian Fierrez, Ignacio Serna, Javier Ortega-Garcia, Íñigo Puente, Jorge Córdova, and Gonzalo Córdova

The Adaptability of a Transformer-Based OCR Model for Historical Documents 34
Phillip Benjamin Ströbel, Tobias Hodel, Walter Boente, and Martin Volk

Using GANs for Domain Adaptive High Resolution Synthetic Document Generation 49
Tahani Fennir, Bart Lamiroy, and Jean-Charles Lamirel

GREC

A Survey and Approach to Chart Classification 67
Anurag Dhote, Mohammed Javed, and David S. Doermann

Segmentation-Free Alignment of Arbitrary Symbol Transcripts to Images 83
Pau Torras, Mohamed Ali Souibgui, Jialuo Chen, Sanket Biswas, and Alicia Fornés

Optical Music Recognition: Recent Advances, Current Challenges, and Future Directions 94
Jorge Calvo-Zaragoza, Juan C. Martinez-Sevilla, Carlos Penarrubia, and Antonio Rios-Vila

Reconstruction of Power Lines from Point Clouds 105
Alexander Gribov and Khalid Duri

KangaiSet: A Dataset for Visual Emotion Recognition on Manga 120
Ruddy Théodose and Jean-Christophe Burie

MuraNet: Multi-task Floor Plan Recognition with Relation Attention 135
Lingxiao Huang, Jung-Hsuan Wu, Chiching Wei, and Wilson Li

Automatic Detection of Comic Characters: An Analysis of Model Robustness Across Domains 151
Javier Lucas, Antonio Javier Gallego, Jorge Calvo-Zaragoza, and Juan Carlos Martinez-Sevilla

FPNet: Deep Attention Network for Automated Floor Plan Analysis 163
Abhinav Upadhyay, Alpana Dubey, and Suma Mani Kuriakose

Detection of Buried Complex Text. Case of Onomatopoeia in Comics Books ... 177
John Benson Louis and Jean-Christophe Burie

Text Extraction for Handwritten Circuit Diagram Images 192
Johannes Bayer, Shabi Haider Turabi, and Andreas Dengel

Can Pre-trained Language Models Help in Understanding Handwritten Symbols? 199
Adarsh Tiwari, Sanket Biswas, and Josep Lladós

CBDAR

On Text Localization in End-to-End OCR-Free Document Understanding Transformer Without Text Localization Supervision 215
Geewook Kim, Shuhei Yokoo, Sukmin Seo, Atsuki Osanai, Yamato Okamoto, and Youngmin Baek

IndicSTR12: A Dataset for Indic Scene Text Recognition 233
Harsh Lunia, Ajoy Mondal, and C. V. Jawahar

IWCP

Reconstruction of Broken Writing Strokes in Greek Papyri 253
Javaria Amin, Imran Siddiqi, and Momina Moetesum

Collaborative Annotation and Computational Analysis of Hieratic 267
Julius Tabin, Mark-Jan Nederhof, and Christian Casey

Efficient Annotation of Medieval Charters 284
Anguelos Nicolaou, Daniel Luger, Franziska Decker, Nicolas Renet, Vincent Christlein, and Georg Vogeler

Greek Literary Papyri Dating Benchmark 296
Asimina Paparrigopoulou, Vasiliki Kougia, Maria Konstantinidou, and John Pavlopoulos

Stylistic Similarities in Greek Papyri Based on Letter Shapes: A Deep Learning Approach 307
Isabelle Marthot-Santaniello, Manh Tu Vu, Olga Serbaeva, and Marie Beurton-Aimar

Author Index 325

VINALDO

Preface

Robust reading, also known as automatic document image processing, is an essential task in various applications areas such as data invoice extraction, subject review, medical prescription analysis, etc., and holds significant commercial potential. Several approaches are proposed in the literature, but datasets' availability and data privacy challenge it.

Considering the problem of information extraction from documents, different aspects must be taken into account, such as (1) document classification (2) text localization (3) OCR (Optical Character Recognition) (4) table extraction (5) key information detection. In this context, machine vision and, more precisely, deep learning models for image processing are attractive methods. In fact, several models for document analysis were developed for text box detection, text extraction, table extraction, etc. Different kinds of deep learning approaches, such as GNN, are used to tackle these tasks. On the other hand, the extracted text from documents can be represented using different embeddings based on recent NLP approaches such as Transformers. Also, understanding spatial relationships is critical for text document extraction results for some applications such as invoice analysis. Thus, the aim is to capture the structural connections between keywords (invoice number, date, amounts) and the main value (the desired information). An effective approach requires a combination of visual (spatial) and textual information.

In this workshop, we aimed to bring together an area for experts from industry, science, and academia to exchange ideas and discuss on-going research in Computer Vision and NLP for scanned document analysis.

The workshop was organized by Rim Hantach (Engie, France) and Rafika Boutalbi (Aix-Marseille University, France). We had one invited speaker: Dr. Matteo Bregonzio the CTO of Datrix and Head of Research and Development at 3rd PLACE, Milano, Italy. The Program Committee was selected to reflect the interdisciplinary nature of the field. For this first edition, we welcomed two kinds of contributions: short and long papers. We received a total of 12 submissions. Each paper was reviewed by two members of the program committee via EasyChair. A double-blind review was used for the short paper submissions, and the authors were welcome to anonymize their submissions. 10 submissions were accepted and were published.

June 2023

Rim Hantach
Rafika Boutalbi

Organization

General Chairs

Rim Hantach	Engie, France
Rafika Boutalbi	Aix-Marseille University, France

Program Committee

Sheuli Paul	Defence Research and Development, Canada
Aravindarajan Subramanian	ENGIE, France
Bill Power	Temple University, USA
Jumanah Alshehri	Temple University
Al Mansour	Boeing, Australia
James Mou	Reed Elsevier, Inc., New York, USA
Swapna Gokhale	University of Connecticut, USA
Felipe Torres	École Centrale Marseille, France
Stanislas Morbieu	Lipade, France

Typefaces and Ligatures in Printed Arabic Text: A Deep Learning-Based OCR Perspective

Omar Alhubaiti[1] and Irfan Ahmad[1,2](✉)

[1] King Fahd University of Petroleum and Minerals, Dhahran 31261, Saudi Arabia
{g201802660,irfan.ahmad}@kfupm.edu.sa
[2] SDAIA–KFUPM Joint Research Center for AI, Dhahran 31261, Saudi Arabia

Abstract. Arabic script is complex, with multiple shapes for the same characters in different positions. Another challenge of the script, in the context of recognition, is ligatures. A combination of a specific two or more character sequence takes a different shape than what those characters normally look like when they appear in a similar position. Deep learning-based systems are widely used for text recognition these days. In this work, we investigate the performance of deep learning systems for two alternative modeling choices: using *characters* as modeling units and using *character shapes* as modeling units. Moreover, we also investigate the impact on text recognition with mixed typefaces, where the training and test sets have samples from multiple typefaces, and discuss the effect of font families on recognition performance. We extend this by studying the effectiveness of the text recognition system in recognizing text from unseen typefaces, i.e., text in the test set is from a typeface not available in training. Finally, we present a methodology to automatically detect ligatures in printed Arabic text. We conducted experiments on the publicly available APTI dataset of printed Arabic text and report the findings and discuss the results.

Keywords: Printed Arabic Text OCR · Omnifont Text Recognition · Unseen Font Text Recognition · Automatic Ligatures Identification

1 Introduction

Arabic text has a complex cursive script that is written from right to left. Characters in Arabic take on different shapes depending on their position in a word (*beginning*, *middle*, *ending*, and *alone*) [7,9]. Not all characters take on all four shapes in different positions, and some of these shapes are dramatically different for some characters than for others, who would look almost identical. Linking between characters also varies between characters, and is a short (or varying length) horizontal dash, known as *Kashida*. Some characters can only be linked to those before or after them, or both. Ligatures are also present in Arabic script,

M. Coustaty and A. Fornés (Eds.): ICDAR 2023 Workshops, LNCS 14194, pp. 5–18, 2023.
https://doi.org/10.1007/978-3-031-41501-2_1

which are unique shapes of specific sequences of characters that would look different if those characters were not in that sequence. Normally, the characters forming a ligature are connected vertically instead of the normal horizontal connections. The only mandatory ligature is *Lam-Alif*, which is hard to read if it were not a ligature. Other ligatures are optional and vary between different typefaces. According to some studies, there are 478 possible ligatures in Unicode for printed Arabic script [1]. In addition to that, Arabic script employs diacritics to account for missing short vowels, since Arabic is an Abjad script [24]. There are four main diacritics, which can appear singly, doubly (*Tanween*/Nunation), or combined with a stress marker (*Shadda*). There are other diacritics like the tilde ˜ـ and *Alif Khanjariyah* ('). It is customary not to write vowels if they are inferable from the context and to only write them when their absence might cause confusion to the reader, with the exception of religious texts and learning materials. All these factors contribute to the difficulty of the text recognition problem in Arabic script.

According to Tahir [23], there are eight main typefaces in the Arabic script: Nashk, Ruqaa, Diwani, Thuluth, Hamayoni, Farsi, Tawqi', and Ijaza. Other typefaces are derived from these and are similar in form to the originals. In modern computer systems, many typefaces can be classified as Naskh variety and come in two forms: Traditional, which is more or less faithful to the rules of writing present in calligraphy books such as [23], and simplified, in which some rules are discarded in favor of readability or simplicity. Other font files such as Diwani or Ruqaa vary very little between implementations and are compliant with calligraphy rules.

In this paper, we present an empirical investigation of the performance of a deep learning-based text recognition system in recognizing Arabic texts in different typefaces. Specifically, we examine the two common modeling choices: using Arabic characters as modeling units and using Arabic character shapes as modeling units to determine which modeling choice is more effective for deep learning-based Arabic OCR. Furthermore, we investigate the impact of mixed typefaces on text recognition, where the training and test sets include samples from multiple typefaces. We also study the effectiveness of the text recognition system in recognizing text from unseen typefaces, i.e., text in the test set is from a typeface not available in the training set, and analyze the effect of font families on recognition performance. Lastly, we introduce a methodology for automatic detection of ligatures in printed Arabic text.

The rest of the paper is organized as follows: In Sect. 2, we discuss the related works in this field. In Sect. 3, we explain the architecture of the deep learning system we employed for Arabic text OCR. In Sect. 4, we present the experiment details and the results. Finally, in Sect. 5 we present the conclusions from our work and some possible future works.

2 Related Work

In this field, recent research employs one of two major classifiers: Hidden Markov Models (HMMs) and deep learning systems based on neural networks. This section first presents the works that use HMMs, followed by those that use deep learning-based systems. Additionally, we describe some of the datasets used in the literature.

In 2008, a study by Prasad et al. [15] utilized HMMs for OCR by detecting the writing line and applying overlapping sliding windows over the line without further segmentation. Hand-crafted features were extracted from the windows. To expand the vocabulary, trimodel units were utilized as HMM models, resulting in a total of 162 HMMs for characters excluding the unigrams. The authors provide detailed information on the HMM construction method. This study was tested on the DAMP dataset and yielded a 15.9% WER for character mode. However, using n-grams improved the performance to 11.5% WER for 5-grams, as there were numerous out-of-vocabulary occurrences during testing.

In [3], Ahmad et al. try reducing HMM models using sub-characters in both printed and hand-written text. The number of HMMs was down to 108 from 178 coming at a slight error cost for handwritten text (e.g. WER of 3.62% up from 3.50%) and slightly better at printed (CER of 5.98% CER down from 6.51%). Authors note that this approach is beneficial when training datasets are imbalanced pertaining to character representation.

Ahmad et al. in [2] performed text recognition using HMMs with an open vocabulary approach. This approach handles multiple typefaces for text-line images. There are three main outputs of this study: a typeface recognizer, a text recognizer for a particular typeface, and unseen typeface recognizer that works by adapting known typefaces. In this work, features are handcrafted (and explained in detailed) and tuned for all the tasks and the recognizer obtains the features using segmented and overlapping sliding windows technique that exploits Arabic scripts' prominent line of writing and trains based on expectation maximization. Results show that splitting typeface and text recognition and adaptation for unseen typefaces provide superior performance. Authors test the framework on their own P-KHATT database which they describe, resulting in 2.89% Character Error Rate (CER) for mono-font, 95% success rate for mixed fonts, and 7.18% CER for unseen fonts. For APTI database, 2.07% CER is the result for mono font recognition. This approach may need testing for all fonts as unseen font in P-KHATT and APTI to further confirm research findings.

An example of work that uses HMMs for printed text recognition while also segmenting text into connected components is [20]. A writing line is established and certain manual features are calculated and HMMs were created without optimizing their number. No results were provided in this work. The dataset for this work is hand-crafted in MS-Word application and is very limited.

An instance of user-friendly systems for OCR is [22] where authors have developed a system for library use to perform OCR on old documents. The system performs image processing (straightening lines etc.), OCR, and translation. This work also describes the interface and API of the system. In terms of OCR, KALDI speech recognition framework is used, which relies on HMMs for language modeling with 165 HMMs used in total. Multiple datasets (ALTEC, IFN/ENIT, KHATT, and HADARA80P) were used for training with special care for ligatures (which are mapped to pronunciation variants in KALDI) and to open the vocabulary. Evaluation is done against three of the four systems mentioned in [6]. The proposed system performed worse than others and authors suggest it is due to difference between training datasets and actual library-setting test (84.6% WER vs FineReader's 100%). This study can benefit greatly from a new dataset that is closer to the target testing domain.

In [13], Jaiem et al. tackle OCR using texture analysis, noting that older works use approaches such as Wavelets and Gabor Filters. This work differentiates between apriori (global) and posterior (context-based) recognition and chooses the former to work on using steerable pyramid (SP) where the image is decomposed recursively to different resolutions by applying filters on it and features extracted at different levels and orientations. A back-propagation Neural Network is used for recognition which is evaluated on APTID/MF yielding results of 99.63% accuracy vs 97.04% using KNN at many typefaces.

When it comes to deep neural networks, many works approach text recognition using RNNs and sometimes CNNs for feature extraction. In [4], Ahmed et al. presents text recognition from scene images. Unconventionally, this work uses CNNs instead of RNNs for this task. For training, input is 50×50 grayscale images augmented to 5 orientations. The architecture is a fully connected forward network using 3×3 and 5×5 filters, Relu for non-linearity, and maxpooling for features. Using EAST dataset which has 27 classes referring to different scene conditions with 2700 characters, authors tested this approach and optimized learning rate to obtain 14.57% error rate. Authors mentioned that low performance was attributed to small training dataset. The specific network architecture is not provided in this study.

Jain et al. in [14] use a CNN(features)-RNN(recognition) architecture for OCR in videos without extracting text from scenes first drawn by the success of such an approach in other similar scripts. The architecture consists of 7 convolutional layers with 4 max-pooling layers in between and 2 batch normalization (CNN), then (RNN) a couple of bidirectional long-short-term-memory layers (blstm) and a soft-max layer, before passing to the CTC layer for sequential recognition. This approach is trained on a hand-crafted dataset from wikipedia then tuned to ACTIV and ALIF datasets. Results vary from (98.17%, 79.67% Character Recognition Rate and Word Recognition Rate, respectively) on some tests to (75.05% and 39.43%), outperforming other systems such as Tesseract in all tests. Accuracy figures are far from the optimum 99% proposed in [5] to approach human-like performance, hence, more wor is needed in this field.

In [17], Rawls et al. were the first to combine deep learning approach for recognition, which uses CNN for feature extraction and LSTM and CTC for recognition with some language modelling in the form of WFST:weighted finite-state transducer using n-gram language models. The architecture composes of 2 convolutional (fully connected + Relu) units (CNN, features are extracted from a sliding window which is hand-optimized), followed by (RNN) (LSTM + fully connected), then a softmax layer, before passing to the CTC objective. This work was tested on the DAMP dataset. Authors specify hardware and parameters, used and report 2.7% and 8.4% CER and WER, respectively, surpassing the benchmark.

A study in 2017 by Alghamdi and Teahan [6] tests four Printed Arabic OCR systems (Automatic Reader, FineReader, Clever Page, and Tesseract) which are first introduced then metrics that measure such systems' performance are discussed, with the authors developing and defining their own suggested metrics (9 in total, including (digit, diacritic, and punctuation), and character (dot, loop, zigzag shaped, and overall) accuracies). Authors discuss several databases settling on KAFD for evaluation purposes. This work then compares the four systems in each of the 9 metrics while giving reasons for each metric's results, where each system excelled at different metrics from others. Authors note low overall accuracy of the systems, while the result being stronger if tested on more datasets. Areas of improvement in the whole field noted by the authors include diacritic and dot performance, and complex typefaces. Qaroush et al. in [16] presented a system for omnifont printed Arabic text recognition. A CNN-based system that relies on explicit character segmentation was developed. The authors report near-perfect segmentation results and high recognition results on APTI dataset of printed Arabic texts.

3 A System for Printed Arabic Text Recognition

The model we use here is adapted from [18]. Figure 1 shows an overview of the model's architecture. This model is a combination of CNN layers for feature detection, and RNN layers for the rest of the tasks which are Recognition and Classification. The CNN part of the model is made of five sets of convolution and max-pooling with the following dimensions: First, we start with 32×128 image with a feature depth of 1, next we have 64×16, 32×8, 32×4, 32×2, and 32×1 layers. The last of which has a feature depth of 256 and is the output of this part of the network.

Next, the RNN input has 32 time steps for a maximum of 32 characters per word and 256 features. Output of this part is a 2D map with 512 hidden cells and 32 time steps and the number of characters is variable (103 when using character shape as a model and 30 when using character as a model). This part is a stacked LSTM (forward and backwards). This is then fed to CTC (Connectionist-Temporal-Classification) to calculate loss for back-propagation. CTC is also used for decoding where we have the number of characters against probabilities for each. Various methods can be used after that like Beam Search or dictionaries but we opt for a simple best path decoding for this work. CTC has the advantages of not needing to segment the image to individual characters.

4 Experiments and Results

4.1 Dataset

We have used the APTI (Arabic Printed Text Image) dataset for conducting teh experiments [21]. APTI is a printed word dataset of more than 110,000 words in 10 different typefaces of which we used seven typefaces. This datasets has

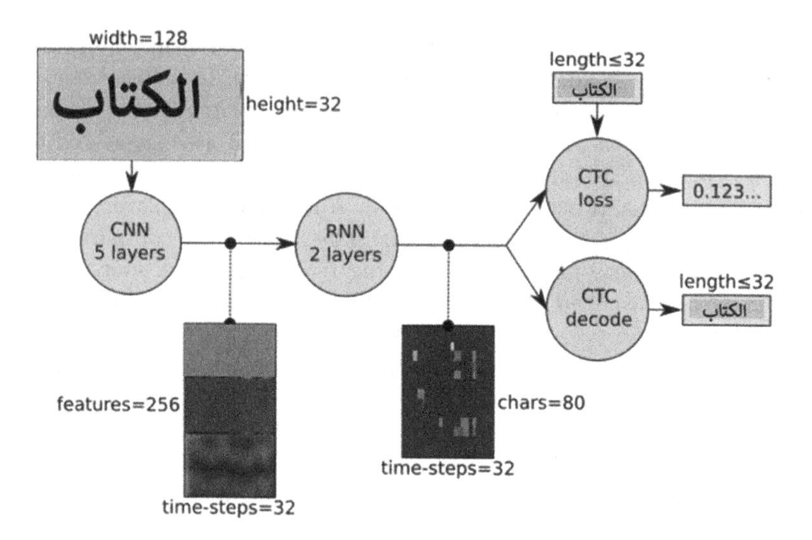

Fig. 1. Arabic text recognition system architecture. (figure adapted from: [19]).

10 sizes per typeface, of which we chose the largest (24 pts) and 4 styles per typeface (Normal, Bold, Italic, old-Italic) of which we selected the Normal. This is a purely computer-rendered dataset without camera or scanning artifacts. This dataset comes in 5 evenly divided sets that have comparable amount of characters in different shapes and positions which can be separated for use in training, validation, and testing. Ground truth for this dataset initially takes character shape into consideration. For use with character as a model approach, we modified that to discard the character shape information. The seven typefaces we used in our experiments from the APTI dataset are shown in Fig. 2.

4.2 Evaluation Measures

In this section, we present the measures we used to evaluate the text recognition results.

Character Error Rate (CER): CER is defined as follows:

$$CER(\%) = \frac{S + I + D}{N} \times 100 \tag{1}$$

where; S, I, and D are substitution, insertion, and deletion errors respectively, and N is the total number of characters in the evaluation set (cf. [12, p. 419–421]).

Word Error Rate (WER): WER is defined as follows:

$$WER(\%) = \frac{S + I + D}{N} \times 100 \tag{2}$$

where; S, I, and D are substitution, insertion, and deletion errors respectively, and N is the total number of words in the evaluation set.

Typeface	Sample Text Images	
Andalus	لآرائهم	فآمالهم
Arabic Transparent	لآرائهم	فآمالهم
Naskh	لآرائهم	فآمالهم
Thuluth	لآرائهم	فآمالهم
Diwani	لآرائهم	فآمالهم
Simplified Arabic	لآرائهم	فآمالهم
Traditional Arabic	لآرائهم	فآمالهم

Fig. 2. Sample text images from the APTI dataset.

4.3 Experiment Setup

We used a machine with an Intel core-i7-7700HQ processor @2.80 GHz with 24 GB of RAM, and an Nvidia GTX 1060-6 GB graphics card over-clocked to 1911 MHz. The system was implemented using GPU accelerated tensor-flow in Python under Microsoft Windows 10 v1809. We used decaying learning rate starting at 0.01 for the first 10 batches, 0.001 for the next 10,000 batches, and 0.0001 for the rest. We keep training until error rate stops improving for 5 epochs. If accuracy did not improve or has declined, that model is discarded and the previous weights are reused for further training.

4.4 Character vs Character Shape as Modeling Unit for Printed Arabic OCR

First we trained the model on two of the seven typefaces, one simple (Arabic Transparent) and one complex (Diwani), to determine whether character as a model (Beginning, middle, and end shapes of a character are treated as one single model) was better or worse than character shape as a model (Beginning, middle, and end are treated as separate models). We did these experiments using 90% of the set-1 of the APTI dataset for training, and the remaining

10% for testing, using 25,000 random images for each epoch. The model only accepts single characters as symbols for shapes. For example, we can represent Alif-Beginning as 'A', Baa-Beginning as 'B', and so on. This poses a problem when having many character shapes as is the case when using character shape as a model. We got around this issue by using additional Unicode characters for extra shapes. We also use this when encoding ligatures as a single unit. When the model validates, it treats all shapes of the same character as one symbol, since when printing this output to a document, the font-system converts a character to its appropriate shape again and thus it makes no sense to penalize the model over this type of mistakes. The results of these experiments are summarized in Table 1. From the results, we can see that using character as a model performs better than using character shape as a model. This is in contrast to what was reported in [10] for handwritten Arabic text recognition using deep learning systems. A possible explanation for this phenomenon is that the character and their different position dependent shapes are quite regular in printed Arabic texts but highly variable in handwritten texts. This, in addition to the fact that we can have more samples per class if we used character as a model than otherwise, and the network is able to select the best features to generalize and use as discriminants for that character.

4.5 Single Typeface, Mixed Typeface, and Unseen Typeface Text Recognition

Next, we ran a set of experiments without involving ligatures at all. These consist of 8 total training: one for each of the seven typefaces and the eighth experiment for all the seven typefaces combined. We trained the model on sets 1–3 of each typeface of size 24-pts with normal style and validated using the set-4. For each epoch, the system selects 30,000 random images from the training set split into 600 batches) except for the last experiment involving all typefaces which has 150,000 random training images per epoch using a 75%–25% training/validation-testing split. Results are present in Table 2. We can see that the model performs quite well even for complex typefaces with lots of overlap. The best results are reported for Simplified Arabic typeface and the worst results are for Naskh typeface. It is also interesting to note that mixed typeface recognition results are as good as single typeface recognition, which are in contrast to what is reported for some HMM-based systems (e.g., [2]). This could possibly be attributed to the high capacity of deep learning models having high number of trainable parameters.

Table 1. Summary of the results for different modeling units on text from two typefaces from the APTI dataset.

Modeling Unit	Typeface	CER	WER
Character	Arabic Transparent	0.82	4.33
	Diwani	0.27	1.56
Character Shape	Arabic Transparent	1.01	5.00
	Diwani	1.54	8.11

Table 2. Summary of the OCR results for different typefaces from the APTI dataset.

Typeface	CER	WER
Andalus	0.26	1.33
Arabic Transparent	0.15	0.73
Naskh	0.28	1.34
Thuluth	0.26	1.28
Diwani	0.24	1.26
Simplified Arabic	0.14	0.67
Traditional Arabic	0.15	0.76
Mixed (All combined)	0.20	1.03

Next, to investigate the performance of the system on unseen typeface and to test whether typeface similarity has an effect on model accuracy, we designed a set of leave-one-out experiments. We train the system on six typefaces and test on the remaining typeface, alternating. Results are summarized in Table 3.

From these results, we can clearly see that for fonts that visually differ from each other, the model was unable to generalize and achieved poor results. For Simplified Arabic and Arabic Transparent typefaces, however, and to a lesser extend, Traditional Arabic, the model was able to generalize very well. The explanation for this is that all these typefaces belong to the same family (Naskh Family). Decotype Naskh was visually more archaic so the model was not able to generalize sufficiently in this case. So we can have the intuition that we can use one typeface from the same family (out of several different Naskh implementations, or several different Andalus implementations, for instance) to have a more general model for that family.

Table 3. Summary of the OCR results for unseen typefaces from the APTI dataset.

Unseen Typeface	CER	WER
Andalus	56.83	98.09
Arabic Transparent	0.18	0.89
Naskh	10.94	39.00
Thuluth	20.00	66.03
Diwani	56.21	98.56
Simplified Arabic	0.29	1.42
Traditional Arabic	5.30	25.73

4.6 An Approach to Automatically Identify Ligatures in Printed Arabic Texts

We performed a pilot study to determine whether our approach will be able to distinguish between a ligature and a non-ligature in a typeface where we do not fully know the ligature (either we do not have the font file or the font file does not contain ligature information, like .ttf fonts). We chose Naskh typeface as a relatively simpler one and Diwani as the complex one. We could not choose a simpler typeface such as Simplified Arabic or Arabic Transparent because these typefaces do not have much ligatures, if at all.

We selected six bigram combinations that are present as ligatures in both typefaces: (*Nuun* or *Miim* in the beginning position with either *Jeem*, *Haa*, or *Khaa* in the middle position). They are the most common ligatures in Arabic (cf. [8,9]). Additionally, We have selected six similar bigram combinations that are non-ligatures (*Saad* or *Daad* in the beginning position either *Jeem*, *Haa*, or *Khaa* in the middle position). Sample ligatures from the APTI dataset is shown in Fig. 3. It should be noted that same character pairs may not form ligatures in other typefaces as illustrated in Fig. 4.

To develop a system to automatically identify ligatures, we train our text recognition system twice. Once treating these 12 character bigrams as ligatures thereby modeling each of them as a single ligature model. This is done by changing the training set annotations. For the second round of training, we treat the 12 character bigrams as individual characters and not as ligatures. Our final ligature identification system then decodes the unseen test sets with the possibility of treating the 12 character bigrams pairs as either ligatures or non-ligatures depending on which has the higher probability.

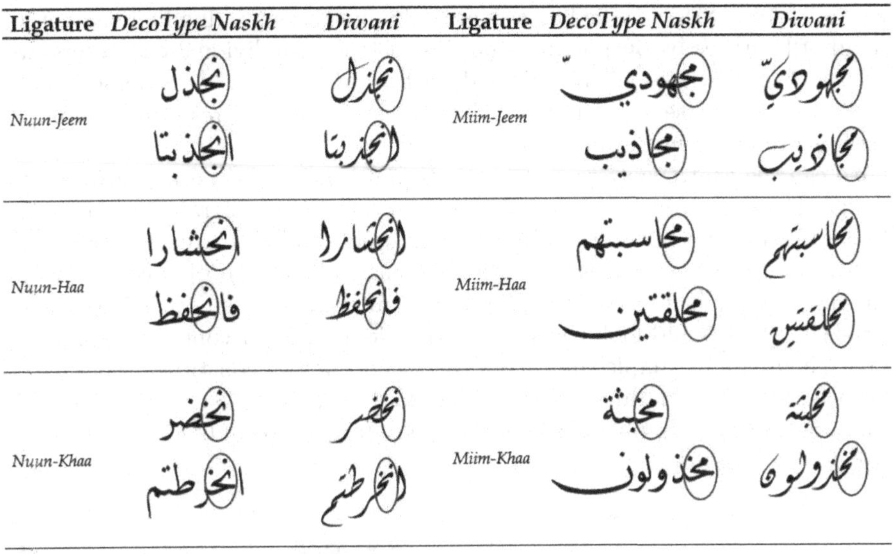

Ligature	*DecoType Naskh*	*Diwani*	Ligature	*DecoType Naskh*	*Diwani*
Nuun-Jeem	نجذل انجذبتا	نجذل انجذبتا	*Miim-Jeem*	مجهودي مجاذيب	مجهودي مجاذيب
Nuun-Haa	انحشارا فانحفظ	انحشارا فانحفظ	*Miim-Haa*	محاسبتهم محلقتين	محاسبتهم محلقتين
Nuun-Khaa	نخضر انخرطتم	نخضر انخرطتم	*Miim-Khaa*	مخبثة مخذولون	مخبثة مخذولون

Fig. 3. Sample ligatures (circumscribed within a blue oval) from the APTI dataset. (Color figure online)

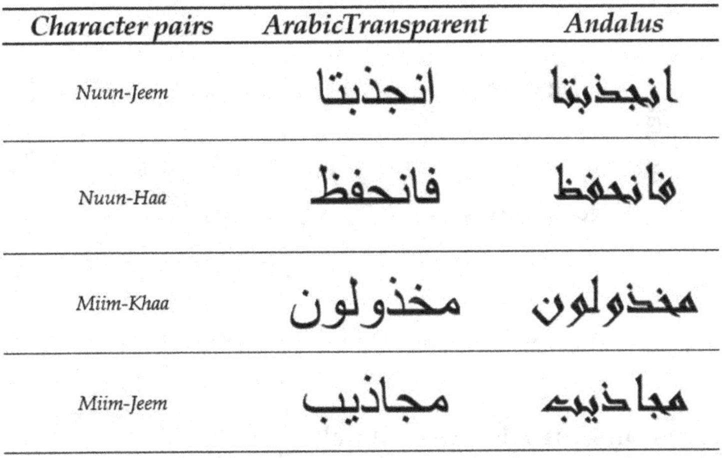

Character pairs	*ArabicTransparent*	*Andalus*
Nuun-Jeem	انجذبتا	انجذبتا
Nuun-Haa	فانحفظ	فانحفظ
Miim-Khaa	مخذولون	مخذولون
Miim-Jeem	مجاذيب	مجاذيب

Fig. 4. Sample text images from the APTI dataset where character combinations do not form ligatures.

If the model detects a ligature correctly, we consider this as a true positive. Additionally, if the model classified a non-ligature pair correctly by outputting individual characters instead of ligatures, this is regarded as a true negative. On the other hand, if a ligature was classified as a non-ligature by outputting individual characters instead of ligatures, this is treated as a false negative. Similarly, if a non-ligature was classified as a ligature by outputting individual characters instead of ligatures by outputting ligatures instead of individual characters, we consider this as a false positive. Based on this setup, we ran our experiments on the unseen test sets of the two selected typefaces. The resulting confusion matrix is presented in Fig. 5.

We can see that the system was able to differentiate between ligatures and non-ligatures effectively. The possible explanation is that a bigram ligature looks visually different from how the original characters would look like in other contexts. Thus the system learns the individual characters from other contexts and ligatures from explicit training on these bigram pairs. On the other hand, non-ligature bigrams would look the same as they look in other contexts. Moreover, we have many more samples in the dataset where the two constituent characters of this non-ligature bigram appear as separate characters. So, the system will lean more towards recognizing the non-ligature bigrams as individual characters instead of as ligatures.

		Predicted	
		Ligatures	**Non-Ligatures**
Actual	**Ligatures**	78%	22%
	Non-Ligatures	16%	84%

Fig. 5. Confusion matrix for Ligature Detection.

5 Conclusions and Future Work

In this paper, we present our work related to printed Arabic text recognition using deep-learning. We investigated two alternative modeling choices and conclude that character as a model perform better than character shape as a model. Further we conducted experiments to investigate the system's performance for single typeface text recognition to observe if different typefaces lead to different recognition performance due to their visual complexity. We observed that the system was able to effectively recognize text in all the seven typefaces which ranges from visual simple to visually complex typefaces. In fact, the system's performance was at times better for visually complex typefaces as compared to some of the simpler ones. Moreover, the system's performance of mixed typefaces text recognition was as good as single typeface text recognition. We also saw that models trained for different typefaces in the same family can generalize to other unseen typefaces from the same family but not to typefaces from different families. Last but not the least, we presented an approach to develop a system to automatically identify ligatures in unseen text. The system, although not perfect, performed reasonably well.

The experiments were conducted on only one dataset and for only seven different typefaces. This clearly is a limitation of the work. Further experiments need to be carried out to confirm the generalizability of the presented conclusions. Furthermore, only six bigram ligatures were experimented. There are many more ligatures, so further work is needed to confirm the effectiveness of the presented approach for automatic identification of ligatures. Accordingly, possible follow-ups for this work include expanding the case study for ligature detection into more ligatures and typefaces, and trying the scheme for additional datasets. Investigating other deep learning models, such as transformer models can also be an interesting future work. This work might also be applicable to handwritten text although the factor that each writer might create their own ligatures can limit the usefulness of this which needs to be investigated. Moreover, deeper investigation into the font-family concept is warranted to determine whether models can generalize within the classical categorization of typefaces as seen in calligraphy books. Last but not the least, generating rare occurring ligatures using systems such as GANs (e.g., [11]) in order to robustly train a text recognition system can be an interesting future work.

Acknowledgment. The authors would like to thank King Fahd University of Petroleum and Minerals (KFUPM) for supporting this work. Irfan Ahmad would like to additionally thank Saudi Data and AI Authority (SDAIA) and KFUPM for supporting him through SDAIA-KFUPM Joint Research Center for Artificial Intelligence grant number JRC–AI–RFP–06.

References

1. Arabic ligature: Graphemica search. https://graphemica.com/search?q=arabic+ligature. Accessed 20 Apr 2023
2. Ahmad, I., Mahmoud, S.A., Fink, G.A.: Open-vocabulary recognition of machine-printed Arabic text using hidden Markov models. Pattern Recogn. **51**, 97–111 (2016)
3. Ahmad, I., Rothacker, L., Fink, G.A., Mahmoud, S.A.: Novel sub-character hmm models for Arabic text recognition. In: 2013 12th International Conference on Document Analysis and Recognition, pp. 658–662. IEEE (2013)
4. Ahmed, S.B., Naz, S., Razzak, M.I., Yousaf, R.: Deep learning based isolated Arabic scene character recognition. In: 2017 1st International Workshop on Arabic Script Analysis and Recognition (ASAR), pp. 46–51. IEEE (2017)
5. Al-Badr, B., Mahmoud, S.A.: Survey and bibliography of Arabic optical text recognition. Signal Process. **41**(1), 49–77 (1995)
6. Alghamdi, M., Teahan, W.: Experimental evaluation of Arabic OCR systems. PSU Res. Rev. **1**(3), 229–241 (2017)

7. Amara, N.: On the problematic and orientations in recognition of the Arabic writing. In: CIFED 2002, pp. 1–10 (2002)
8. Elarian, Y., Ahmad, I., Awaida, S., Al-Khatib, W., Zidouri, A.: Arabic ligatures: analysis and application in text recognition. In: 2015 13th International Conference on Document Analysis and Recognition (ICDAR), pp. 896–900. IEEE (2015)
9. Elarian, Y., Ahmad, I., Zidouri, A., Al-Khatib, W.G.: Lucidah ligative and unligative characters in a dataset for Arabic handwriting. Int. J. Adv. Comput. Sci. Appl. **10**(8) (2019)
10. Eltay, M., Zidouri, A., Ahmad, I.: Exploring deep learning approaches to recognize handwritten Arabic texts. IEEE Access **8**, 89882–89898 (2020)
11. Eltay, M., Zidouri, A., Ahmad, I., Elarian, Y.: Generative adversarial network based adaptive data augmentation for handwritten Arabic text recognition. PeerJ Comput. Sci. **8**, e861 (2022)
12. Huang, X., Acero, A., Hon, H.W., Reddy, R.: Spoken Language Processing: A Guide to Theory, Algorithm, and System Development, vol. 1. Prentice Hall PTR, Upper Saddle River (2001)
13. Jaiem, F.K., Kanoun, S., Eglin, V.: Arabic font recognition based on a texture analysis. In: 2014 14th International Conference on Frontiers in Handwriting Recognition, pp. 673–677. IEEE (2014)
14. Jain, M., Mathew, M., Jawahar, C.: Unconstrained scene text and video text recognition for Arabic script. In: 2017 1st International Workshop on Arabic Script Analysis and Recognition (ASAR), pp. 26–30. IEEE (2017)
15. Prasad, R., Saleem, S., Kamali, M., Meermeier, R., Natarajan, P.: Improvements in hidden Markov model based Arabic OCR. In: 2008 19th International Conference on Pattern Recognition, pp. 1–4. IEEE (2008)
16. Qaroush, A., Awad, A., Modallal, M., Ziq, M.: Segmentation-based, omnifont printed Arabic character recognition without font identification. J. King Saud Univ.-Comput. Inf. Sci. **34**(6), 3025–3039 (2022)
17. Rawls, S., Cao, H., Sabir, E., Natarajan, P.: Combining deep learning and language modeling for segmentation-free OCR from raw pixels. In: 2017 1st International Workshop on Arabic Script Analysis and Recognition (ASAR), pp. 119–123. IEEE (2017)
18. Scheidl, H.: Handwritten text recognition in historical documents. In: Tuwien Faculty of Informatics, Technische Universität Wien, Diplomarbeit. Tuwien (2018)
19. Scheidl, H.: Build a handwritten text recognition system using tensorflow. Medium, 09-Aug-2020 (2019)
20. Shaker, A.S.: Recognition of off-line printed Arabic text using hidden Markov models. Ibn AL- Haitham J. Pure Appl. Sci. **31**(2), 230–238 (2018). https://doi.org/10.30526/31.2.1952. http://jih.uobaghdad.edu.iq/index.php/j/article/view/1952
21. Slimane, F., Ingold, R., Kanoun, S., Alimi, A.M., Hennebert, J.: A new arabic printed text image database and evaluation protocols. In: 2009 10th International Conference on Document Analysis and Recognition, pp. 946–950. IEEE (2009)
22. Stahlberg, F., Vogel, S.: QATIP-an optical character recognition system for Arabic heritage collections in libraries. In: 2016 12th IAPR Workshop on Document Analysis Systems (DAS), pp. 168–173. IEEE (2016)
23. Tahir, K.M.: Tarikh al-khatt al-Arabi wa-adabuh: huwa kitāb tārīkhī ijtimāī adabī muzayyan bi-al-uwar al-khaīyah wa-al-rusāum al-fūtughrāfīyah. al-Matbaah al-Tijariyah al-Hadīthah (1982)
24. Zitouni, I.: Natural Language Processing of Semitic Languages. Springer, Heidelberg (2014). https://doi.org/10.1007/978-3-642-45358-8

Leveraging Knowledge Graph Embeddings to Enhance Contextual Representations for Relation Extraction

Fréjus A. A. Laleye(✉), Loïc Rakotoson, and Sylvain Massip

Opscidia, Paris, France
{frejus.laleye,loic.rakotoson,sylvain.massip}@opscidia.com

Abstract. Relation extraction task is a crucial and challenging aspect of Natural Language Processing. Several methods have surfaced as of late, exhibiting notable performance in addressing the task; however, most of these approaches rely on vast amounts of data from large-scale knowledge graphs or language models pretrained on voluminous corpora. In this paper, we hone in on the effective utilization of solely the knowledge supplied by a corpus to create a high-performing model. Our objective is to showcase that by leveraging the hierarchical structure and relational distribution of entities within a corpus without introducing external knowledge, a relation extraction model can achieve significantly enhanced performance. We therefore proposed a relation extraction approach based on the incorporation of pretrained knowledge graph embeddings at the corpus scale into the sentence-level contextual representation. We conducted a series of experiments which revealed promising and very interesting results for our proposed approach. The obtained results demonstrated an outperformance of our method compared to context-based relation extraction models.

Keywords: Relation extraction · Knowledge Graph Embeddings · Contextual representation

1 Introduction

In recent years, there has been growing interest among researchers in transforming text into structured information for data mining purposes. The use of pre-trained language models, both general and domain-specific, has been shown to greatly improve performance on information extraction tasks [1,6,7,18]. However, for more specialized domains with limited data, information extraction remains a challenge. Although these models can model contextual information in text, they may not be as effective in specialized fields due to a lack of domain-specific knowledge. To address this issue, some researchers have proposed injecting external knowledge into pre-trained models, including well-supplied ontologies [15–17]. However, this approach is only viable if well-annotated ontologies or knowledge graphs are available in the target domain.

M. Coustaty and A. Fornés (Eds.): ICDAR 2023 Workshops, LNCS 14194, pp. 19–31, 2023.
https://doi.org/10.1007/978-3-031-41501-2_2

The idea of combining contextual representations from large pretrained models such as BERT with semantic representation in knowledge graphs has been widely proven to improve classification tasks, as demonstrated in [11,25]. This same hypothesis holds true for relation extraction tasks, but only when there is a large knowledge graph available to build rich knowledge graph embeddings. However, what if such a knowledge graph is not available? In this paper, we propose an approach that leverages the global knowledge available in the training corpus to enrich contextual representations of documents and improve the performance of relation extraction based on BERT. Our approach only relies on local knowledge from the training data, making it useful for specialized domains without large specific or humanly annotated knowledge bases.

2 Related Work

2.1 Pretrained Languages Models

Several contextual representation models, such as BERT [4], GPT-3 [22] or Galactica [19], have been developed to encode words in context and have shown significant improvements in various natural language processing tasks. However, these models may not be capable of representing multiple senses of a word or capturing relational knowledge between words in context. For example, the word *geometry* can have different meanings depending on the context, such as in mathematics or chemistry [19], and such differences may be difficult for these models to capture. To address this limitation, pre-trained contextual models have been developed for specific domains, such as BioBERT, which was trained on biomedical documents (PubMed abstracts and PMC full-text articles) [13]; ClinicalBert, which was trained on clinical notes in the MIMIC-III database [9], and PubMed-BERT, which was trained on abstracts from PubMed and full-text articles from PMC [6]. These domain-specific models have shown promising results in classification and relation extraction tasks, indicating the importance of incorporating domain-specific knowledge in pre-training contextual models.

2.2 External Knowledge Injection

In order to address the aforementioned limitation, a common approach is to enhance contextual representations by incorporating external expert knowledge. Several recent works, including [16,17,26], have utilized entity and relation knowledge from external knowledge bases such as Wikidata to improve language representation. In particular, [16] demonstrated that the addition of metadata features and knowledge graph embeddings leads to improved document classification results. Similarly, [17] incorporated knowledge base graph embeddings pretrained on Wikidata with transformer-based language models to improve performance on the sentential Relation Extraction task. Another study [15] experimented with injecting specialized knowledge from CN-dbpedia [23] and HowNet [5], which are large-scale language knowledge bases, into the language representation on twelve NLP tasks. Their findings highlight the usefulness of knowledge graphs for tasks requiring expert knowledge. However, unlike these methods, our proposed approach dynamically learns the global and hierarchical structure of the graph built on a corpus for the relation extraction task, rather than incorporating embeddings pretrained on external knowledge graphs directly into the contextual representation. Additionally, we incorporate relational information between entities learned dynamically from their hierarchical structure in the datasource, rather than entity embeddings.

It is worth noting that all state-of-the-art methods presented in the literature require access to knowledge contained in large knowledge graphs. In these approaches, the embeddings are first pretrained on a broad range of information derived from the knowledge graph, which is separate from the graph constructed from the original data, before being incorporated into contextual representations of sentences or documents. In contrast, our proposed approach is motivated by the desire to dynamically learn the global and hierarchical structure of the graph built from the data corpus for the relation extraction task. It is important to emphasize that we do not inject entity embeddings into the contextual representation, but rather incorporate relational information dynamically learned from the hierarchical structure of the data source between two entities.

3 Proposed Approach

In this section, we describe each component that composes the improved semantic relation extraction system we propose.

3.1 Model Architecture

We propose an approach that leverages global representations constructed from local knowledge graphs using training data independent of domain and data specialization. As illustrated in Fig. 1, our model architecture consists of two main modules. The first module focuses on creating these global representations, while the second module jointly learns relational and contextual representations. Both modules are detailed in the following sections.

In Fig. 1, the collection of triples $\{e_i, r_k, e_j\}$ designates the set of entity-relations extracted from the training documents where e_1 and e_2 represent the entities and r the semantic relation to be predicted in a given document d. Each triple corresponds to a document. We denote a document $d = \{w_1, w_2, w_3, ..., w_n\}$ as a sequence of tokens, where n is the length of this document.

3.2 Pretrained Knowledge Graph Embeddings

The primary goal of this module is to generate embeddings from a hierarchical knowledge graph that captures local and structural information of all entity pairs in the documents. The knowledge graph is designed to represent relational information between entities and properties between relations, allowing for the

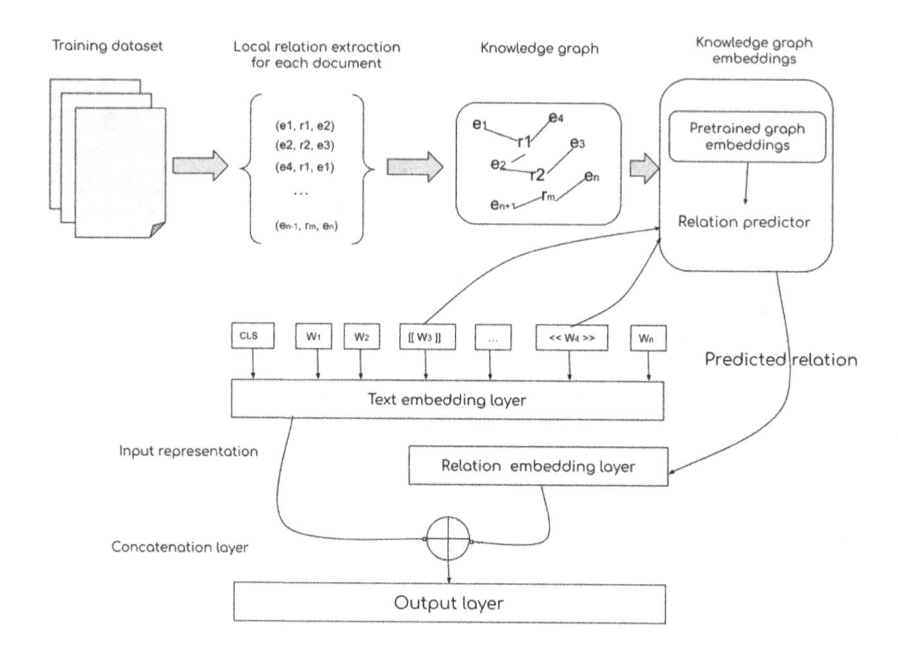

Fig. 1. Architecture of our proposed approach.

encoding and learning of relational semantic representations necessary for semantic relation prediction. To achieve this, we constructed a set of triples consisting of entities and relations found in the training data.

Given the set of documents D, we constructed the graph $KG = \{h, r, t\} \subseteq E \times R \times E$ with E and R respectively the set of entities and the set of relations. Although the set D only contains sentences and does not explicitly specify the knowledge graph KG, one can be generated by extracting the set of all available entities E in the documents and all relations R to form KG whose edges are represented by $r \subseteq R$ and nodes by $\{h, t\} \subseteq E$.

The relational representations are therefore computed by optimizing the function $f(h, t) = hM_r t$ over the set of triples where M_r is a relation-specific diagonal matrix that captures the properties of the relation between two entities. Dense vector representations (i.e., knowledge graph embeddings) are generated for each entity by training the knowledge graph, such that the distances between these vectors predict the presence of edges in the graph. Using this model, we can predict a relation r given a question of the form $(e_1, ?, e_2)$, which is essentially a link prediction task in a knowledge graph.

Our method for relation extraction does not involve masking entity types in documents, as we wish to preserve both the semantic and relational information between entity mentions. Additionally, we have not included entity types in the entity vocabulary, in line with our motivation to use only data available in the training set. It is worth noting that the knowledge graph embeddings used in previous works [10,14,27] are obtained from a wide range of entity properties and information drawn from external knowledge graphs, rather than directly from the graph structure provided by the training data. Our approach involves pre-training a model of knowledge graph embeddings to infer relation representations from two entities, and subsequently using this model during the training of the final semantic relation extraction system.

3.3 BERT-Based Document Contextuel Representation

To obtain the distributed representation of a document and encode the context for the relation extraction task, we opted for a pre-trained BERT-like model. Following the success of Transformer [21], BERT models have been experimented with in most NLP tasks and adapted in several domains. In this work, we have been interested in BERT models such as SpanBERT [10], a replica of BERT designed to represent and predict spans of text, SciBERT [1], a pretrained language model for scientific text based on BERT and PubMedBERT [6], a BERT model pre-trained with a collection of PubMed abstracts. These are all designed to learn a sequential contextual representation of the input sequence through a multi-layer, multi-head self-attention mechanism. Specifically, the pre-trained BERT model, in our text embedding layer, is used to produce the token representation which represent the context-sensitive word embeddings. Given a document d, we insert the following four markers [[,]], <<, and >> at the beginning and end of the two entities (e_1, e_2). An example from chemprot dataset is shown in Fig. 2

```
<< Androgen >> antagonistic effect of estramustine phosphate (EMP) metabolites
   on wild-type and mutated [[ androgen receptor ]].
```

Fig. 2. Document with special tokens.

3.4 Input Representation

We built the input representation by summing the representation from BERT and relation representation obtained with the knowledge graph embeddings. The pretrained knowledge graph embedding model has the ability to assign a dense vector representing the relation that may exist between the two entities present in the text. Given a sequence of words $S = \{w_1, w_2, w_3, ...w_j\}$ as a document where w_i is the $j - th$ token, including the special tokens, we built the input representation to aggregate both the contextual information and the learned relational structure.

First, in text embedding layer, we used BERT-based model to produce tokens embeddings H_c from S:

$$H_c = BERT(\{[CLS]\} \cup \{w_i \in S\}) \tag{1}$$

where $[CLS]$ is a special token that is put in the first position of each document. We obtained the contextual representation of the document by deriving its final state.

Secondly, to obtain the relation representation H_r, we construct the query $(e1, ?, e2)$, from the document, as input to the knowledge graph embedding model from which it infers the vector representation of r:

$$H_r = h * M_r * t \tag{2}$$

where h and t are respectively the embeddings of $e1$ and $e2$ and M_r the relation embedding matrix learned during pre-training. It's worth mentioning that a limitation of our approach of using only training data to build the knowledge graph embedding model is that, with its inability to incorporate context, it can't infer an unseen relation during training. This can be a problem for the incorporation of the relation representation in the contextual vector. A simple solution that we employed is tu use a zero vector as embedding vector if no relation is found given a query. This has the advantage of always ensuring the concatenation operation with the BERT context vector.

Finally, in the concatenation layer, we construct the final representation which integrates both the contextual representation H_c and the learned relation representation H_r:

$$\nu_d = \underbrace{BERT(\{[CLS]\} \cup \{w_i \in S\})}_{H_c} + W^* \underbrace{h * M_r * t}_{H_r} \tag{3}$$

where $\nu_d = H_c + W^* H_r$ if the model of knowledge graph emneddings infers a relation and $\nu_d = H_c$ otherwise. The learnable weights W^* are composed of one linear layer and dropout ($p = 0.1$) for regularization. Intuitively, ν_d aggregates and can fuse contextual and relational information across document tokens.

4 Experiments

4.1 Dataset

Although our approach may be domain agnostic, in this paper we focus our study of biomedical relation extraction on predicting relations between two biomedical entities given a sentence. We therefore chose to carry out experiments with the relation extraction datasets from BLURB benchmark [6]. Table 1 presents the datasets and their statistics on the various training, development and testing sets.

ChemProt. ChemProt is a disease chemical biology dataset consisting of 1820 PubMed abstracts with annotated chemical-protein interactions [12]. The dataset contains the protein and chemicals entity types and about 10, 031 relations.

DDI. DDI dataset consists of 1025 documents from two DrugBank database and MedLine which are labeled with drug entities and drug-drug interactions [8]. All interactions are categorized into four true and one vacuous relation.

GAD The GAD (Genetic Association Database) is an archive of published genetic association studies which consists of scientific excerpts and abstracts distantly annotated with the presence or absence gene-disease associations [3].

Table 1. Dataset statistics

Dataset	Entity type	Train	Dev	Test
Chemprot	Chemical-Protein	18,035	11,268	15,745
DDI	Drug-Drug	25,296	2,496	5,716
GAD	Gene-Disease	4261	532	534

4.2 Baselines

We compared our approach, which is based on the incorporation of a learned relation representation into contextualized document representations, to the following state-of-the-art context models for the semantic relation extraction task.

1. **SciBERT** [1] leverages unsupervised pretraining on a large multi-domain corpus of scientific publications to improve performance on downstream scientific NLP tasks.
2. **SpanBERT** [10] is a self-supervised pre-training method designed to better represent and predict spans of text.
3. **BioBERT** [13] is a domain-specific language representation model pre-trained on large-scale biomedical corpora.
4. **PubMedBERT** [6] a pretraining model using abstracts from PubMed and full-text articles from PubMedCentral that achieved state-of-the-art performance on many biomedical NLP tasks.

These models were chosen for their application in the biomedical field and because they have proven their effectiveness in the extraction of biomedical relationship. They have all been finetuned on the 3 benchmarked datasets according to author recommendations for result reproduction. It should be mentioned that the recommendations were not enough to achieve the same performance claimed in their paper even when using their open-source code. For example, with Scibert[1], we obtained F1-macro score of 74.2% on the ChemProt dataset while the authors mentioned 83.64% in their paper. In this paper, we reported the scores from our own experiments.

[1] https://github.com/allenai/scibert.

4.3 Overall Performance

Once the knowledge graph is created using all entity pairs and their relations, we build the knowledge graph embedding model by using several popular Knowledge Graph Embedding approaches such as a translation-based model TransE [2], DistMult [24] and ComplEx [20] which are the semantic matching models. The effectiveness of each model was tested and evaluated on the three datasets and on the link prediction task. Two standard measures are used as evaluation metrics: Mean Reciprocal Rank (MRR) and Hits at N. For a given entity pair, we perform a ranking of all the triples to calculate $hits@N$ and Mean Reciprocal Rank. $Hits@10$ denotes the fraction of actual triples that are returned in the top 10 predicted triples We report MRR, and Hits at 1 and 10 in Table 2 for the evaluated models. Among these KGE methods, ComplEx performed the best on the link prediction task. As a result, we only use knowledge graph embeddings from ComplEx in the following experiments.

Table 2. Comparing the KGE models on all the 3 datasets used in our study. We conclude an outperformance of the ComplEx model compared with the TransE and DistMult models.

Model	ChemProt			DDI			GAD		
	Hits@1	Hits@10	MRR	Hits@1	Hits@10	MRR	Hits@1	Hits@10	MRR
TransE	0.44	0.51	0.49	0.51	0.56	0.58	0.31	0.45	0.38
DistMult	0.52	0.72	0.59	0.64	0.77	0.71	0.37	0.47	0.45
ComplEx	0.67	0.84	0.72	0.71	0.88	0.78	0.44	0.52	0.51

The accuracy and F1 scores for the relation extraction models on the 3 benchmark datasets are shown in Table 3. The performances of relation extraction based on contextual representations are presented at the top of the table.

The results show a very significant performance of the PubMedBERT model on all 3 datasets compared to SciBERT, BioBERT and SpanBERT. We can also notice that, although SpanBERT is designed to adapt to the recognition of participating entities in a relation extraction task, it is the model that performed less well ($AVG_{F1} = 62.70\%$; $AVG_{Acc} = 77.88\%$) from our experiments. BioBERT and PubMedBERT gained on average, respectively +3 points and +5 points on SciBERT model. This is due to the aggregated biomedical domain knowledge in the BioBERT and PubMedBERT models compared to SciBERT. SciBERT remains a domain specific model but general for the scientific domain.

Table 3. Evaluation of binary classifiers on test data. The best results in each column are highlighted in bold font. This table provides a comparison of relation extraction performance between our proposed approach, which incorporates kGE-based relation representation, and the SOTA pretrained approach.

Model	ChemProt		DDI		GAD	
	Acc	F1	Acc	F1	Acc	F1
SOTA pretrained models						
SciBERT	86.9	74.2	87.55	76.10	76.75	50.60
BioBERT	88.15	77.13	91.05	77.10	**78.70**	**55.45**
PubMedBERT	**89.5**	**79.63**	**91.65**	**78.80**	78.69	54.60
SpanBERT	77.20	65.33	80.10	70.95	76,35	51.80
Incorporating kGE-based relation representation using our proposed approach						
SciBERT+ComplEx	90.35	79.45	90.65	80.10	86.45	59.20
BioBERT+ComplEx	92.25	84.33	93.65	**86.10**	87.05	61.40
PubMedBERT+ComplEx	**93.55**	**86.83**	**93.70**	85.85	**89.66**	**62.82**
SpanBERT+ComplEx	88.05	75.33	86.55	77.75	84.75	58.88

In the second part of the Table 3, we have reported the performance results of our approach for incorporating relational knowledge into the contextual representation for the semantic relation extraction task. The findings demonstrate the ability of our approach to produce a textual representation that can be enriched with contextual and relational knowledge in order to increase the performance of a relation extraction model. With our approach, we significantly improved the F1 scores of the baseline models: +9 points for SciBERT and +8 points for PubMedBERT on GAD dataset, +9 points for BioBERT on DDI dataset and +10 points for SpanBERT on ChemProt dataset. These improved performances are shown in Figs. 3 and 4 which respectively present the average of the F1 scores and accuracies on all of the 3 datasets.

In order to illustrate the limitations of relying solely on textual context for relation extraction, we conducted an evaluation to assess the ability of the model to generalize to documents where the relation representation provided by the Knowledge Graph Embedding (KGE) model does not align with the ground truth relations and is semantically distant from the context. To achieve this, we proposed a setting in which the test data did not include any relationships from the training set of the KGE model. Intuitively, this setting forces the relation extraction model to rely on more than just the entities, since they lack explicit links to the predicted relation. We present the evaluation results in Table 4.

Based on the results presented in Table 4, we can confirm that our approach consistently enhances the effectiveness of relation extraction models, even when the relation representation is out of context. Our proposed method outperforms baseline approaches and validates the generalization capability of our approach to incorporate both context and relational structure to improve relation extraction models.

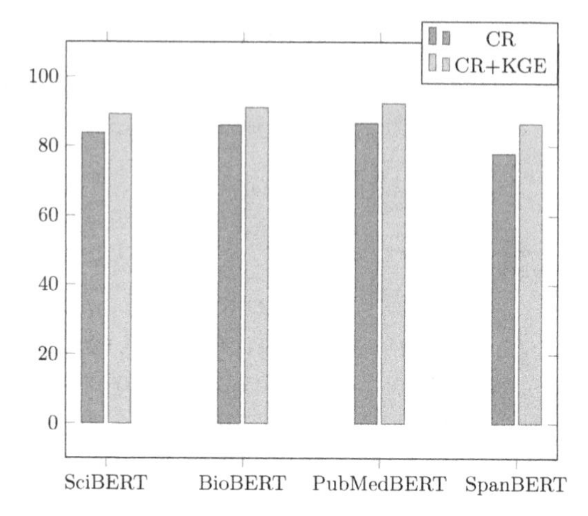

Fig. 3. Average of model accuracies on all 3 datasets. CR (Contextual Representation) denotes relation extraction using only contextual document embeddings and CR+KGE denotes relation extraction using our approach of incorporating knowledge graph embeddings into contextual representation.

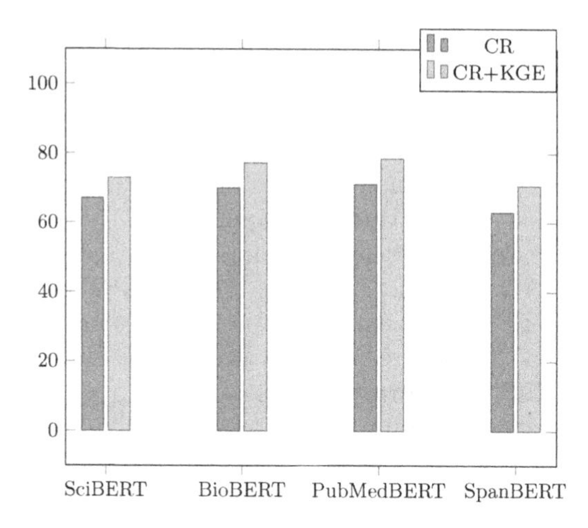

Fig. 4. Average of F1 scores on all 3 datasets. CR (Contextual Representation) denotes relation extraction using only contextual document embeddings and CR+KGE denotes relation extraction using our approach of incorporating knowledge graph embeddings into contextual representation.

Furthermore, our findings demonstrate the ability of our model to capture implicit and missing relations between entities in the training set, which is made possible by leveraging the knowledge information provided by the hierarchical graph constructed in the first step.

Table 4. Experimental result of relation extraction on the test sets not including relations from training set.

Model	Acc	F1
SciBERT+ComplEx	86.7	69.15
BioBERT+ComplEx	87.16	71.25
PubMedBERT+ComplEx	89,15	75.75
SpanBERT+ComplEx	81.25	65.4

5 Conclusion

In this paper, we proposed an approach for enriching contextual representations in semantic relation extraction task by leveraging the structural and relational knowledge provided by a hierarchical knowledge graph. Our proposed pre-training task of the knowledge graph incorporates knowledge information into language representation models, resulting in a better fusion of relational and contextual information. The experimental results on the BLURB benchmark demonstrate that our approach outperforms the contextual representation models cited in the related works section by a significant margin of about 10 points in F1 scores. Moreover, our approach is able to infer semantic relations that are not included in the pre-training dataset of the knowledge graph embeddings model, which is a significant advantage over the existing works. Our proposed approach focuses on leveraging the knowledge available in the training data and does not require pre-existing knowledge graphs, making it particularly useful for specialized domains with limited external resources. Based on our findings, pre-training the knowledge graph embeddings model enables capturing implicit knowledge related to entities, resulting in enriched contextual representations of semantic relations. These results provide a promising direction for future research in the field of semantic relation extraction.

References

1. Beltagy, I., Lo, K., Cohan, A.: Scibert: a pretrained language model for scientific text. In: Conference on Empirical Methods in Natural Language Processing (2019)
2. Bordes, A., Usunier, N., Garcia-Duran, A., Weston, J., Yakhnenko, O.: Translating embeddings for modeling multi-relational data. In: Burges, C., Bottou, L., Welling, M., Ghahramani, Z., Weinberger, K. (eds.) Advances in Neural Information Processing Systems, vol. 26. Curran Associates, Inc. (2013). https://proceedings.neurips.cc/paper/2013/file/1cecc7a77928ca8133fa24680a88d2f9-Paper.pdf
3. Bravo, À., González, J.P., Queralt-Rosinach, N., Rautschka, M., Furlong, L.I.: Extraction of relations between genes and diseases from text and large-scale data analysis: implications for translational research. BMC Bioinform. **16**, 55:1–55:17 (2015). https://doi.org/10.1186/s12859-015-0472-9
4. Devlin, J., Chang, M., Lee, K., Toutanova, K.: BERT: pre-training of deep bidirectional transformers for language understanding. In: Burstein, J., Doran, C., Solorio,

T. (eds.) Proceedings of the 2019 Conference of the North American Chapter of the Association for Computational Linguistics: Human Language Technologies, NAACL-HLT 2019, Minneapolis, MN, USA, 2–7 June 2019, Volume 1 (Long and Short Papers), pp. 4171–4186. Association for Computational Linguistics (2019). https://doi.org/10.18653/v1/n19-1423

5. Dong, Z., Dong, Q., Hao, C.: Hownet and its computation of meaning. In: International Conference on Computational Linguistics (2010)
6. Gu, Y., et al.: Domain-specific language model pretraining for biomedical natural language processing. ACM Trans. Comput. Healthcare **3**(1) (2021). https://doi.org/10.1145/3458754
7. Gutierrez, B.J., et al.: Thinking about GPT-3 in-context learning for biomedical IE? think again. arXiv abs/2203.08410 (2022)
8. Herrero-Zazo, M., Segura-Bedmar, I., Martínez, P., Declerck, T.: The DDI corpus: an annotated corpus with pharmacological substances and drug-drug interactions. J. Biomed. Inform. **46**(5), 914–920 (2013). https://doi.org/10.1016/j.jbi.2013.07.011. https://www.sciencedirect.com/science/article/pii/S1532046413001123
9. Huang, K., Altosaar, J., Ranganath, R.: Clinicalbert: modeling clinical notes and predicting hospital readmission. arXiv abs/1904.05342 (2019)
10. Joshi, M., Chen, D., Liu, Y., Weld, D.S., Zettlemoyer, L., Levy, O.: Span-BERT: improving pre-training by representing and predicting spans. Trans. Assoc. Comput. Linguist. **8**, 64–77 (2020). https://doi.org/10.1162/tacl_a_00300. https://aclanthology.org/2020.tacl-1.5
11. Karl, F., Scherp, A.: Transformers are short text classifiers: a study of inductive short text classifiers on benchmarks and real-world datasets (2022). https://doi.org/10.48550/ARXIV.2211.16878. https://arxiv.org/abs/2211.16878
12. Krallinger, M., Rabal, O., Akhondi, S.A., Pérez, M.P., Santamaría, J., Intxaurrondo, A., et al.: Overview of the biocreative VI chemical-protein interaction track (2017)
13. Lee, J., et al.: BioBERT: a pre-trained biomedical language representation model for biomedical text mining. Bioinformatics **36**(4), 1234–1240 (2019). https://doi.org/10.1093/bioinformatics/btz682
14. Lin, C., Miller, T., Dligach, D., Bethard, S., Savova, G.: EntityBERT: entity-centric masking strategy for model pretraining for the clinical domain. In: Proceedings of the 20th Workshop on Biomedical Language Processing, pp. 191–201. Association for Computational Linguistics, Online (2021). https://doi.org/10.18653/v1/2021.bionlp-1.21. https://aclanthology.org/2021.bionlp-1.21
15. Liu, W., et al.: K-bert: enabling language representation with knowledge graph. In: AAAI Conference on Artificial Intelligence (2019)
16. Ostendorff, M., Bourgonje, P., Berger, M., Schneider, J.M., Rehm, G., Gipp, B.: Enriching bert with knowledge graph embeddings for document classification. In: Proceedings of the 15th Conference on Natural Language Processing, KONVENS 2019, Erlangen, Germany, 9–11 October 2019 (2019). https://corpora.linguistik.uni-erlangen.de/data/konvens/proceedings/papers/germeval/Germeval_Task1_paper_3.pdf
17. Papaluca, A., Krefl, D., Suominen, H., Lenskiy, A.: Pretrained knowledge base embeddings for improved sentential relation extraction. In: Proceedings of the 60th Annual Meeting of the Association for Computational Linguistics: Student Research Workshop, Dublin, Ireland, pp. 373–382. Association for Computational Linguistics (2022). https://doi.org/10.18653/v1/2022.acl-srw.29. https://aclanthology.org/2022.acl-srw.29

18. Peng, Y., Yan, S., Lu, Z.: Transfer learning in biomedical natural language processing: an evaluation of BERT and ELMo on ten benchmarking datasets. In: Proceedings of the 18th BioNLP Workshop and Shared Task, Florence, Italy, pp. 58–65. Association for Computational Linguistics (2019). https://doi.org/10.18653/v1/W19-5006. https://aclanthology.org/W19-5006
19. Taylor, R., et al.: Galactica: a large language model for science (2022). https://doi.org/10.48550/ARXIV.2211.09085. https://arxiv.org/abs/2211.09085
20. Trouillon, T., Dance, C.R., Gaussier, É., Welbl, J., Riedel, S., Bouchard, G.: Knowledge graph completion via complex tensor factorization. J. Mach. Learn. Res. **18**, 130:1–130:38 (2017)
21. Vaswani, A., et al.: Attention is all you need. In: Guyon, I., et al. (eds.) Advances in Neural Information Processing Systems, vol. 30. Curran Associates, Inc. (2017). https://proceedings.neurips.cc/paper/2017/file/3f5ee243547dee91fbd053c1c4a845aa-Paper.pdf
22. Winata, G.I., Madotto, A., Lin, Z., Liu, R., Yosinski, J., Fung, P.: Language models are few-shot multilingual learners. In: Proceedings of the 1st Workshop on Multilingual Representation Learning, Punta Cana, Dominican Republic, pp. 1–15. Association for Computational Linguistics (2021). https://doi.org/10.18653/v1/2021.mrl-1.1. https://aclanthology.org/2021.mrl-1.1
23. Xu, B., et al.: CN-DBpedia: a never-ending Chinese knowledge extraction system. In: International Conference on Industrial, Engineering and Other Applications of Applied Intelligent Systems (2017)
24. Yang, B., tau Yih, W., He, X., Gao, J., Deng, L.: Embedding entities and relations for learning and inference in knowledge bases. CoRR abs/1412.6575 (2014)
25. Ye, Z., Jiang, G., Liu, Y., Li, Z., Yuan, J.: Document and word representations generated by graph convolutional network and bert for short text classification. In: European Conference on Artificial Intelligence (2020)
26. Zhang, Z., Han, X., Liu, Z., Jiang, X., Sun, M., Liu, Q.: ERNIE: enhanced language representation with informative entities. In: Proceedings of the 57th Annual Meeting of the Association for Computational Linguistics, Florence, Italy, pp. 1441–1451. Association for Computational Linguistics (2019). https://doi.org/10.18653/v1/P19-1139. https://aclanthology.org/P19-1139
27. Zhou, W., Chen, M.: An improved baseline for sentence-level relation extraction. In: Proceedings of the 2nd Conference of the Asia-Pacific Chapter of the Association for Computational Linguistics and the 12th International Joint Conference on Natural Language Processing (Volume 2: Short Papers), pp. 161–168. Association for Computational Linguistics, Online only (2022). https://aclanthology.org/2022.aacl-short.21

Extracting Key-Value Pairs in Business Documents

Eliott Thomas[1(✉)], Dipendra Sharma Kafle[1], Ibrahim Souleiman Mahamoud[1,2], Aurélie Joseph[2], Mickael Coustaty[1], and Vincent Poulain d'Andecy[2]

[1] Université de La Rochelle, L3i, Avenue Michel Crépeau, 17042 La Rochelle, France
{eliott.thomas,dipendra.kafle,ibrahim.mahamoud, mickael.coustaty}@univ-lr.fr

[2] Yooz, Immeuble le sequoia, Parc d'Andron, 30470 Aimargues, France
{ibrahim.mahamoud,aurelie.joseph,vincent.poulaindandecy}@getyooz.com

Abstract. Key-value extraction is a challenging task in document AI, particularly in business documents such as invoices. Accurately extracting key-value pairs from such documents is crucial for downstream tasks like accounting, analytics, and decision-making. In this paper, we propose a method for grouping and linking key-value pairs in business documents using a combination of rule-based methods and pretrained transformers. Our method comprises three steps: grouping words into word groups using rule-based methods that rely on layout information obtained via OCR, classifying the groups of words into key, value, or other using a pretrained BERT multilingual model, and linking the predicted key-value pairs based on their relative positions in the document, leveraging the layout information.

Furthermore, we present a comparative study of our approach against state-of-the-art methods for key-value extraction in business documents. Our approach is designed to be adaptable and requires minimal semantic and language-specific knowledge, making it suitable for a wide range of business documents. This flexibility allows our method to be easily applied to real-world scenarios, where documents may vary in layout and language. Additionally, our method is computationally efficient, allowing for fast processing of large volumes of documents. Our results demonstrate the effectiveness of combining rule-based methods with pretrained transformers for key-value extraction in business documents, providing a strong foundation for developing more accurate and robust methods in the future.

Keywords: Key-value extraction · Business documents · Pretrained transformers

1 Introduction

Businesses deal with vast amounts of information every day, ranging from invoices and receipts to purchase orders and contracts. Extracting key informa-

M. Coustaty and A. Fornés (Eds.): ICDAR 2023 Workshops, LNCS 14194, pp. 32–46, 2023.
https://doi.org/10.1007/978-3-031-41501-2_3

tion from these documents is crucial for efficient data processing and decision-making. Key-value extraction is a common technique used to identify and extract important information from business documents. The process involves identifying key entities (such as "date of document:", "name:", etc.) and their corresponding values (such as "12/07/1983", "John", etc.) within the document.

While key-value extraction is essential, it can be a challenging and time-consuming task, especially when dealing with large volumes of documents. Manual extraction is often error-prone, time-consuming, and expensive, which is why automated extraction methods are becoming increasingly popular. However, automated key-value extraction poses its own set of challenges, such as dealing with variations in document layouts and languages.

In this paper, we present a full pipeline for Machine-Learning-based (ML-based) key-value extraction from business documents. Our approach is composed of three main steps: Grouping, Semantic Entity Recognition, and Relation Extraction. You can see the structure of the pipeline in Fig. 1. We evaluate our method using the XFUND and FUNSD datasets, covering various types of form documents in different languages, as well as an unannotated private dataset for testing purposes. As globalization continues to drive international business, there is a growing need for natural language processing models that can effectively handle multiple languages and dialects, making multilingual models an increasingly important area of research. Our proposed method offers a comprehensive solution for key-value extraction in business documents, with competitive results in some languages and tasks. We also discuss the limitations and areas for improvement, such as incorporating semantic and contextual information to enhance the extraction process. Overall, our method offers a promising approach to automate key-value extraction and improve data processing efficiency in businesses.

In this paper, we provide an overview of the existing literature on key-value extraction from business documents (Sect. 2). We then describe the datasets used to evaluate our proposed method, including the XFUND [8] and FUNSD [4] datasets, as well as our own unannotated private dataset (Sect. 3). Our proposed method consists of three main steps: Grouping, Semantic Entity Recognition, and Relation Extraction, which we detail in Sect. 4. We present the results of our experiments and compare our approach to existing state-of-the-art methods in Sect. 5. Finally, we discuss the limitations of our method and suggest possible avenues for future research in Sect. 6.

2 Related Work

In recent years, there has been a growing interest in developing multilingual natural language processing models. One prominent example is Multilingual BERT (mBERT) [3], a pre-trained neural network that can handle multiple languages in a single model. The advantage of using mBERT is that it has already learned general features of language from a large corpus of text in multiple languages, making it more versatile and efficient than training separate models for each language. This is particularly useful in a business context where documents may contain multiple languages.

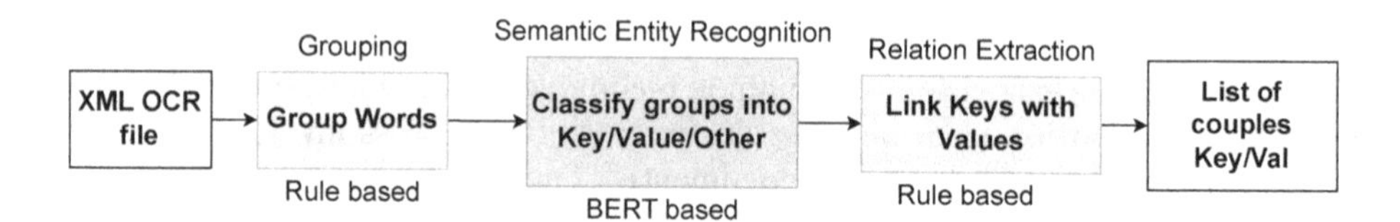

Fig. 1. Overview of the proposed pipeline for key-value extraction from business documents, including the three main steps of Grouping, Semantic Entity Recognition, and Relation Extraction

Prior research has explored the use of deep learning techniques for key-value extraction from semi-structured documents using only text, as demonstrated in previous work such as [6]. The paper proposes a deep learning-based pipeline for extracting information from semi-structured documents without considering the document layout. It defines "information" as entities with labels and values, forms relational triplets using relationships between entities, and achieves up to 96% test accuracy on real-world documents from the Ghega dataset [5]. The Ghega dataset consists of 136 patents and 110 datasheets documents, and while limited to a small set of pre-defined classes and only using text, not layout information, to extract key-value pairs, it serves as a valuable benchmark for the proposed method. It's important to note that the methodology and dataset used in this paper differ from those in the other three papers mentioned.

Recent works have shown that pretraining multilingual language models at scale can lead to significant performance improvements for cross-lingual transfer tasks. For instance, Facebook AI presented XLM-R [2], a Transformer-based masked language model pretrained on 100 languages with over 2TB of filtered CommonCrawl data, which outperforms mBERT [3] on various cross-lingual benchmarks. Notably, XLM-R performs particularly well on low-resource languages. The authors also provided a detailed empirical analysis of the factors that contribute to achieving these performance gains. This includes examining the trade-offs between positive transfer and capacity dilution, as well as studying the performance of high and low resource languages at scale.

Microsoft Corporation proposed an information-theoretic framework, INFOXLM [1], that formulates cross-lingual language model pre-training as maximizing mutual information between multilingual-multi-granularity texts. The framework provides a unified view of existing methods for learning cross-lingual representations and inspired the authors to propose a new pre-training task based on contrastive learning. They regard a bilingual sentence pair as two views of the same meaning and encourage their encoded representations to be more similar than the negative examples. Joint training on monolingual and parallel corpora is done to improve the cross-lingual transferability of pre-trained models. The experimental results on various benchmarks show that the approach achieves significantly better performance.

Recently, multimodal pre-training with text, layout, and image has achieved state-of-the-art performance for visually-rich document understanding tasks.

LayoutXLM [7] is a multimodal pre-trained model presented by the authors to address language barriers for visually-rich document understanding. To evaluate the model, the authors introduced the XFUND [8] dataset, which includes form understanding samples in 7 languages, with manually labeled key-value pairs for each language. Experimental results showed that LayoutXLM significantly outperformed existing state-of-the-art cross-lingual pre-trained models on the XFUND dataset.

These three pre-trained models, XLM-R, INFOXLM, and LayoutXLM, were used as baselines for comparison by the authors of LayoutXLM in their paper. We will use them as our own baselines.

3 Datasets

Many natural language processing (NLP) tasks, such as named entity recognition, sentiment analysis, and text classification, require annotated datasets to train and evaluate models. However, annotated datasets for business documents are often proprietary and not publicly available, which makes it difficult for researchers to develop and evaluate models for form analysis tasks. The FUNSD [4] dataset fills this gap by providing a publicly available dataset in english, for analysis of forms, which have layouts similar to those of business documents. With the increasing globalization of businesses, the ability to analyze forms in multiple languages is becoming increasingly important. The extension of the FUNSD dataset to the multilingual XFUND [8] dataset allows for the development of models that can effectively analyze forms in various languages, making it more suitable for real-world applications.

The XFUND dataset contains 199 forms (149 for training and 50 for testing) in each of the seven languages: Chinese, Japanese, Spanish, French, Italian, German, and Portuguese. Each form in the dataset consists of a single page and is annotated with information about the location and content of text blocks, tables, and other visual elements. The annotations include bounding boxes for individual words and groups of words, as well as labels for different field types such as questions, answers, headers, and others.

The annotation process of the XFUND dataset involved human annotators using the Microsoft Read API to generate OCR tokens with bounding boxes. An in-house GUI annotation tool was then used to group the tokens into entities and assign predefined labels to them. Key-value pairs were also annotated by linking related entities together. The annotations include bounding boxes for individual words and groups of words, as well as labels for different field types such as questions, answers, headers, and others.

The XFUND dataset is a valuable resource for researchers interested in developing and evaluating models for cross-lingual form analysis. Its annotations provide rich information about the content and layout of forms in different languages, and its small size and simplicity make it a convenient dataset for experimentation and development.

In the following sections of the paper, we will treat both the XFUND and FUNSD datasets as a single dataset, despite the difference in languages. This is because both datasets share the same document format and distribution of text blocks, tables, and other visual elements. Figure 2 shows a sample document from the FUNSD dataset, and the documents in the XFUND dataset follow the same structure. Therefore, we can leverage the annotations from both datasets to train and evaluate our model for form analysis tasks.

Fig. 2. Visualization of the FUNSD dataset, showing the distribution of the four classes. Purple represents 'Other', orange represents 'Question', blue represents 'Header', and green represents 'Answer'. (Color figure online)

Mode de paiement : cheque
Da'ⁿ d'échéance : 03/05/2019

Référence	Description	Quantité	Prix unitaire	Total
	Objet;Intervention sur évacuation des eaux usées du bâtiment C.			
AR0001	Main d'oeuvre Forfait intervention et déplacement ; désengorgement de l'évacuation principale du réseau d'évacuation des eaux usées.	1,00	555,00	555,00

Fig. 3. Visualization of the 20k dataset, showing the distribution of the four classes. Purple represents 'Other', orange represents 'Question', blue represents 'Header', and green represents 'Answer'. (Color figure online)

In addition to the public XFUND dataset, which is annotated and publicly available, we utilize a private collection of business documents for testing purposes. This industrial dataset includes invoices, purchase orders, and delivery notes in French and English, obtained from various channels within our company. While not annotated for our specific task, this diverse dataset of 10 different types of invoices, such as orders, quotations, and statements, offers a valuable resource for manual testing and evaluation of our key-value extraction system. The documents in this dataset exhibit a wide range of layouts and structures, reflecting the varied nature of the business documents our system processes.

Since the dataset contains about 20k documents, we will call it 20k in the rest of the paper. Figure 3 shows a sample document from the 20k dataset.

4 Methodology

As explained in the introduction before, the methodology for extracting key-value couples is made in three steps: Grouping, Semantic Entity Recognition and Relation Extraction. To the best of our knowledge, no research has adapted key-value Extraction to invoices until now.

4.1 Grouping

Accurately grouping words is crucial for automated key-value pair extraction from business documents, but the diverse and complex layouts of these documents can pose significant challenges for rule-based methods. Invoices, receipts, and purchase orders are just a few examples of business documents that often have multiple columns, tables, and non-uniform font sizes. This can make it difficult to accurately group words together and extract meaningful information.

One common challenge is splitting words belonging to the same key-value pair across different columns or tables. In some cases, the key and value may be separated by multiple lines, making it challenging to group them together. Noise in the document, such as logos, images, and headers/footers, can also disrupt the grouping process and lead to incorrect groupings.

The grouping strategy is flexible enough to handle different document layouts and formats, allowing for accurate and efficient extraction of key-value pairs. In the case of the XFUND dataset, the words are already grouped, but for most datasets, grouping the words is a critical step to ensure accurate extraction.

By overcoming the challenges posed by complex and diverse layouts in business documents, automated key-value pair extraction can significantly reduce the time and effort required for data entry and analysis.

Optical Character Recognition (OCR), is a technology that converts scanned images or handwritten text into machine-encoded text. The output of an OCR system typically includes a hierarchy of elements such as blocks, lines, words, and characters. *Blocks* are the highest level of organization and represent areas of text that are visually distinct from other areas on the page, such as tables, images, or headers. *Lines* are the next level down and represent a sequence of words that appear horizontally in the same line. *Words* are the smallest unit of text and represent a group of characters that are separated by spaces or punctuation. Finally, characters are individual units of text that make up words. The OCR output provides the necessary text for automated key-value extraction from business documents. Here are the main steps used to group the words in a document, in order:

1. Join the words if they are on the same *Line*. We will see in Sect. 5 why it was the preferred base for our grouping

2. If a word has 0 confidence (according to the OCR) we ignore it
3. If 2 words are too far away, we split the line. Too far meaning further than the longest word of the line
4. We then split the lines when there is a colon. This anticipates demarcation between keys and values
5. We finally remove useless remaining lines: no text, or text with mostly special characters (>50%)

There are 2 examples in Fig. 4, with the top ones being raw OCR and the bottom ones after processing.

(a) Split on the colon

(b) Split when too far

Fig. 4. Before-and-after illustration of the entity extraction process using the line (top) and processed (bottom) methods

4.2 Semantic Entity Recognition

Semantic Entity Recognition (SER) is a crucial step in the key-value extraction process, as it involves identifying and classifying the different components of a document into *key*, *value*, *header*, and *other* categories. In our case, we are using the XFUND dataset to train our model on this task. To ensure a fair comparison with the baseline system, we evaluate our approach against the XFUND dataset with the original four classes. This allows us to compare the performance of our model to the baseline using the same evaluation metrics. However, in practical usage, our goal is to accurately extract key-value pairs from forms, and we are less interested in the headers and other classes. Therefore, when using the complete pipeline, we group the *header* and *other* classes together and focus solely on the key and value classes, which allows us to simplify the problem into a three-class classification task. Another reason is the very rare presence of the class *header* as you can see in Table 1. There are then 2 scenarios, 4 class output for the comparison and 3 class output for the real world usage.

We use a BERT multilingual model [3] to perform the classification task, which has been pre-trained on large amounts of text data in different languages. To further enhance the performance of our model, we apply various preprocessing techniques such as:

- Lowercasing the text
- Replacing some of the words by a corresponding token. For example, emails are replaced by *$EMAIL$*. We also replace phone numbers, times, dates, numbers, and geopolitical entities (GPEs) with their corresponding tokens.

Overall, these steps ensure that our SER model is accurate and robust enough to handle the complex and varied layouts of different types of business documents.

Table 1. Distribution of the classes in the XFUND dataset (train and val)

Class	Support
Answer	0.39
Other	0.31
Question	0.28
Header	0.02

The BERT model we used in our SER system is pre-trained on a large corpus of text, which allows it to learn general features of language. This provides a major advantage for our task, as it can improve performance even with limited training data. This is especially useful for business documents, where data availability can be a challenge. To make the classification decision, we used the CLS token output. The CLS token is a special token added to the beginning of the input sentence in BERT models. The output of the CLS token represents the entire input sequence and is used as a vector representation of the sentence to feed into a feed-forward neural network. The hidden layer size of this network was 768, the learning rate of 2e-5 and we used the Adam optimizer for training. We also used a weighted cross-entropy loss function, based on the class distribution of the train set, to handle class imbalance and give more importance to minority classes during training. You can see that the minority class here is clearly the *header* class, as shown in Table 1.

For the language of our SER model, we opted for BERT-multilingual, which can handle multiple languages in a single model. This choice was made to ensure the model's versatility for handling different languages in the future. However, we still trained a separate model per language to obtain the best performance. One alternative approach would have been to train the model on all languages together and test it on all languages, but this would have resulted in longer training times and potentially less accurate results.

4.3 Relation Extraction

In the linking step of our methodology, we aim to connect the predicted keys with their corresponding predicted values. This process is based on the output of the Semantic Entity Recognition step where we classify each group of tokens into one of the three classes: key, value or other. It is important to note that in our problem setting, the questions serve as keys and the answers as values.

To link a predicted key with its corresponding value, we utilize the layout information in the form of OCR coordinates of each group of words. Specifically, we use a distance-based approach, where we calculate the distance between the bounding boxes of each predicted key-value pair. We take the minimum distance between the two bounding boxes with the two closest corners or the distance between the two centroids. We also take into account the alignment of the key-value pair, considering the angle between the boxes to ensure that they are horizontally or vertically aligned. The angle is calculated as the angle between the line connecting the two bounding boxes and the horizontal line. You can see an example of the of distance and alignment, from the french part of XFUND, in Fig. 5.

The key-value pair with the highest score is then selected, resulting in an accurate linking of the pairs and the desired structured output. We experimented with different settings to optimize the scoring mechanism, including exploring different distance metrics and weightings for the angle metric. The details of these experiments and their corresponding results will be presented in Sect. 5.

Although the linking process is effective in connecting predicted keys with their corresponding values based on layout information, it should be noted that this method only considers the physical location of key-value pairs and does not take into account their semantic meaning. While this approach works well for most documents, it may face limitations in more complex settings where unstructured data or non-standard layouts are present. In such cases, additional methods that incorporate semantic understanding may be necessary to improve the accuracy and reliability of the extraction process. Nevertheless, our proposed method is still efficient and effective, and provides a viable solution for structured data extraction from various document types. Additionally, it has the advantage of being faster than semantic-based methods.

(a) Minimum distance between 2 bounding boxes

(b) Angle between 2 bounding boxes

Fig. 5. This figure shows the minimum distance (in pixels) and angle (in degrees) between two bounding boxes for entity extraction. The specific bounding boxes shown here are [175, 673, 490, 716] and [587, 703, 726, 744], with a distance of 98 pixels and an angle of 5°.

5 Results

5.1 Grouping

In this subsection, we evaluate the performance of the grouping task on two datasets: the original XFUND dataset and XFUND_OCR, which we obtained by applying OCR extraction to the images in the original dataset. The XFUND dataset provides annotated data for cross-lingual document analysis of business forms, where the words are already grouped into entities based on their spatial proximity. Specifically, we compare the entities present in the original XFUND dataset with those obtained from XFUND_OCR followed by a grouping process.

After loading the XFUND_OCR dataset, we must choose a first step to join the words together: *Blocks*, *Lines* or *Words* ? The experimental protocol goes as follows:

1. Load the XFUND and the XFUND_OCR datasets
2. Form the entities (group the texts and bounding boxes together) based on the level of extraction chosen: *Blocks*, *Lines* or *Words*
3. For each document in the original dataset, compare the entities of the corresponding document in XFUND_OCR based on their bounding boxes
4. We iterate over all the entities of the ground truth. If we can find an entity in the OCR output where the IOU is higher than 0.55, we consider the ground truth entity detected correctly by the OCR method
5. Calculate the precision, recall, and F1-score
6. Compare the performance of the grouping obtained from each level of extraction

$$IOU = \frac{Area\_bbox\_ground\_Truth \cap Area\_bbox\_predicted}{Area\_bbox\_ground\_Truth \cup Area\_bbox\_predicted} \quad (1)$$

Table 2. Proportion of the entities correctly grouped together (all languages)

Level	Answer	Question	Header	Other
Block	0,47	0,56	0,67	0,38
Line	**0,60**	**0,60**	**0,97**	**0,77**
Word	0,28	0,23	0,07	0,12

We can see in Table 2 that the extraction based on *Lines* is consistently better than the one based on *Blocks* or *Words*, so we keep that as a base.

Now that we decided to base the entities on the *Lines* extracted by the OCR, we need to measure how good are the new modifications (presented in Sect. 4) compared to the lines unmodified. The experimental protocol is the same, except step 2 where we can add the modifications. We try with and without the modifications, and we try in different languages.

Table 3. F1 score showing the good correspondence between entities

Option/F1 Score	FR	ES	DE
(a) Line	0.67	0.74	**0.68**
(b) Line & modifications	**0.68**	**0.76**	**0.68**

The results in Table 3 show that using the OCR output modified by the grouping algorithm improves F1-score slightly: precision similar but recall is consistently higher due to the numerous splits, which enable the detection of more entities grouped together in a single line. While the difference in performance between the two methods is not substantial, it is worth noting that the steps involved in this process are crucial when applying it to a business document, which may have more complex layouts and structures than a simple form document (notably tables).

5.2 Semantic Entity Recognition

In this section, we present the experimental protocol we followed to evaluate the performance of our proposed approach for SER on the XFUND and FUNSD datasets in multiple languages. We then present the results against the baselines. Our experimental protocol consists of the following steps:

1. Load the train and evaluation sets of the XFUND and FUNSD datasets
2. Split the datasets into training and evaluation according to the original split of the authors
3. Preprocess the text according the methodology
4. Train the transformer-based model on the training sets using the input sequences and output labels
5. Evaluate the trained model on the evaluation sets and record the performance metrics, such as precision, recall, and F1 score
6. Repeat steps 2 to 5 for all languages in the datasets (EN, FR, DE, ES)
7. Compare the performance of the models trained (per language) to the baselines

We evaluated our method against three baselines (LayoutXLM [7], XLM-RoBERTa [2], InfoXLM [1]) on four languages (English, French, German, Spanish). Across all languages, our method consistently outperforms XLM-RoBERTa and InfoXLM in terms of F1-score, as shown in Table 4. In addition, our method outperforms LayoutXLM in French, achieving the highest scores among all four languages. These results demonstrate the effectiveness of our method for semantic entity recognition in multilingual business documents.

Regarding the superior performance of our method in French, one possible explanation is that our private dataset 20k was composed mostly of French documents, which may have influenced the preprocessing steps or other aspects of the method, leading to improved performance on French text. Further investigation is needed to confirm this hypothesis.

It is worth noting that the evaluation was carried out using Language-specific Fine-tuning, meaning that the model was trained on each language separately and then tested on each language separately. This approach differs from Multitask Fine-tuning, where one trains the model on all languages at once and then tests it on separate languages. Although we did not try Multitask Fine-tuning in this study due to the length of training and the risk of losing performance, it could be interesting to explore this approach in future research.

Table 4. F1 scores of OUR method against the baselines in 4 different languages

Language	OUR	xlm-roberta-base	infoxlm-base	layoutxlm-base
EN	*0.77*	0.67	0.69	**0.79**
FR	**0.81**	0.67	0.70	*0.79*
DE	**0.82**	0.68	0.70	**0.82**
ES	*0.72*	0.61	0.62	**0.76**

5.3 Relation Extraction

In addition to evaluating our method for semantic entity recognition, we also assess its performance for relation extraction, which involves identifying and linking entities in a document. For this task, we use a rule-based method that relies solely on the layout information of the documents. Specifically, we experiment with different distance and alignment measures to determine the most effective approach for our method. In the following section, we present the protocol for our relation extraction experiments and compare the performance of our method with three baselines on the same four languages. Here is the experimental protocol we followed for this part of the process:

1. We use the FUNSD and XFUND datasets (from the eval split) as our test sets for relation extraction
2. We do not perform any training as this is a zero-shot relation extraction task
3. We evaluate our method based on the accuracy of the predicted links
4. We experiment with different distance and alignment measures to determine the best approach for our method
5. We compare our method against the three same baselines as before: LayoutXLM, InfoXLM, and XLM-RoBERTa
6. We compare the performance of our method and baselines on the same four languages as before: English, French, German, and Spanish

We compared two distance metrics for our rule-based method, **DIST-C** and **DIST-M**. **DIST-C** is the distance between the centroids of the bounding boxes of two entities, while **DIST-M** is the minimum distance between the bounding boxes. Additionally, we experimented with three alignment measures: no angle (**NO-ANG**), the angle (**ANG**) and the squared angle (**ANG**^{2}). We used six combinations of these metrics that you can find in Table 5.

We compare the F1 scores obtained by different distance and alignment methods for relation extraction. The distances are measured in pixels and the angles in degrees. Our results showed that **DIST-M** consistently outperformed **DIST-C**, and that incorporating the angle (squared or not) improved performance compared to using only the distance metric. Overall, we found that **DIST-M+ANG** provided the best performance among the six combinations, and we used this approach to compare against the three baselines.

Table 5. Comparison of F1 scores for relation extraction using different distance and alignment methods.

Language	DIST-C			DIST-M		
	NO-ANG	ANG	ANG^2	NO-ANG	ANG	ANG^2
FR	0.48	0.50	0.66	0.60	0.64	**0.74**
EN	0.39	0.45	0.52	0.52	**0.58**	0.56
DE	0.40	0.42	0.45	0.47	**0.50**	0.48
ES	0.49	0.51	0.55	0.58	**0.59**	**0.59**

We compare the performance of our proposed method against the baseline approaches on the XFUND and FUNSD datasets in Table 6. Overall, our rule-based method achieved competitive performance compared to the three baselines across the four languages. Notably, we achieved the highest accuracy in English and were as good as LayoutXLM in French. While our method was second in Spanish and equal to last with XLM-RoBERTa in German, it's important to note that our method is a fast and simple approach that relies only on layout information. Our results suggest that, with further development and integration of semantic information, our approach could form the basis of a more powerful system for relation extraction.

Table 6. F1 scores comparison between our proposed method and baseline approaches on the XFUND and FUNSD datasets.

LANG	OUR	xlm-roberta-base	infoxlm-base	layoutxlm-base
FR	**0.64**	0.50	0.49	**0.64**
EN	**0.58**	0.27	0.29	0.55
DE	0.50	0.50	0.53	**0.66**
ES	0.59	0.53	0.55	**0.69**

6 Conclusion

This paper presents a full pipeline for key value extraction from business documents. The pipeline is composed of three main steps: Grouping, Semantic Entity Recognition and Relation Extraction. To develop and evaluate our approach, we used the XFUND and FUNSD datasets which covers a variety of forms documents in different languages. We also used an unannotated private dataset, mostly for testing purposes in all 3 tasks. Our approach achieves competitive results in some languages and tasks, such as beating state-of-the-art LayoutXLM in French for Semantic Entity Recognition. Our results for Grouping are particularly noteworthy as this task has received little attention in the existing literature, making our approach a novel contribution in this area. Despite the lack of prior research, our method was able to achieve reasonably good results, demonstrating its potential for extracting key information from business documents.

Overall, our proposed method provides a comprehensive solution for key-value extraction in business documents that outperforms existing methods in some scenarios.

While our proposed method shows promise, there are still areas for improvement and limitations to consider. One such limitation is that Grouping and Relation Extraction rely heavily on layout information, which could lead to difficulties with unseen document layouts. To improve the results on these tasks, incorporating semantic and contextual information could be beneficial. In contrast, the Semantic Entity Recognition task may benefit from using layout information since text alone is not enough. Addressing these limitations and making improvements to our proposed method will help advance the field of key-value extraction even further.

Our proposed method provides a promising approach for key-value extraction from business documents, with competitive results in some languages and tasks. The method has limitations and areas for improvement, such as incorporating semantic and contextual information. Future research could explore these directions to further improve key-value extraction in business documents and enable efficient data extraction from a wider range of sources.

Acknowledgment. This work was supported by the French government in the framework of the France Relance program and by the YOOZ company under the grant number ANR-21-PRRD-0010-01. We also would like to thank Guénael Manic, Mohamed Saadi, Jonathan Ouellet and Jérôme Lacour from YOOZ for their support.

References

1. Chi, Z., et al.: InfoXLM: an information-theoretic framework for cross-lingual language model pre-training. In: Proceedings of the 2021 Conference of the North American Chapter of the Association for Computational Linguistics: Human Language Technologies, pp. 3576–3588. Association for Computational Linguistics, Online (2021). https://doi.org/10.18653/v1/2021.naacl-main.280. https://aclanthology.org/2021.naacl-main.280
2. Conneau, A., et al.: Unsupervised cross-lingual representation learning at scale. CoRR abs/1911.02116 (2019). http://arxiv.org/abs/1911.02116
3. Devlin, J., Chang, M., Lee, K., Toutanova, K.: BERT: pre-training of deep bidirectional transformers for language understanding. CoRR abs/1810.04805 (2018). http://arxiv.org/abs/1810.04805
4. Jaume, G., Ekenel, H.K., Thiran, J.P.: FUNSD: a dataset for form understanding in noisy scanned documents. In: Accepted to ICDAR-OST (2019)
5. Medvet, E., Bartoli, A., Davanzo, G.: A probabilistic approach to printed document understanding. Int. J. Doc. Anal. Recogn. (IJDAR) **14**(4), 335–347 (2011). https://doi.org/10.1007/s10032-010-0137-1
6. Shehzad, K., Ul-Hasan, A., Malik, M.I., Shafait, F.: Named entity recognition in semi structured documents using neural tensor networks. In: Bai, X., Karatzas, D., Lopresti, D. (eds.) DAS 2020. LNCS, vol. 12116, pp. 398–409. Springer, Cham (2020). https://doi.org/10.1007/978-3-030-57058-3_28

7. Xu, Y., et al.: LayoutXLM: multimodal pre-training for multilingual visually-rich document understanding (2021)
8. Xu, Y., et al.: XFUND: a benchmark dataset for multilingual visually rich form understanding. In: Findings of the Association for Computational Linguistics: ACL 2022, Dublin, Ireland, pp. 3214–3224. Association for Computational Linguistics (2022). https://doi.org/10.18653/v1/2022.findings-acl.253. https://aclanthology.org/2022.findings-acl.253

Long-Range Transformer Architectures for Document Understanding

Thibault Douzon[1,2(✉)], Stefan Duffner[1], Christophe Garcia[1], and Jérémy Espinas[2]

[1] INSA Lyon, LIRIS, Lyon, France
{thibault.douzon,stefan.duffner,christophe.garcia}@insa-lyon.fr
[2] Esker, Lyon, France
{thibault.douzon,jeremy.espinas}@esker.com

Abstract. Since their release, Transformers have revolutionized many fields from Natural Language Understanding to Computer Vision. Document Understanding (DU) was not left behind with first Transformer based models for DU dating from late 2019. However, the computational complexity of the self-attention operation limits their capabilities to small sequences. In this paper we explore multiple strategies to apply Transformer based models to long multi-page documents. We introduce 2 new multi-modal (text + layout) long-range models for DU. They are based on efficient implementations of Transformers for long sequences. Long-range models can process whole documents at once effectively and are less impaired by the document's length. We compare them to LayoutLM, a classical Transformer adapted for DU and pre-trained on millions of documents. We further propose 2D relative attention bias to guide self-attention towards relevant tokens without harming model efficiency. We observe improvements on multi-page business documents on Information Retrieval for a small performance cost on smaller sequences. Relative 2D attention revealed to be effective on dense text for both normal and long-range models.

Keywords: Document Understanding · Long-range Transformers · Relative Attention

1 Introduction

Digital documents are everywhere around us, in the form of born digital PDF or scanned paper, they carry much information and can be easily exchanged. They can be used to exchange information from an issuer to its recipient or to archive its content. Information is generally structured in some way depending on the document type. Invoices and scientific articles, for example, do not follow the same structure because their objective is different. Both are codified to carry information very efficiently such that most invoices and most articles look the same but differ in content. Document Understanding is a field gaining increasing attention in the past years, as automating document-related processes can

M. Coustaty and A. Fornés (Eds.): ICDAR 2023 Workshops, LNCS 14194, pp. 47–64, 2023.
https://doi.org/10.1007/978-3-031-41501-2_4

drastically improve efficiency of information processing in a wide range of fields and industries. Recent advances in Neural Network architectures allowed better document understanding and enabled tackling more complex tasks: Question Answering [20], Layout Segmentation [18] and Information Extraction [11].

In particular, models based on Transformer architectures have led to a breakthrough in these domains since their first release in 2017 [28]. They have been widely used on Natural Language Processing tasks and their performance is unequaled [29]. The implementation of Transformer models in DU [32] was swift, revealing the potential of attention-based models in this domain. More recently, the trend is towards multi-modal models that combine multiple different information representations such as text, image, sound, layout. Those models have shown great results on short, single page documents, but are difficult to apply to long, multi-page or dense documents. This is because the self-attention time and space computational complexity is $O(N^2)$ where N is the length of the sequence. It effectively limits the usage of Transformer models on long sequences due to either long training or lack of GPU memory.

In this work, we explore several approaches and architectures in order to use Transformer models on long documents. For simplicity, we limit our study to text and layout (i.e., text position in the page) modalities, and chose to focus on document length to evaluate the model efficiency. We compare various encoder-only models on Sequence Tagging tasks with business and academic documents. We also study the impact of relative attention based on document layout instead of a linear token position, and its implementation for long-range Transformers.

2 Related Work

2.1 From NLP to Document Understanding

This work derives from both long-range Transformers proposed in NLP tasks, trying to process longer sequences at once and Transformer architectures adapted to DU. Before the proposal of Transformers, the *de facto* architecture for NLP has been Recurrent Neural Networks. Multiple improvements have been proposed, for example to tackle vanishing gradients like Long-Short Term Memory cells [9]. Coupled with Conditional Random Fields, bidirectional LSTM encoders were then capable at most text understanding task [16]. For more complex Information Retrieval, where target information can span multiple tokens, `BIESO` tags allow better decoding by precisely locating the beginning and end of the information. Although long sequences can be processed with Recurrent Neural Networks, longer input negatively affects the performance of encoder-decoder architectures [2]. Hence, the attention mechanism was quickly adopted for those architectures as an "information highway" between the encoder and the decoder.

In addition to these new architecture developments, large progress has been made in the past years on how to learn usable word representations. Before, word embeddings were trained at the same time as the other model's parameters. Then, approaches like Word2Vec and GloVe [21,22] showed that self-supervised learning improves finetuning on all tasks. Major improvements came from contextual embeddings, first introduced by Elmo [23]. Contrary to static embeddings, contextual embeddings can better represent words with multiple meanings in adequation with their surroundings.

This is where Transformer models rose, heavily relying on (self-)attention and pre-training giving unprecedented performance at the time. Most NLP challenges leaderboards were monopolized by BERT like models, growing bigger and deeper by the day [4,5,19].

In parallel to those quick improvements, the DU community developed alternatives to bi-LSTM, using multiple modalities to provide more useful information to the model. Some used convolutions over a grid mixing image and text [6,12], others proposed graph-based models [33] to represent a document.

The revolution in DU came from Transformer architectures. Pre-trained models able to leverage large document collections outperformed all previous approaches. LayoutLM [32], for example, only introduced 2D positional embeddings over BERT and was pre-trained on the RVL-CDIP [17] collection. It opened the way to many other models applying Transformers to previous design [6], leveraging end-to-end capacities of encoder-decoder models [24], or providing image and text to the models like a visual Transformer [15]. Because the Transformer output is independent of the sequence order, positional embeddings are classically added to the input. It is also possible to introduce relative bias to the self-attention mechanism to promote local interactions inside the self-attention.

Most recent models for DU propose to leverage as much information as possible by using multiple modalities: text, layout and image. Either by combining Convolutional Neural Networks with Transformers [24,31] or mixing visual with classical Transformers [10,13]. Even though those approches provide superior results, we chose to not include image information to our architectures.

2.2 Long Range Transformers

Since the introduction of BERT [7] and GPT [26], Transformers have demonstrated their capacity to understand and model language [29]. Their ability to manipulate words can be visualised through the amount of attention each token allows to other tokens. However, dot-product attention computation involves a $O(N^2)$ time and memory complexity where N is the sequence length. It limits the capacity of Transformer-based models in dealing with long sequences as they need too much GPU memory and/or take too long to process.

Many modifications have been proposed to replace the attention layer with some efficient approximation that can be computed in $O(N)$ or $O(N \log(N))$. They have been developed and tested with NLP tasks where long sequences are most likely to be found like long text summarization and translation. Some models use attention patterns [1,3,34] to limit token attention to a fixed number of other tokens. Some combination of sliding window, global and random patterns provide a simple but efficient attention. A balance needs to be found between more attention context and attention complexity. It is also possible to learn the attention pattern by limiting attention to tokens that share some locality sensitive hash [14]. Others proposed to replace the $N \times N$ attention matrix with a low-rank approximation. Empirical observations on multiple NLP tasks show that the attention matrix can be replaced with a lower rank approximation without harming the attention process too much [30].

However, long range Transformer architectures have not yet been used on DU tasks, mostly due to datasets not containing lengthy documents.

3 Datasets

We used 2 document datasets, where our choice was mainly made based on document length and the task itself. We wanted a NLP task that can be represented as Sequence Tagging in order to test the whole encoder with long inputs. Both datasets consist of English-only documents with close to perfect OCR extraction. They provide word-level axis aligned bounding boxes in the form that can be fed to the model as layout information. We use the OCR provided order for the input sequence and do not further analyze documents to extract their structure.

3.1 Business Documents

The first dataset consists of Customer Orders submitted to a commercial platform between 2018 and 2021. Due to privacy concerns, these documents cannot be shared. It contains 80k documents that can be divided in 9000 different issuers with no more than 50 documents from the same issuer. Usually, an issuer only emits documents with the same template for convenience. About 55% of documents can be tokenized into a sequence of 512 tokens which fit into classical Transformer default maximum length. Only 5% of documents are longer than 2048 tokens, following a long tail of distribution. In order to evaluate the models' generalization abilities, we split into train, validation and test sets such that templates in the test set have not been seen by the model during training.

The task consists of Information Extraction on multiple known classes: *document number*, *date*, *total amount*, *item ID numbers* and *item quantities*. Some information only appears once in the document (e.g., *document number*, *date* and *total amount*) while others are repeated for each line item in the business order. We call *header* fields those only occurring once and *table* fields others as they are most of the time structured in a table layout. There could be between 1 and 50 items present in any document, their number is not known in advance. Figure 1 shows the labelling of a multi-page document. Even though header field are sometimes repeated on each page, it is only labeled once in order to stay consistent acrosstemplates. Labels are provided at the word level based on manual customer document extraction. We also controlled labelling quality and rejected from the dataset documents with missing mandatory fields or wrong number of line items.

A superset of this dataset was used for pre-training models on business documents. It consists of 300k Customer Orders and 100k Invoices from the same

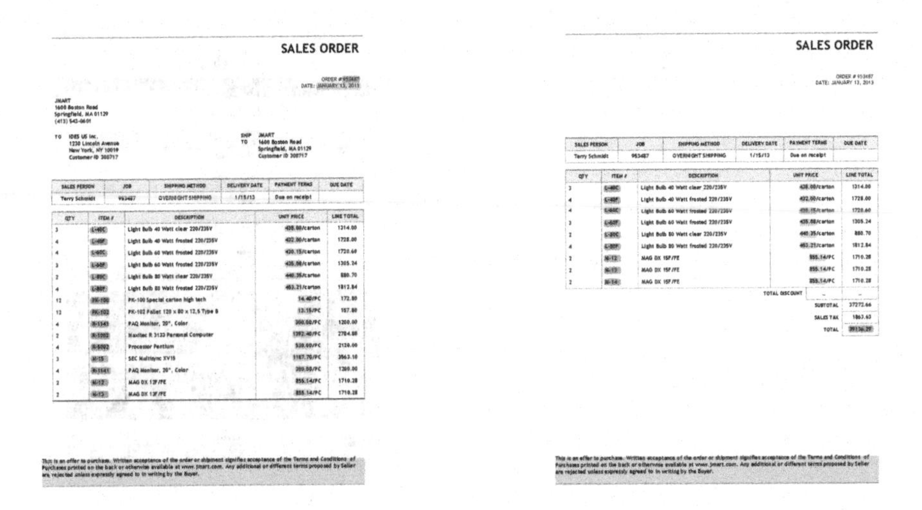

SALES ORDER

ORDER # 953487
DATE: JANUARY 13, 2013

JMART
1600 Boston Road
Springfield, MA 01129
(413) 543-0601

TO IDES US Inc.
1230 Lincoln Avenue
New York, NY 10019
Customer ID 300717

SHIP TO JMART
1600 Boston Road
Springfield, MA 01129
Customer ID 300717

SALES PERSON	JOB	SHIPPING METHOD	DELIVERY DATE	PAYMENT TERMS	DUE DATE
Terry Schmidt	953487	OVERNIGHT SHIPPING	1/15/13	Due on receipt	

QTY	ITEM #	DESCRIPTION	UNIT PRICE	LINE TOTAL
3	L-40C	Light Bulb 40 Watt clear 220/235V	438.00/carton	1314.00
4	L-40F	Light Bulb 40 Watt frosted 220/235V	432.00/carton	1728.00
4	L-60C	Light Bulb 60 Watt frosted 220/235V	430.15/carton	1720.60
3	L-60F	Light Bulb 60 Watt frosted 220/235V	435.08/carton	1305.24
2	L-80C	Light Bulb 80 Watt clear 220/235V	440.35/carton	880.70
4	L-80F	Light Bulb 80 Watt frosted 220/235V	453.21/carton	1812.84
12	PK-100	PK-100 Special carton high tech	14.40/PC	172.80
12	PK-102	PK-102 Pallet 120 x 80 x 12,5 Type B	13.15/PC	157.80
4	R-1343	PAQ Monitor, 20", Color	300.00/PC	1200.00
2	R-1002	Maxitec R 3133 Personal Computer	1392.40/PC	2784.80
4	R-5002	Processor Pentium	530.00/PC	2120.00
3	M-15	SEC Multisync XV15	1187.70/PC	3563.10
4	R-1341	PAQ Monitor, 20", Color	300.00/PC	1200.00
2	M-12	MAG DX 13F/FE	855.14/PC	1710.28
2	M-13	MAG DX 13F/FE	855.14/PC	1710.28

This is an offer to purchase. Written acceptance of the order or shipment signifies acceptance of the Terms and Conditions of Purchases printed on the back or otherwise available at www.jmart.com. Any additional or different terms proposed by Seller are rejected unless expressly agreed to in writing by the Buyer.

SALES ORDER

ORDER # 953487
DATE: JANUARY 13, 2013

SALES PERSON	JOB	SHIPPING METHOD	DELIVERY DATE	PAYMENT TERMS	DUE DATE
Terry Schmidt	953487	OVERNIGHT SHIPPING	1/15/13	Due on receipt	

QTY	ITEM #	DESCRIPTION	UNIT PRICE	LINE TOTAL
3	L-40C	Light Bulb 40 Watt clear 220/235V	438.00/carton	1314.00
4	L-40F	Light Bulb 40 Watt frosted 220/235V	432.00/carton	1728.00
4	L-60C	Light Bulb 60 Watt frosted 220/235V	430.15/carton	1720.60
3	L-60F	Light Bulb 60 Watt frosted 220/235V	435.08/carton	1305.24
2	L-80C	Light Bulb 80 Watt clear 220/235V	440.35/carton	880.70
4	L-80F	Light Bulb 80 Watt frosted 220/235V	453.21/carton	1812.84
2	M-12	MAG DX 15F/FE	855.14/PC	1710.28
2	M-13	MAG DX 15F/FE	855.14/PC	1710.28
2	M-14	MAG DX 15F/FE	855.14/PC	1710.28
			TOTAL DISCOUNT –	–
			SUBTOTAL	37272.66
			SALES TAX	1863.63
			TOTAL	39136.29

This is an offer to purchase. Written acceptance of the order or shipment signifies acceptance of the Terms and Conditions of Purchases printed on the back or otherwise available at www.jmart.com. Any additional or different terms proposed by Seller are rejected unless expressly agreed to in writing by the Buyer.

Fig. 1. Sample pages with colored labels similar to those in the Business Documents dataset. Both pages come from the same document, the first page in on the left and the last page on the right. Some information are repeated across pages of a document. (Color figure online)

commercial platform. All documents were submitted and processed by the platform but later rejected due to labelling errors or bad habits. Fortunately, this does not impact the OCR quality and allows us to pre-train our models on a large collection of recent documents. We chose to use it for pre-training instead of RVL-CDIP [8] for the OCR quality difference.

3.2 DocBank

DocBank [18] is a dataset containing 500k public research article pages. It contains English documents spanning various research fields. Documents were obtained on arXiv and were annotated with PDFPlumber, a PDF parser that accurately extracts item bounding boxes. The task consists in document layout analysis. Li et al. [18] provide both pixel and word-level annotations for CV and NLP models. The order of words is defined from top-to-bottom and left-to-right, except for multicolumn documents where whole columns are ordered left-to-right. In this work we will only use textual information along the word 2D positions.

Docbank segmentation task contains 12 categories (e.g. *title*, *paragraph*, *figure* etc.) representing semantic parts of a research article. Because articles contain dense paragraphs, most pages are longer than 512 tokens once tokenized. In fact only 11% of the test documents contains less than 512 tokens and 84% contains between 512 and 2048 tokens (Fig. 2).

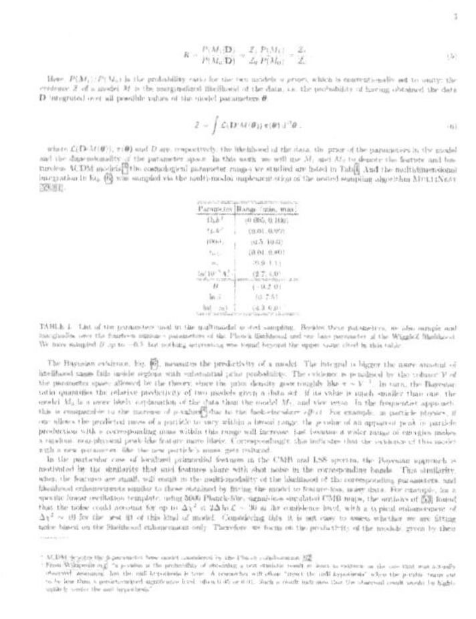

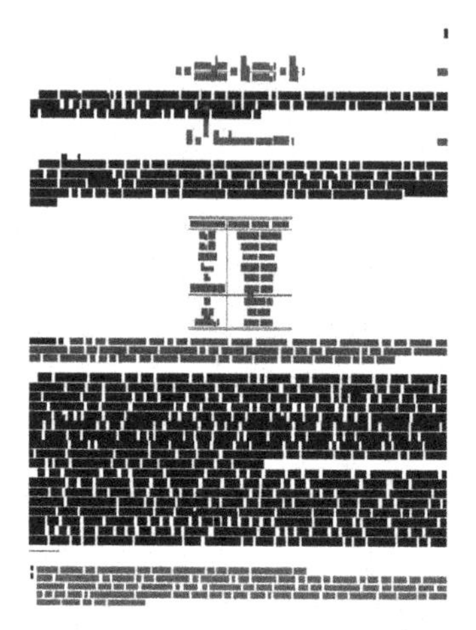

Fig. 2. DocBank sample image on the left and its corresponding segmentation on the right. Each color represents one class (black for *paragraph*, purple for *equation*, ...). (Color figure online)

4 Models

We compared LayoutLM, a Transformer for DU which is our baseline, with our long range contributions LayoutLinformer and LayoutCosformer[1]. They only differ by their implementation of self-attention: LayoutLM uses full self-attention like BERT, LayoutLinformer uses a low-rank approximation first proposed by [30] and LayoutCosformer uses a kernel-based method introduced in [25] as a replacement. We further detail how they work in the subsequent subsections.

We chose those models over other efficient Transformers based on the convenience to adapt them from linear text to 2-dimensional documents. Efficient attention based on sliding windows [3,34] does not transpose nicely to 2D documents because the sliding window mechanism is deeply linked to the linear order of words. Even though our approach tries to provide words in a natural order, in some documents it does not reflect the human reading order – for example for table content. To mitigate this issue, we preferred to rely on global attention or 2D local attention.

Similarly to how LayoutLM was adapted from BERT, we adapt Linformer and cosFormer models to process documents by adding a 2D positional embedding and a page embedding to the input. We chose to use learned embeddings to simplify weight transfer from LayoutLM to our long-range models.

[1] Models implementation and weights available at https://github.com/thibaultdouzon/long-range-document-transformer.

4.1 LayoutLM

LayoutLM [32] has proven its capacities on most tasks related to documents since its release. It reuses BERT [7] encoder and tokenizer, and only modifies the positional encoding by introducing a 2D encoding for word boxes boundary and size. This modification allows the model to leverage layout information provided by the OCR. LayoutLM's computational bottleneck is the self-attention layer. In Transformers, self-attention [28] takes *queries* Q, *keys* K and *values* V and computes a weighted average of *values* for each input. The weights are given by the dot product between each pair of *queries* and *keys*. It can be formulated $\texttt{softmax}(QK^\top)V$, where $Q, K, V \in \mathbb{R}^{N\times d}$ and N represents the sequence length and d the model hidden size. Figure 3 describes the self-attention operation. The matrix $\texttt{softmax}(QK^\top)$ is the attention matrix containing the intensity of attention between each pair of tokens. Xu et al. [32] pre-trained the model on RVL-CDIP [8] which contains 7 millions scanned documents released in the 90' from the tobacco industry. Two versions of LayoutLM was have been released: base and large, and it outperforms all preceding text-only language models on classification and information retrieval tasks.

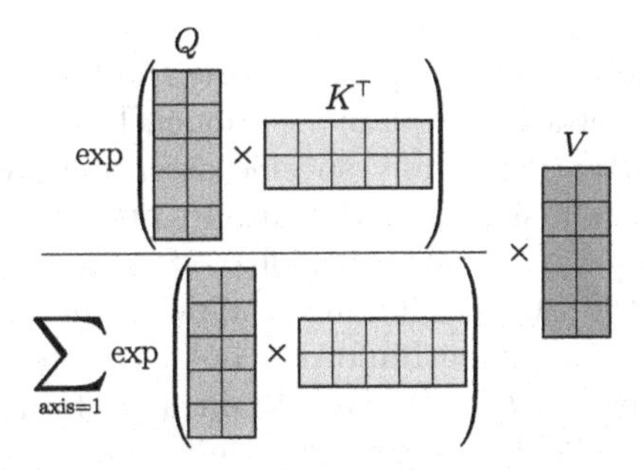

Fig. 3. Illustration of the attention mechanism used in LayoutLM, normalization and multiple heads aside. In this example, $N = 5$ and $d = 2$. Due to the softmax operator, the product $QK^\top$ must be computed, resulting in $O(N^2)$ complexity.

In our experiments, we only use the base model with maximum sequence length $N = 512$ and hidden size $d = 768$. For longer documents, we split the tokenized sequence into chunks of maximum length and process them separately.

4.2 LayoutLinformer

Our first contribution, LayoutLinformer is based on the Linformer architecture [30] and adapted to document processing by adding 2D positional encodings and using LayoutLM pre-trained weights. Although true self-attention can only be computed in $O(N^2)$, it can be approximated very efficiently by leveraging the low rank of the attention matrix $QK^\top$. In Fig. 4, we illustrate LayoutLinformer's attention mechanism. Keys and values sequence length dimension is projected on a smaller space of size k through a linear transformation: $K' = P_K K$ where $P_k \in \mathbb{R}^{k \times N}$ is the learned projection matrix (respectively $V' = P_V V$ where $P_V \in \mathbb{R}^{k \times N}$). This means the size of the new attention matrix $Q(P_K K)^\top$ is $N \times k$, reducing the complexity of self-attention to $O(Nk)$.

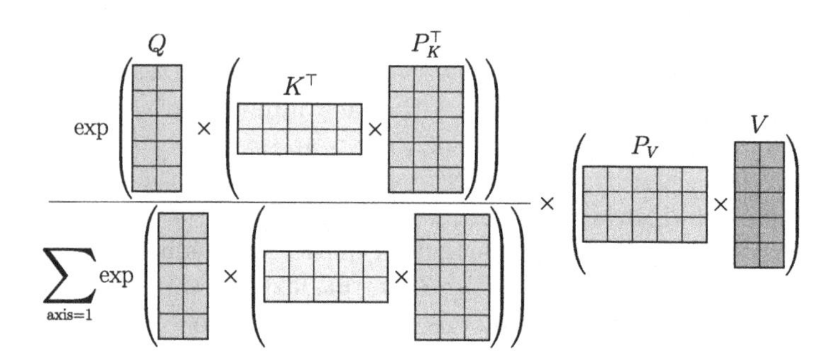

Fig. 4. LayoutLinformer attention mechanism. In this example, $N = 5$, $d = 2$ and $k = 3$. Efficient matrix multiplication ordering reduces the complexity to $O(Nk)$.

An immediate drawback of this projection is the loss of ability to visualize the attention matrix in order to explain the model. It is also no longer possible to implement causal attention or any specific attention pattern. On the other hand, Linformer provides a simple modification to the Transformer in order to make it manage longer sequences with global attention. Most model weights are identical between the two architectures, allowing us to transfer LayoutLM pre-trained weights into DocumentLinformer before further pre-training.

Wang et al. [30] showed that it can obtain a performance comparable to Roberta [19] on multiple NLP benchmarks. He brang evidence that its performance is mostly determined by the projection dimension k, and that increasing sequence length N did not degrade results. Therefore, we chose to apply LayoutLinformer with $N = 2048$ and $k = 512$ in order to compare its performances with LayoutLM.

4.3 LayoutCosformer

Our second contribution, called LayoutCosformer, is based on the cosFormer [25] model which is another efficient alternative to the original Transformer. Similarly to LayoutLinformer, we transferred pre-trained weights from LayoutLM to DocumentCosFormer thanks to the similarities between architectures. It achieves linear complexity by replacing the non-linear similarity computation between Q and K with a linear operation. More specifically, Qin et al. [25] proposed to replace $\exp(QK^\top)$ with $\Phi(Q)\Phi(K^\top)$ where Φ is a nonlinear function. Figure 5 illustrates in more detail how LayoutCosformer attention works. In order to keep values of the similarity matrix positive, a good choice is $\Phi = \texttt{ReLU}$. Computations can then be reordered to decrease the complexity to $O(N)$.

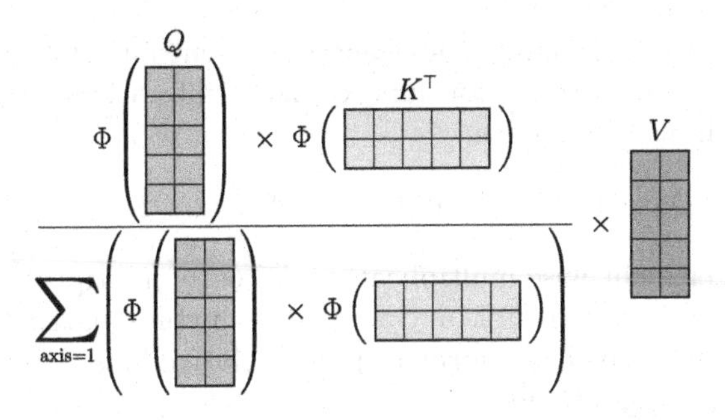

Fig. 5. LayoutCosformer efficient attention mechanism with $N = 5$ and $d = 2$. The linear similarity enable computing first $\Phi(K^\top)V$ and factorize $\Phi(Q)$ out of the summation.

In addition to its linear self-attention complexity, Qin et al. [25] include a relative self-attention bias towards nearby tokens. They cannot simply add the bias to the $N \times N$ similarity matrix before multiplying with values because it would mean a quadratic complexity. Their solution is to use functions that can be decomposed into a sum of products: $f(x, y) = \sum_n g_n(x) \times h_n(y)$. If we call B the bias matrix where $B_{i,j} = f(i, j)$, their biased similarity matrix can be written $\Phi(Q)\Phi(K^\top) \odot B$ where $\odot$ is the element-wise product. Then when looking at the attention from token i to token j we obtain:

$$
\begin{aligned}
s_{i,j} &= \Phi(Q_i)\Phi(K_j^\top)B_{i,j} \\
&= \Phi(Q_i)\Phi(K_j^\top)\sum_n g_n(i) \times h_n(j) \\
&= \sum_n \Phi(Q_i)\Phi(K_j^\top)g_n(i)h_n(j) \\
&= \sum_n (\Phi(Q_i)g_n(i)) \times (\Phi(K_j^\top)h_n(j))
\end{aligned}
$$

Using this trick, they proposed to use a cosine bias $B_{i,j} = \cos(\frac{\pi}{2M}(i - j))$ which can be decomposed into $B_{i,j} = \cos(\frac{\pi}{2M}i)\cos(\frac{\pi}{2M}j) + \sin(\frac{\pi}{2M}i)\sin(\frac{\pi}{2M}j)$. With the normalization constant M set to the maximum sequence length, they ensure $0 < B_{i,j} < 1$ with a maximum when $i = j$. In the next subsection, we demonstrate how it can also be applied to 2D relative attention.

4.4 2D Relative Attention

Global self-attention is a powerful tool for capturing long-range dependencies. However, although distant dependencies can be relevant, most attention should be toward close neighbors. Relative attention [24,27] selectively focuses on specific parts of the input by biasing the base self-attention. This was proven useful on text which that can be represented as a linear sequence, but due to complex layouts, the sequence order is suboptimal to determine locality. In order to better capture local context in documents, we introduced 2D relative attention based on the token positions inside the document.

In LayoutLM, we pre-compute for each document an attention bias matrix B and modify the self-attention formula to take it into account. More precisely, we replace the self-attention with:

$$\text{RelativeAttention}(Q, K, V, B) = \left(\text{softmax}(QK^\top) \odot B\right) V$$

where $\odot$ denotes element-wise multiplication. Directly multiplying the attention matrix by some bias is very flexible and allows for any bias matrix to be chosen. It also matches the way LayoutCosformer applies relative bias to its self-attention, thus allowing to compare them.

On the other hand, it is nontrivial to implement relative attention for global long-range Transformers. Because LayoutLinformer compresses the sequence dimension of the Key matrix, it is not possible to apply custom 2D attention bias to LayoutLinformer. For LayoutCosformer it is possible to reuse the same trick as in the 1D version with another bias function.

Because the function must remain separable into a sum of products, a good choice is to use exponentials and trigonometric functions. We first prove that the product of two separable functions is also itself separable. Let $f^1 = \sum_n g_n^1(x) \times h_n^1(y)$ and $f^2 = \sum_m g_m^2(x) \times h_m^2(y)$ be two functions separable into sum of products, then:

$$\begin{aligned} f^1(x,y) \times f^2(x,y) &= \left(\sum_n g_n^1(x) \times h_n^1(y)\right) \times \left(\sum_m g_m^2(x) \times h_m^2(y)\right) \\ &= \sum_n \sum_m \left(g_n^1(x) \times h_n^1(y) \times g_m^2(x) \times h_m^2(y)\right) \\ &= \sum_{n,m} (g_n^1(x) g_m^2(x)) \times (h_n^1(y) h_m^2(y)) \end{aligned}$$

Which can also be separated into a sum of products.

We chose to compare 2 different attention biases. The first one is simply the product cosine bias along both X and Y axis. It captures local context in every direction with variations close to euclidean distance. We define B^{squircle}[2] the following:

$$B_{i,j}^{\text{squircle}} = \cos\left(\frac{\pi}{2M}(x_i - x_j)\right) \times \cos\left(\frac{\pi}{2M}(y_i - y_j)\right)$$

where x_i and y_i (resp. x_j and y_j) are positions of token i (resp. j) along X and Y axis. In practice we used the coordinates of the center of each token bounding box.

Although this bias correctly captures 2D locality, documents complex layout sometimes implicitly calls for other definition of proximity in order to understand it. For instance, Fig. 6 shows a table from a purchase order.

In this configuration, in order to grasp correctly the meaning of a cell in the table, the model needs to make the connection with the table header positioned at the beginning of the page. When multiple line items are spanning the whole page,

[2] Squircle are intermediate shape between square and circle, see https://en.wikipedia.org/wiki/Squircle. Contours of the surface described by B^{squircle} is not actually a squircle but also range from square to circle.

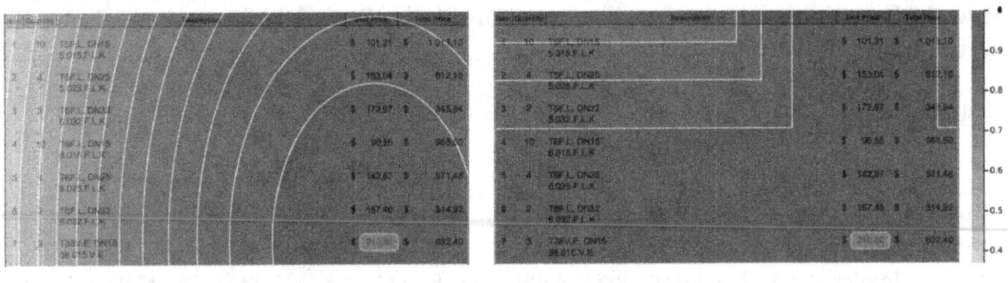

(a) Squircle relative attention bias. (b) Cross relative attention bias.

Fig. 6. Contour plots for squircle and cross relative attention bias applied to token "210, 80" (bottom-right corner). Because token positions are normalized between 0 and 1000, tokens along the same line cannot fully attend to each other on the left while they are unaffected on the right.

we hypothesize that this relative attention might hurt the performance due to the long-distance separating tokens. To deal with this issue, we propose another bias pattern. Its objective is to allow attention to tokens that are aligned with each other along the X or Y axis. To this end, we define $B^{\text{cross}}_{i,j} = \max\{\cos(\frac{\pi}{2M}(x_i - x_j)), \cos(\frac{\pi}{2M}(y_i - y_j))\}$. We illustrate the differences with an example shown in Fig. 6. With cross relative attention bias, the highlighted token (the price of an item) can better attend to the column header "Unit Price" and to its related line. In general, tokens inside a table can fully attend to their corresponding column header and line. This should prove helpful for understanding tables by guiding the model attention towards semantically related tokens.

5 Experiments

Our models are pre-trained on our Business Documents collection for 200k steps using Masked Visual-Language Modeling [32]. They are then finetuned on each dataset. For both tasks, we use `BIESO` tags to help the model decode predictions spanning multiple tokens. We performed our experiments on two RTX A6000 for pre-trainings and single RTX A6000 for fine tunings. LayoutLM models runs with a batch size of 48 and sequence length of 512 while long-range models (LayoutLinformer and LayoutCosFormer) can only get to a batch size of 16 with sequence length of 2048 on a single device. We accumulate gradient for 96 data samples before updating model's weights. We use Adam with learning rate $lr = 2 \cdot 10^{-5}$ and linear warmup for 5% of the training steps followed by linear decrease.

5.1 Long-Range

Theoretical results on models architectures hints towards LayoutLinformer and LayoutCosformer being much more efficient the longer the sequence. We use a dummy inference task with increasing sequence lengths and compare our 2 models with LayoutLM base architecture. The results are available in Table 1. They reveal how the computational complexity of full self-attention disables LayoutLM when dealing with sequence longer than 1024. Its memory consumption

Table 1. Duration and memory consumption of the 3 models for various sequence lengths on an inference task.

Model name	Time in seconds/*Memory in GiB*											
	Sequence length											
	512		1024		2048		4096		8192		16384	
LayoutLM	1.41	*1.25*	2.83	*2.50*	7.39	*5.01*	23.43	*13.69*	-		–	
LayoutLinformer	1.18	*1.35*	1.92	*2.26*	3.54	*3.28*	6.90	*5.19*	13.08	*8.96*	25.65	*16.78*
LayoutCosformer	2.03	*1.36*	2.50	*2.37*	4.68	*3.38*	9.00	*5.38*	17.23	*9.59*	33.96	*17.59*

limits our tests with LayoutLM up to sequence length of 4096, longer sequences couldn't fit into a single GPU. On the other hand, LayoutLinformer and LayoutCosformer performed as predicted, with LayoutCosformer being slightly slower and more memory hungry than LayoutLinformer.

It turns out document's length also greatly impacts models metrics performance on the Customer Order dataset. For better visualization, we group documents into 3 length categories: short (document fits into 512 tokens), medium (between 513 and 2048) and long (2049 or more tokens). LayoutLM models can process short documents in a single sequence but need to split other documents into multiple independent sequences. Short and medium documents fit into LayoutLinformer and LayoutCosformer sequence length but not long documents. When a model cannot process a document in a single sequence, we split the document into multiple sequences and process them separately.

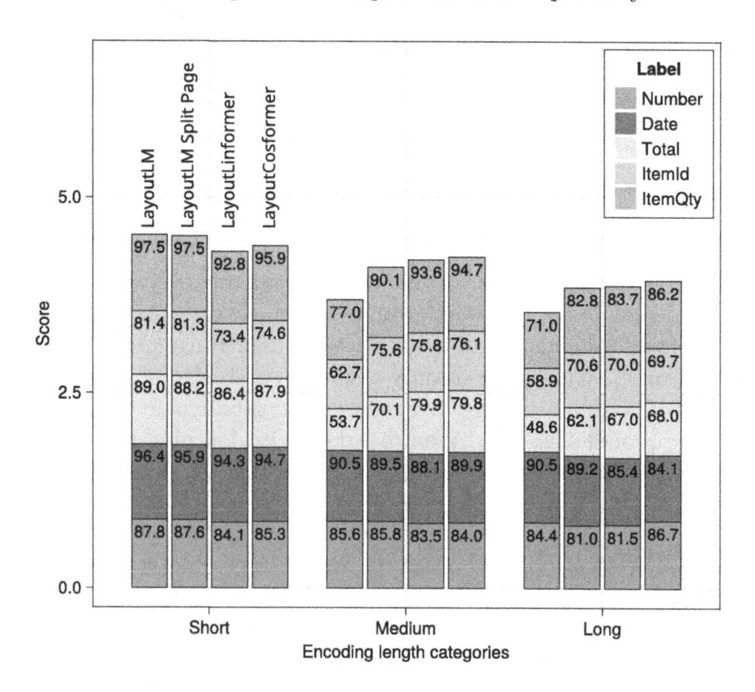

Fig. 7. F1-score stacked bar plot of multiple models on the Business Orders dataset. In each document length categories, models are in the same order.

In Fig. 7, we compare our pre-trained LayoutLM models with LayoutLinformer and LayoutCosformer. First, we discovered LayoutLM is very sensitive to the split position for medium and long documents. Introducing a sequence split when a new page is started greatly improves performance, we call this model LayoutLM SplitPage. It performs better on *total amount* (from 53.7% to 70%), item ID number (from 62.7% to 75.6%) and *quantity* (from 77.0% to 90.1%) recognition for medium and long documents. The repetitive structure of multipage documents combined with the fact that most pages fit in a 512 tokens sequence allow the model to not get lost. *Document number* and *date* are mostly not affected because they almost always occur at the beginning of the document, which is not affected by the splitting strategy.

Although LayoutLinformer and LayoutCosformer perform slightly worse than LayoutLM for short documents on all classes (around 74% F1 score on *item ID number* versus 81% for LayoutLMs), their performance decreases less than LayoutLM's on medium documents. On those medium documents, even LayoutLM SplitPage drops from 88.2% to 70.1% F1 score on the *total amount* while both long-range models only reduce performance from roughly 87% to 80%. We also noticed date recognition performance degrades across all models with longer documents which is not expected because *dates* are usually at the top of the first page. The same can be noted for the *order number* at a smaller scale. It might be due to a correlation between document's length and layout: short and medium / long documents do not share layouts. And because there are twice more short documents than longer ones, it is harder to generalize to new layouts. Overall, the performance of long-range models is more consistent across wide variety of document lengths.

Table 2. F1 weighted average for each model and document length categories. All models were first pre-trained on the Business Documents collection.

Model name	F1 weighted macro average		
	Short	Medium	Long
LayoutLM	95.36	95.84	91.42
LayoutLinformer	95.20	96.49	91.41
LayoutCosformer	94.03	95.91	91.40

We performed the same experiments on Docbank dataset, except for the page-splitting part as all documents are single page. At first we compared models performance for each document length categories in Table 2. It contains average F1 score across all labels weighted by the support of each label. It turns out length categories introduce bias in the composition of pages, with labels being very sparsely represented in some categories. This bias implicitly selects more first page in short pages (with lower text density), and medium sized pages contain a lot of paragraphs.

We observe the same drop in performance for long-range models on short documents, with LayoutLinformer providing better results across the board than LayoutCosformer. But we notice LayoutLM perform slightly better on medium documents than on short. Long-range models follow the same pattern with a greater difference between short and medium pages, LayoutCosformer almost gaining 2 average F1 percentage points. There is almost 20 times fewer long documents than medium, which could explain part of the global performance loss. Unfortunately, due to those biases, it is difficult to draw conclusions on model's performance.

Table 3. Results on Docbank dataset for LayoutLMs and long-range models. Models with a asterisk (*) are ours. They were pre-trained on the Business Documents collection before finetuning on Docbank.

Model name	2D Relative attention	F1 score												Macro average
		Abst.	Author	Caption	Equa.	Figure	Footer	List	Para.	Refe.	Section	Table	Title	
LayoutLM (Li et al. [18])	–	98.1	85.9	95.9	89.4	**100.0**	89.5	**89.4**	**97.8**	93.3	**95.9**	86.3	**95.7**	93.1
LayoutLM (Xu et al. [32])	–	98.3	89.6	96.0	89.0	99.7	91.6	88.2	97.5	**93.5**	94.3	87.4	90.4	93.0
LayoutLM (*)	–	97.8	87.5	94.9	87.2	99.7	90.5	84.0	97.1	93.2	92.8	85.7	88.6	91.6
LayoutLM (*)	Squircle	**98.4**	90.2	**96.1**	89.7	99.8	92.0	88.9	97.6	93.4	94.6	87.7	90.3	**93.2**
LayoutLM (*)	Cross	**98.4**	**90.3**	96.0	89.6	99.8	**92.1**	88.7	97.6	93.4	94.6	87.5	90.7	**93.2**
LayoutLinformer (*)	–	97.9	88.9	93.7	90.0	99.5	91.1	87.9	97.5	93.2	91.3	87.6	88.7	92.3
LayoutCosformer (*)	–	97.2	87.2	91.0	88.1	99.3	90.6	87.4	97.1	93.2	81.4	87.0	88.3	90.7
LayoutCosformer (*)	Squircle	97.0	85.4	92.4	89.2	98.8	90.7	84.2	97.2	93.2	85.6	87.9	86.8	90.7
LayoutCosformer (*)	Cross	97.4	86.9	93.8	**91.2**	98.9	91.7	87.5	97.5	93.1	87.4	**89.0**	88.1	91.9

Table 3 compiles results for LayoutLM and long-range models for all labels. First, we can make sure our training pipeline performs on par with what Docbank authors reported for LayoutLM base model by comparing their results and the ones we obtained by using public LayoutLM weights. Except for *author* and *title* labels, both results are very close, and the macro average is almost identical. Secondly, pre-training on business documents negatively impacts LayoutLM performances on all labels, losing 1.4 F1 percentage points on average. This advocates for pre-training data crucial role in later model finetuning results and its composition. Finally, long-range models performed on the same level as LayoutLM. LayoutLinformer even being more performant than our pre-trained LayoutLM. Overall, even though LayoutCosformer seems less performant on this task, both long-range models performed better than our pre-trained LayoutLM on *table* and *equation*. Those two labels might beneficiate from long-range references, giving the model hints of their presence in the current sequence.

5.2 Relative Attention

We conduct the same experiments on models with 2D relative attention and compare their performance with their flat attention counterpart. On the business order dataset, Table 4 shows slight gains when using squircle attention with LayoutLM. For all document lengths, information retrieval is improved a few percentage points of F1 score over our previous LayoutLM Split Page implementation. Though, we do not observe the same improvement with the cross shaped attention pattern. This might indicate focusing on very local neighbors helps LayoutLM making the right decision. Overall, relative attention improves results in some circumstances but not as much as splitting every page did. However, when combined with LayoutCosformer, we observe a significant degradation in performance for all labels with the squircle attention while the cross pattern provides similar results as the raw LayoutCosformer.

Table 4. Macro average F1 score on the Business Orders dataset with 2D relative attention.

Model name	2D Relative attention	Macro average F1 score		
		Short	Medium	Long
LayoutLM Split Page	–	90.0	82.2	77.2
LayoutLM Split Page	Squircle	**90.4**	83.0	77.8
LayoutLM Split Page	Cross	90.0	82.0	77.6
LayoutCosformer	–	87.6	85.0	**79.0**
LayoutCosformer	Squircle	85.8	82.2	73.4
LayoutCosformer	Cross	87.6	**85.2**	77.2

On Docbank task, relative attention provides noticeable performance gains for both LayoutLM and LayoutCosformer. We provide all results in Table 3. LayoutLM with relative attention is standing out, going from 91.6% F1 score to 93.2% for both squircle and cross patterns. Most improvements are made on *author*, *equation* and *list*, each gaining at least 2 F1 score points. Both resulting models even beat Docbank's authors version by a thin margin. This is impressive knowing those models were pre-trained on the same business order dataset as our base LayoutLM which suffered a 1.5 F1 score performance drop as a consequence. It turns out *author*, *equation* and *list* were also the fields where our LayoutLM performance dropped the most compare to stock LayoutLM. Applying cross shaped relative attention to LayoutCosformer also improves performance across most labels. It even outperforms all other models on *equation* and *table* fields which benefit most from very long attention.

6 Conclusion

In this work, we showed the impact of document length on Transformer-based models applied to Document Understanding. Depending on the document's type and the task, model's performance on longer documents can be negatively impacted with F1 score dropping 20% for the most impacted. We explored several alternatives including another sequence split strategy and long-range layout-aware models based on Linformer and cosFormer architectures. They all proved to successfully reduce the performance gap between short and long documents (down to only 10% performance drop), sometimes at a small cost on short document's metrics. We also introduce relative attention based on 2D textual layout instead of the classical sequence order. It produces better results on dense text, significantly improving both LayoutLM and LayoutCosformer on the Docbank layout segmentation task.

In addition to other efficient Transformer architectures, we plan to investigate other ways to use longer sequences for DU. For example, in multi-modal models, this may allow fitting the whole text and visual patches of a document in a single sequence without needing more compute capabilities.

References

1. Ainslie, J., et al.: ETC: encoding long and structured inputs in transformers. In: Proceedings of the 2020 Conference on Empirical Methods in Natural Language Processing (EMNLP), pp. 268–284. Association for Computational Linguistics, Online (2020). https://doi.org/10.18653/v1/2020.emnlp-main.19. https://aclanthology.org/2020.emnlp-main.19
2. Bahdanau, D., Cho, K., Bengio, Y.: Neural machine translation by jointly learning to align and translate (2016). http://arxiv.org/abs/1409.0473. Type: article
3. Beltagy, I., Peters, M.E., Cohan, A.: Longformer: the long-document transformer (2020). http://arxiv.org/abs/2004.05150. Type: article
4. Brown, T., et al.: Language models are few-shot learners. In: Advances in Neural Information Processing Systems, vol. 33, pp. 1877–1901. Curran Associates Inc. (2020). https://papers.nips.cc/paper/2020/hash/1457c0d6bfcb4967418bfb8ac142f64a-Abstract.html
5. Chowdhery, A., et al.: PaLM: Scaling Language Modeling with Pathways (2022)
6. Denk, T.I., Reisswig, C.: BERTgrid: contextualized embedding for 2D document representation and understanding. arXiv preprint arXiv:1909.04948 (2019)
7. Devlin, J., Chang, M.-W., Lee, K., Toutanova, K.: BERT: pre-training of deep bidirectional transformers for language understanding. In: Proceedings of the 2019 Conference of the North American Chapter of the Association for Computational Linguistics: Human Language Technologies, Minneapolis, Minnesota (Volume 1: Long and Short Papers), pp. 4171–4186. Association for Computational Linguistics (2019). https://doi.org/10.18653/v1/N19-1423. https://aclanthology.org/N19-1423
8. Harley, A.W., Ufkes, A., Derpanis, K.G.: Evaluation of deep convolutional nets for document image classification and retrieval. In: International Conference on Document Analysis and Recognition (2015). http://arxiv.org/abs/1502.07058. Type: article
9. Hochreiter, S., Schmidhuber, J.: Long short-term memory. Neural Comput. **9**(8), 1735–1780 (1997). https://doi.org/10.1162/neco.1997.9.8.1735. ISSN 0899-7667
10. Huang, Y., Lv, T., Cui, L., Lu, Y., Wei, F.: LayoutLMv3: pre-training for document AI with unified text and image masking (2022). http://arxiv.org/abs/2204.08387. Type: article
11. Huang, Z., et al.: ICDAR2019 competition on scanned receipt OCR and information extraction. In: 2019 International Conference on Document Analysis and Recognition (ICDAR), pp. 1516–1520 (2019). https://doi.org/10.1109/ICDAR.2019.00244. http://arxiv.org/abs/2103.10213

12. Katti, A.R., et al.: Chargrid: towards understanding 2D documents. In: Proceedings of the 2018 Conference on Empirical Methods in Natural Language Processing, Brussels, Belgium, pp. 4459–4469. Association for Computational Linguistics (2018). https://doi.org/10.18653/v1/D18-1476. https://aclanthology.org/D18-1476
13. Kim, G., et al.: OCR-free document understanding transformer (2022). http://arxiv.org/abs/2111.15664. Type: article
14. Kitaev, N., Kaiser, L., Levskaya, A.: Reformer: the efficient transformer (2020). http://arxiv.org/abs/2001.04451. Type: article
15. Kolesnikov, A., et al.: An image is worth 16 × 16 words: transformers for image recognition at scale. In: International Conference on Learning Representations (2021)
16. Lample, G., Ballesteros, M., Subramanian, S., Kawakami, K., Dyer, C.: Neural architectures for named entity recognition. In: Proceedings of the 2016 Conference of the North American Chapter of the Association for Computational Linguistics: Human Language Technologies, San Diego, California, pp. 260–270. Association for Computational Linguistics (2016). https://doi.org/10.18653/v1/N16-1030. https://aclanthology.org/N16-1030
17. Lewis, D., Agam, G., Argamon, S., Frieder, O., Grossman, D., Heard, J.: Building a test collection for complex document information processing. In: Proceedings of the 29th Annual International ACM SIGIR Conference on Research and Development in Information Retrieval, SIGIR 2006, pp. 665–666. Association for Computing Machinery, New York (2006). https://doi.org/10.1145/1148170.1148307. ISBN 9781595933690
18. Li, M., et al.: DocBank: a benchmark dataset for document layout analysis. In: Proceedings of the 28th International Conference on Computational Linguistics, Barcelona, Spain, pp. 949–960. International Committee on Computational Linguistics (Online) (2020). https://doi.org/10.18653/v1/2020.coling-main.82. https://aclanthology.org/2020.coling-main.82
19. Liu, Y., et al.: RoBERTa: a robustly optimized BERT pretraining approach (2019). http://arxiv.org/abs/1907.11692. Type: article
20. Mathew, M., Karatzas, D., Jawahar, C.V.: DocVQA: a dataset for VQA on document images. In: Winter Conference on Applications of Computer Vision, pp. 2200–2209 (2021). https://openaccess.thecvf.com/content/WACV2021/html/Mathew_DocVQA_A_Dataset_for_VQA_on_Document_Images_WACV_2021_paper.html
21. Mikolov, T., Sutskever, I., Chen, K., Corrado, G.S., Dean, J.: Distributed representations of words and phrases and their compositionality. In: Advances in Neural Information Processing Systems, vol. 26. Curran Associates Inc. (2013). https://papers.nips.cc/paper/2013/hash/9aa42b31882ec039965f3c4923ce901b-Abstract.html
22. Pennington, J., Socher, R., Manning, C.: GloVe: global vectors for word representation. In: Proceedings of the 2014 Conference on Empirical Methods in Natural Language Processing (EMNLP), Doha, Qatar, pp. 1532–1543. Association for Computational Linguistics (2014). https://doi.org/10.3115/v1/D14-1162. https://aclanthology.org/D14-1162
23. Peters, M.E., et al.: Deep contextualized word representations. In: Proceedings of the 2018 Conference of the North American Chapter of the Association for Computational Linguistics: Human Language Technologies, New Orleans, Louisiana (Volume 1: Long Papers), pp. 2227–2237. Association for Computational Linguistics (2018). https://doi.org/10.18653/v1/N18-1202. https://aclanthology.org/N18-1202

24. Powalski, R., Borchmann, L., Jurkiewicz, D., Dwojak, T., Pietruszka, M., Pałka, G.: Going full-TILT boogie on document understanding with text-image-layout transformer (2021). http://arxiv.org/abs/2102.09550. Type: article
25. Qin, Z., et al.: cosFormer: rethinking softmax in attention. In: International Conference on Learning Representations (2022). https://doi.org/10.48550/arXiv.2202.08791. Type: article
26. Radford, A., Wu, J., Child, R., Luan, D., Amodei, D., Sutskever, I.: Language models are unsupervised multitask learners (2019)
27. Shaw, P., Uszkoreit, J., Vaswani, A.: Self-attention with relative position representations. In: Proceedings of the 2018 Conference of the North American Chapter of the Association for Computational Linguistics: Human Language Technologies, New Orleans, Louisiana (Volume 2: Short Papers), pp. 464–468. Association for Computational Linguistics (2018). https://doi.org/10.18653/v1/N18-2074. https://aclanthology.org/N18-2074
28. Vaswani, A., et al.: Attention is all you need. In: Advances in Neural Information Processing Systems, vol. 30. Curran Associates Inc. (2017). https://papers.nips.cc/paper/2017/hash/3f5ee243547dee91fbd053c1c4a845aa-Abstract.html
29. Wang, A., Singh, A., Michael, J., Hill, F., Levy, O., Bowman, S.: GLUE: a multi-task benchmark and analysis platform for natural language understanding. In: Proceedings of the 2018 EMNLP Workshop BlackboxNLP: Analyzing and Interpreting Neural Networks for NLP, Brussels, Belgium, pp. 353–355. Association for Computational Linguistics (2018). https://doi.org/10.18653/v1/W18-5446. https://aclanthology.org/W18-5446
30. Wang, S., Li, B.Z., Khabsa, M., Fang, H., Ma, H.: Linformer: self-attention with linear complexity (2020). http://arxiv.org/abs/2006.04768. Type: article
31. Xu, Y., et al.: LayoutLMv2: multi-modal pre-training for visually-rich document understanding. In: Proceedings of the 59th Annual Meeting of the Association for Computational Linguistics and the 11th International Joint Conference on Natural Language Processing (Volume 1: Long Papers), pp. 2579–2591. Association for Computational Linguistics, Online (2021). https://doi.org/10.18653/v1/2021.acl-long.201. https://aclanthology.org/2021.acl-long.201
32. Xu, Y., Li, M., Cui, L., Huang, S., Wei, F., Zhou, M.: LayoutLM: pre-training of text and layout for document image understanding. In: Proceedings of the 26th ACM SIGKDD International Conference on Knowledge Discovery & Data Mining, KDD 2020, pp. 1192–1200. Association for Computing Machinery, New York (2020). https://doi.org/10.1145/3394486.3403172. ISBN 9781450379984
33. Yu, W., Lu, N., Qi, X., Gong, P., Xiao, R.: PICK: processing key information extraction from documents using improved graph learning-convolutional networks (2020). http://arxiv.org/abs/2004.07464. Type: article
34. Zaheer, M., et al.: Big bird: transformers for longer sequences. In: Advances in Neural Information Processing Systems, vol. 33, pp. 17283–17297. Curran Associates Inc. (2020). https://proceedings.neurips.cc/paper/2020/hash/c8512d142a2d849725f31a9a7a361ab9-Abstract.html

KAP: Pre-training Transformers for Corporate Documents Understanding

Ibrahim Souleiman Mahamoud[1,2](✉), Mickaël Coustaty[1], Aurélie Joseph[2], Vincent Poulain d'Andecy[2], and Jean-Marc Ogier[2]

[1] La Rochelle Université, L3i Avenue Michel Crépeau, 17042 La Rochelle, France
mickael.coustaty@univ-lr.fr

[2] Yooz, 1 Rue Fleming, 17000 La Rochelle, France
{ibrahim.souleimanmahamoud,aurelie.joseph, vincent.poulaindandecy}@getyooz.com, jean-marc.ogier@univ-lr.fr

Abstract. Pre-trained models have proven their efficiency. Despite their good performance, these models require a lot of data and resources to allow state-of-the-art results. In this paper, we propose KAP a pre-trained model adapted for the domain specificity for corporate documents. KAP takes into account the domain specificity of corporate documents and proposes a model that integrates the local context of each word (i.e the words at the top, bottom, and right of the word). That is useful to help a better understand the 2D context of a document. We also propose a co-attention for a better understanding of the correlation between image and text. In addition to this, we integrate a special tokenizer to focus on the essential words in a corporate document in order to reduce the vocabulary size. KAP achieves state-of-the-art results on several datasets despite being pre-trained on only 72,000 documents.

Keywords: NLP · Attention mechanism · Visqual Question Answering · Document classification · Document Analysis and Recognition

1 Introduction

In the last few years, several pre-trained models have been proposed [12,13,16, 17]. These pre-trained models contribute greatly to improving the performance of the state-of-the-art. They contributed to the improvement of many tasks, we can quote the translation, classification, Visual Questions Answering, etc. In particular, we find pre-training models on documents. The objective of pre-trained models is to build up a prior understanding of the documents in order to be finetuned on the final tasks like automatic document understanding (i.e. document classification, token classification, visual question answering, etc).

The first pre-trained models that were developed, such as Bert [16] and Roberta [17], only used the textual modality. It learns bidirectional representation by predicting the original vocabulary of the randomly masked words.

M. Coustaty and A. Fornés (Eds.): ICDAR 2023 Workshops, LNCS 14194, pp. 65–79, 2023.
https://doi.org/10.1007/978-3-031-41501-2_5

These models were adapted to specific tasks by fine-tuning only the 1D text alignment.

Unlike sequential text, corporate documents have a strong 2D structure. Some models were defined by including layout information and visual information such as LayoutLMv3, LayoutLMv2 and LAMBERT [18]. These models have provided state-of-the-art (SOTA) results on several Document Analysis and Recognition (DAR) tasks.

Despite their contributions, these models have many limitations. These limitations can be summarised in three points. (1) The pre-training carried out by these models allowed to have representations of the words (i.e embedding) adapted to be reused or finetuned. These representations are highly depending on the pre-trained data. This pre-trained data is generally very diverse (i.e. it concerns all types of documents). This diversity makes them good at performing on as many documents as possible. But if for a specific domain, these models do not perform well with fine-tuning, what are the possible options? Should we use the pre-trained model with corporate documents or create a new pre-trained model for a specific domain?. (2) To answer this question, we will focus on the second limitation. These models does not take into account all the specificities of a given context. In order to achieve suitable results with corporate documents, some or all of these architectures would have to be changed. (3) Finally, the third point is that these models are very cumbersome and require a lot of resources (time, data) to converge. They have a high environmental impact and a high financial cost.

To address these limitations we propose KAP. We can summarise these contributions in the following way.

- We propose a pre-trained model on a specific type of dataset. This model can be used on other types of datasets to allow everyone to have their own pre-trained model.
- KAP proposes a model able to better use the local context of each word and the visual features.
- We combine the best of the deep learning and domain expert strategies to achieve a pre-trained model self-supervised.
- KAP Achieves state-of-the-art performance in experiments conducted. The proposed model is lightweight as it reduces the vocabulary size with a tokenizer able to focus on the essential words.

The type of documents that we will be interested in this study concerns corporate documents. More specifically, documents such as invoices and purchase orders. These documents are very often used in companies. Finding an efficient model for these types of documents is both a scientific challenge and a real opportunity for better automation of document processing.

2 Related Works

In this section, we will focus on pre-trained models more specifically on their model base (i.e. models with a low number of parameters).

We can categorize these models into three parts. The first part corresponds to the models using only the textual modality, then the models using in addition the 2D position modality and finally the multimodal models.

Among the models only pre-trained on the textual modality, we can quote Bert [16], Roberta [17]. Bert is basically a multi-layer bidirectional Transformer encoder. The objective function used in Bert are two, the masked language model (MLM) and the next sentence prediction (NSP). The MLM objective allows to randomly mask a token and to retrieve the masked token. NSP is a binary classification task taking a pair of sentences as inputs and classifying whether they are two consecutive sentences. With these contributions, Bert has been able to use pre-entrained features to be finetuning on Natural Language Processing (NLP) tasks. To improve some of the weaknesses of BERT, the RoBERTa model was proposed. RoBERTa has essentially made three contributions to BERT. (1) RoBERTa is trained with dynamic masking (i.e. randomly masked tokens each time) unlike BERT which fixes the masked token at the preprocessing stage. RoBERTa showed that dynamic masking was better. (2) They showed that using all the words in a single document was better than using a pair of sequences as opposed to BERT. (3) Finally RoBERTa allowed the use of a larger batch size to converge faster but also to have better performance. Despite these contributions, these models have several gaps. BERT and RoBERTa are not trained on 2D structural information.

LAMBERT [18] has contributed by integrating this 2D positional information into the word representation. They introduced the Layout-Aware Language Model, which allowed them to integrate this 2D information into their architecture. Their experimentation showed on several datasets using documents for example on CORD [8], SROIE [19] that it improved over RoBERTa. The result present by LAMBERT was more than 2% of f1-score compared to SOTA on almost all the tested datasets. However, LAMBERT does not use visual information from the documents, which contain descriptive visual information about a document

We can also mention the paper [21], which proposes a method for classifying tokens. In addition to taking into account word positions like LAMBERT, this method also adds the context surrounding each word. They define a neighborhood zone all the way to the left of the page and about 10% of the page height above. In addition to adding this context, this method also takes into account the type of token, for example, if the token is a date, it looks for all the dates in the document as candidates and thus limits the number of candidate words. Despite the relevance of this method, it is nevertheless limited to token classification and cannot be easily used for Visual Question Answering. In addition, this method does not use the visual information in the document, which can sometimes be useful when the text is not clearly visible.

To better take into account the image and layout of the documents the LayoutLMv3 [13] has been proposed. This model allows the extraction of visual features with transforms based on the VIT [7] and not convolution. They proposed linear projection features of image patches before feeding them into the multimodal Transformer. They also proposed a new objective Word-Patch Alignment (WPA) feature in addition to the existing Mask Language Model (MLM) and Masked Image Modeling (MIM). LayoutLMv3 also performed well and confirms the hypothesis that the use of images makes a real contribution to the overall understanding of the document.

Despite the good performance of LayoutLMv3, when we conducted tests on a specific domain dataset such as CHIC [5] (i.e. a dataset for Visual Question-Answering on corporate documents) the results are not as great as expected. Based on this observation, we hypothesized the following. (1) The training data has a strong impact on the word representation that will be produced. We assume that the use of corporate documents in the pre-training stage would have a positive impact. (2) The model used in the pre-training stage is often cumbersome and resource intensive. We would like to propose a model that can minimize the pre-training time and still perform well so that it can be reused more easily.

3 KAP

Based on the limitations identified in the Sect. 2, we have proposed KAP which is an answer to the following questions. (1) How to minimize the vocabulary size?, because there is a correlation between the size of the vocabulary and the weight of the model. (2) How to better exploit the features present in the data? and finally (3) What would be the objective function to automatically pre-train the model?

3.1 Model Architecture

KAP is a pre-trained model on a corporate documents dataset in order to learn the representation of a word based on the image and its context (see Fig1).

The model input contains text embedding and their 1D and 2D positions. In addition to these inputs we have added two new inputs to better represent a word according to the surrounding context and image features (see Sect. 3.2).

The encoder allows to transform these inputs into hidden outputs. The encoder used and the basic architecture is from RoBERTa. At the final stage, we have the Mask Language Modeling (MLM) objective function which masks the tokens using a new strategy (see Sect. 3.3).

3.2 Co-attention

We propose a co-attention mechanism to make better use of the surrounding context of each word (i.e. the words at the top, bottom, and right of the word) and also the feature of the global image. To this end, we have chosen to use a similar approach to the co-attention mechanism used in [10,11]. This co-attention mechanism is described in Fig. 2, and it allowed adapting the representation of a word to its surrounding context or to the feature of the overall image.

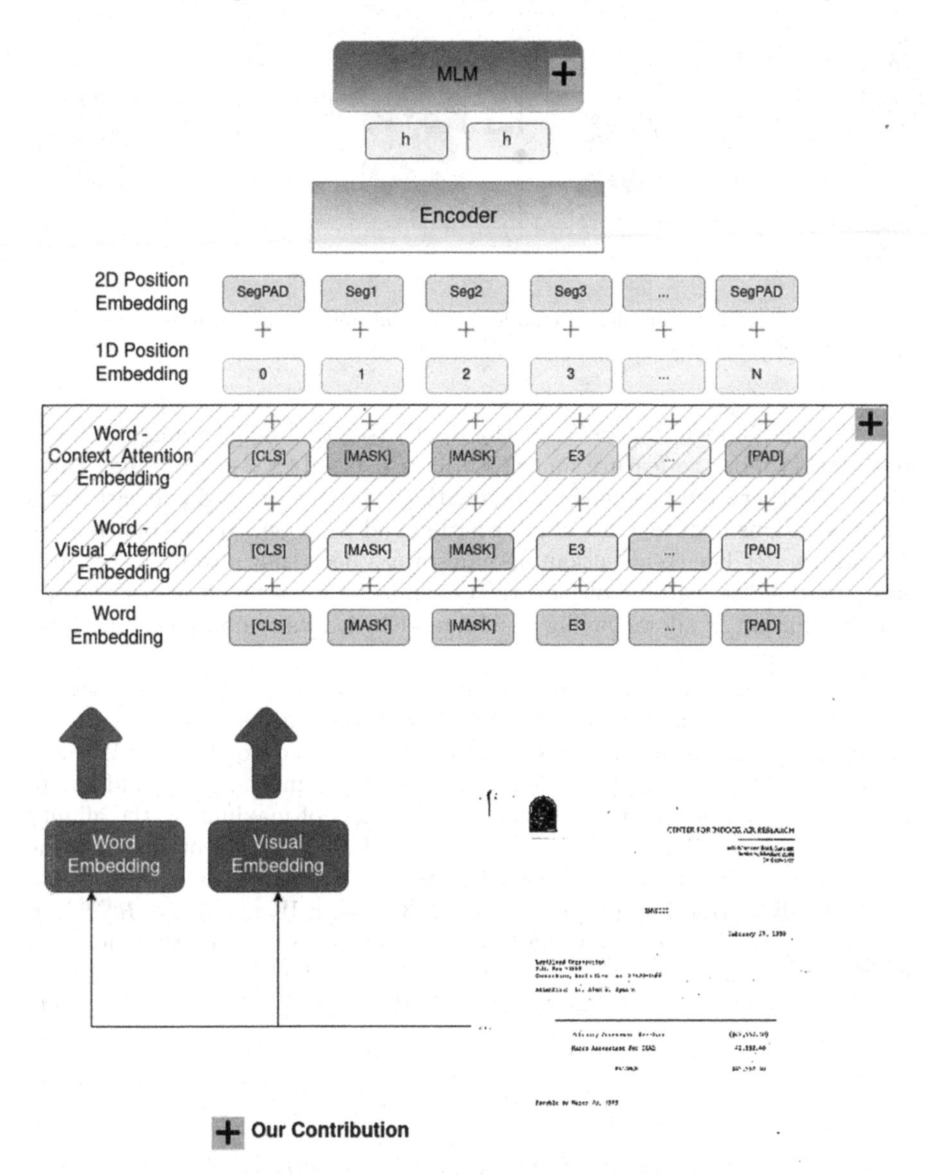

Fig. 1. KAP architecture and pretraining objective. KAP is a pre-trained model for multimodal understanding of corporate documents. Its purpose is to provide context-sensitive representations of words (i.e. top, bottom and right of the word) but also features of the overall image.

More specifically, this co-attention mechanism has two main advantages: (1) The size of the word features and the CI features (the surrounding context feature and the image feature) are not of the same size. For the surrounding context we took 5 words for each direction, the global size is therefore $5 * 3 * d$

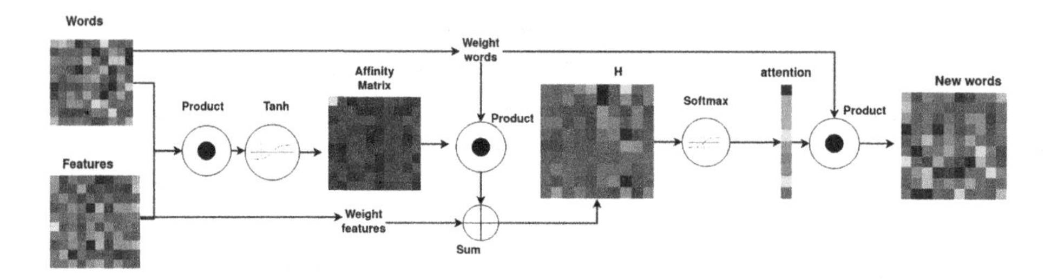

Fig. 2. The co-attention model to enable contextual and visual features to be better taken into account.

where d is the hidden dimension of the model (see the model configuration in Appendix A). For the feature image, we have a $14 * 14 * d$ size. Adding these features together would mean compressing the data or taking a single part. Both of these solutions involve a loss of information and we prefer the use of attention mechanisms. (2) The co-attention mechanism has been shown to be effective on tasks requiring the learning of joint relationships between two features. Beyond a simple addition, it allows having a small model with its own weight to perform this learning automatically.

In order to adapt this representation of words, we have to go through several steps. The first step is to calculate the Affinity Maxtrix (Af) between the CI features (i.e. either the context or the image feature) see Eq. 1 where W_a contains the weight $\in R^{d \times d}$. After calculating the affinity matrix, we would like to calculate the attention matrix which simply consists of maximizing the affinity over the location of CI features $a[n] = max_i(Afi, n)$. Instead of choosing the max activation, we will instead consider the affinity matrix as a feature and learn to predict attention a with the Eqs. 2, 3. Where $W_{word}, W_{ci} \in R^{d \times d}$ and $w_{hi} \in R^{d \times 1}$. Finally, we adapt the features of the words by multiplying them with the attention a.

$$Af = tanh(CI^T W_a Words) \tag{1}$$

$$H = tanh(W_{word}\ Words + (W_{ci}\ CI)\ C) \tag{2}$$

$$a^i = softmax(w_{hi}^T\ H_i) \tag{3}$$

$$Word\_CI\_Attention = \sum_{m=1}^{N} a_n^i\ CI \tag{4}$$

In the following, we will explain the contextual features and the visual features that are involved in the co-attention mechanism.

Contextual Features: In corporate documents, the words respect a 2D alignment. This alignment allows the human to read a document easily and to know quickly which word is linked with others.

We will exploit these alignments to retrieve the context surrounding each word. This context allows in most cases to retrieve the closest but also the most significant words for each word in three directions (i.e. top, bottom, and right of the word). To retrieve this surrounding context we used a rule-based expert algorithm (see Appendix A).

In this algorithm, we have described the pseudo-code of a recursive function that allows us to retrieve this surrounding context. We will first browse the set of words in documents and for each word W we will call the function $GET\_CONTEXT$.

In this function, we first retrieve the indexes of the words aligned to a chosen direction. From these indexes, we line up in rows if necessary (i.e. this concerns the context up and down as otherwise, we would have mixed words). Once the words are ordered in line, we retrieve the first word in the index list, from this word we calculate the distance $dist1$ with the source word W. Then if the function is called a second time we calculate the distance $dist2$ to the previously selected word. If $dist1 < dist2$ then we stop the recursion and return the set of selected indexes. This allows us to stop when a word is far from the other words selected previously.

Once these indexes are retrieved we retrieve for each word the context around it (top, bottom, and right) of the word. For the numeric and alphanumeric words, we take the context which contains character strings in order to associate each numeric value with the associated keyword.

For each document, we recovered approximately 3000 tokens which correspond to the whole of the surrounding context. To avoid mixing these tokens, we added special tokens to our tokenizer (see Sect. 3.2) "[RIGHT]", "[BOTTOM]", "[TOP]" and "[LEFT]" to separate these tokens according to the direction of the context. In the end, we use co-attention with the surrounding context.

Image Feature: For the automatic comprehension of documents, it is important to know how to exploit the visual feature. This visual feature contains information that is not found in the text of the document or in the position of the words. It allows having a global vision of the document and of the elements that compose it.

In the first step, we standardize the size of all our images, we reshape these images in size (224, 224). Then we used the linear projection features of the image proposed by LayoutLMv3 [13]. We chose this model because it was light and allowed us to recover these visual features very quickly. This choice was reinforced by the good results obtained with the LayoutLMv3 model compared to other models using other features. For each image, we have recovered the following feature size $14 \times 14 \times d$. We then used these features in our co-attention to allow each word to change its representation according to the global image.

Although the co-attention architecture remains the same when using feature images and surrounding context. However, in KAP we have co-attention-visual and co-attention-context which do not share the same weight. This removes confusion in the model between visual and contextual features.

Tokenizer: The tokenizer is used during pre-processing and allows words to be transformed into unique numbers. The set of unique tokens and associated values is called a vocabulary. To build this vocabulary the tokenizer trains on the text set. The size of the vocabulary is chosen in advance before the tokenizer is trained. In general, this size depends on the size of the dataset and the diversity of the text that composed it. The choice of vocabulary size has a strong impact on the final weight of the pre-trained model.

The larger the vocabulary size, the larger the data needed for training and the longer the time needed to converge. Finding a good compromise between vocabulary size and performance is not easy. However, in KAP we propose a solution to minimize the vocabulary size while keeping the same performance. We propose to introduce new special tokens. Special tokens allow the association of a significant element to a number understandable by the model.

We have added new special tokens to our tokenizer to allow us to focus only on information with important semantics and group the rest on special tokens. In the context of corporate documents, there are two groups of words. The first are words that each bring different semantics and others in which only their syntax counts and have no real semantic differences between them. For example, the numbers in a document, whether it is the number 10, 100, or 1000 that is next to the keyword "reference number" do not change the overall information of the document. From this observation we have added the following tokens "[numeric]", "[float]", "[alphanumeric]", "[IBAN]", "[Email]", "[date]" , "[email]", "[date]" and "[currency]". We will call the words with these types of syntax Value-Words. We will use "[numeric]" for the token consisting of numeric values. The special token "[float]" will be used for all float numeric values (i.e. this is used to separate for example amounts that are float from other numbers).

We associate each Value-Words with its special token through two types of models. The first one uses the characters of the word and another one is based on rules. The first model allows us to know if a word is numeric, float or alphanumeric. This is easy to do and features exist for most programming languages. Then other models based on rules allow us to determine the syntax type, for example, for IBAN we took the standard [9] to detect if a character string corresponds well to an IBAN. To see how many percent these Value-Words represent in a vocabulary, we trained a tokenizer on the text of 150 000 corporate documents and built a large vocabulary of 25000 words. We found that these Value-Words represent 9000 tokens in the vocabulary which is about 36% of the vocabulary.

From this observation, we conducted an ablation study to see the impact of adding these special tokens and we found that these tokens improved the results while greatly reducing the computation time (see Sect. 6)

To build this tokenizer, we used Byte-Pair Encoding (BPE) [4], a hybrid between character- and word-level representations, that allows handling the large common vocabularies from natural language corpora.

3.3 Pre-training Objective

KAP is pre-trained with the MLM objectives to learn multimodal representation in self-supervised learning. In the related works section, we have seen the masking strategy used by Bert and the improvement made by RoBERTa. Our proposed objective function is a Mixed-strategy.

The Mixed-strategy is composed of a static part (we will call this part Key-Value tokens) and of a dynamic part which masks randomly some tokens of the documents.

In a corporate document, some information is more important than others, and this information is often organized as key-value pairs (i.e. Total Amount (key) - 1250 (value)). We will call it "Key-Value" and the rest of the words will be called "Other-Token".

With the expert algorithm, we have extracted for each document the set of values and their associated key. To achieve this extraction, we first selected the Value-Words (i.e. numeric, alphanumeric, etc). From these words, we used the algorithm $GET\_CONTEXT$ (see Appendix A) to retrieve the surrounding context. Then we took from the 3 contexts (i.e. top, bottom and right of the word) the closest to the word Value-Words to become its key. This allowed us to have a self-supervised model to automatically annotate relevant Key-value (or Question Answers) in a document. After extracting these Key-Values we will use it in the masking strategy.

The classical masking strategy consists in masking some percentage of the tokens of the document. This strategy is relevant but is very resource intensive and requires a lot of data to converge on an acceptable solution. We propose with KAP to guide the masking of the words on what have the most important for the corporate documents. We will mask the words extracted from the Key-Value as much as possible (we will call this guiding masking) and also add a specific percentage of the Other-Token in the documents.

To determine the best percentage of masking we did an ablation study to determine the right compromise between guided masking and dynamic masking. In the end we randomly masked 15% of the tokens in the document. In this 15%, 70% of the tokens where masked with the guided strategy and 30% using the dynamic strategy. For guide masking we have masked either the key or the value but never both at the same time. For the words masked with the guide strategy, we have tried to mask as much as possible the keys (i.e. this presents about 80% of the words masked by the guide strategy).

After describing our mixed strategy, we will describe the MML objective function formula. The pre-training objective consists in maximizing the log-likelihood of correct masked words based on the contextual tokens (i.e. words_context and words_visual), as well as the element of the question-answer pair that is not masked. The masked text tokens is $y_{t'}$. The correct text token Y^t and the other features CIP^d (positions features, context features and visual features). We use cross-entropy loss as follows. Where $M^{'}$ and represent the positions masked positions. We denote the Transformer model parameters by θ and minimize the cross-entropy loss described in Eq. 5

$$L_{MLM}(\theta) = -\sum_{m=1}^{M^{'}} \log p\theta \left(y^{t^{'}} \| CIP^{d}, Y^{t} \right) \tag{5}$$

3.4 Pre-training Data

To achieve a good understanding of corporate documents, it is necessary to have a dataset that best fit to a general vision of corporate document on which the model will be used later (like in a finetuning process). To build this pre-training dataset, we have chosen three document sources. The first one corresponds to the public dataset RVL-CDIP [20], we used only the invoice class of dataset which contains 20 000 documents. Thereafter, we chose to also include a public dataset (DocVQA [15]) that contains 12,000 documents and 50,000 questions-answer. These questions are very diverse and can concern any type of token in the document. This Docvqa dataset will allow us to know the impact of using other types of documents (i.e. emails, advertisements, etc) in addition to corporate documents. In addition to these unilingual datasets we used XFUND [2] a multilingual dataset. This contains labeled forms with key-value pairs in 7 languages (Chinese, Japanese, Spanish, French, Italian, German, and Portuguese). We used the XFUND dataset which represents about 1500 documents.

Finally we also added private data which contain 36 500 documents, most of which are in French. For all this dataset, a document consists of one page.

4 Experiments

4.1 Fine-Tuning on Multimodal Tasks

To test the efficiency of KAP, we confront our model with various datasets. These datasets are used in several tasks concerning documents. They are widely used in the SOTA to compare the performances on automatic document processing models. We fine-tuning KAP for the DAR tasks (i.e. document analysis and comprehension) on publicly available benchmarks. The results are presented in Table 1.

Document Visual Question Answering. The Visual Question Answering (VQA) task is an important task in the NLP domain. It allows to estimate the level of language understanding but also requires a better analysis of the images by the state-of-the-art models. We can formalize this task as an extraction task, the model will predict the start and end position of the response using the word set of the document. Several datasets exist for the VQA task and in this study, we will focus on CHIC [5], DOCVQA [15].

The CHIC dataset is a dataset that focuses on corporate documents. The CHIC questions and answers concern mainly the words numeric, alphanumeric, date, etc which are important in the automatic processing of corporate documents. This dataset contains about 3800 question-answers on 700 documents in several languages (i.e. English, French, German, Italian, Spanish).

The final results show us that KAP outperforms the state-of-the-art models by about 2% of ANLS score on Visual Question Answering datasets.

Document Form Understand. The task of understanding form and receipt requires the extraction of information from structured forms. The task of this dataset is classification of token into key, value, other and heeader.

XFUND [2] is a dataset containing forms from 7 languages (Chiness, Japanese, Spanish, French, Italian, German, Portuguese) Each language includes 199 forms. In this dataset there are 1,393 fully annotated forms (i.e. to annotate the type of word and the word with which it is associated).

The results show that KAP pre-training on this dataset has largely contributed to improving performance by about 1.

Document Image Classification
Document classification predicts the category of the document according to the age and text of the document. We have chosen to use RVL-CDIP in this task. RVL-CDIP dataset contains 400,000 document images, among them 320,000 are training images, 40,000 are validation images, and 40,000 are test images. For document classification, KAP is less efficient than SOTA models. This is mainly due to the different types of documents present in RVL-CDIP. The LayoutLM have been trained on documents similar to the one present in RLV-CDIP. In contrast to KAP which performs most of its pre-training on document corporate.

Table 1. This table contains the results of the proposed KAP model and the results of the SOTA model (LayoutLMv3, LayoutLMv2, Bert). The three modalities first Text (T), the second modality Text-Image-Layout (TIL), and the last is Text, Image, Layout, and Context of each word TILC.

Modality			CHIC	DOCVQA	RVL-CDIP	XFUND
Method		Param	ANLS	ANLS	Accuracy	F1
Human			88.09	85.17	94.25	98.11
T	*Bert*	~110M	15.36	32.89	41.45	45.57
	KAP	~40M	**43.32**	**56.10**	**61.11**	**48.60**
TIL	$LayoutLMv2_{Base}$ [14]	~200M	69.69	75.63	95.25	74.21
	$LayoutLMv3_{Base}$ [13]	~133M	68.01	78.76	**95.44**	78.76
	KAP	~41M	**70.56**	**74.11**	79.28	**71.93**
TILC	**KAP**	~41M	**75.77**	**81.87**	89.75	**81.28**

5 KAP Strengths and Weaknesses

The model we propose, like all other models, has its strengths and weaknesses. The contributions of KAP are both its strength and its weakness in some cases. We can summarise this in three points. The first is that the vocabulary used in KAP is essentially based on the words of corporate documents. This implies that in documents containing these words, KAP perform well, otherwise, it will depend on whether these words are present in the training. This limitation was especially noted for data not seen in pre-training such as RVL-CDIP. In future work we will study the impact of the size of the vocabulary and the importance of the words it contains.

The second point concerns the use of local contextual information of each word. This has led to many improvements, but there are still cases where the model can be confused. This confusion arises when a word is associated with a context that does not concern it. For example, if we have a number 2825 and around we find at the top a keyword Reference number (i.e. being a key): 256 (i.e. its associated value). If we ask the question reference number the answer could be 2825. This is due to the wrong association of the number 2825 with the word reference number. While the associated keyword of 2825 is found in its diagonal axis. The improvement of these cases is possible, which could be done if in GET_CONTEXT algorithm included also other contexts than the current context (i.e. left, top and right of word).

Finally, the last point concerns the use of visual features. The strategies we have used, despite being light and simple, nevertheless have their limits. We have noted a better understanding of the overall 2D visual feature of the document which could correct errors of at least 12.6% on the CHIC dataset.

6 Ablation Studies

We will first analyse the impact of the masking strategy. We tested three strategies see the result in Table 2. Firstly we tested the static strategy, in this test we took all the Key-Values (described in the Sect. 3.3) in a document. Then in second time we tested a dynamic strategy which consists in masking 10% of the tokens of the documents, and any type of words in a document. Then finally we tested the Mixed strategy described in Sect. 3.3 which is a mixture of the two preceding strategies. When we analyse the results on the CHIC data we see that the static strategy converges faster than the others. The Mixed Strategy although, although slower than the static one, still gave a score of +1.70 ANLS. The dynamic strategy was also good but gave a slightly worse performance than the mixed strategy.

Secondly, we will analyse the usefulness of the tokenizer used in KAP, show the result in Table 3. We first tested a tokenizer with a vocabulary of size 5000 containing all the words without any special token. Then we added some special tokens (described in the Sect. 3.2) and compared the two results. Although the two results are close, the model using the tokenizer with the special token converges faster.

Table 2. Performance of impact masking

Method	Pre-training duration (hours)	CHIC		
		F1	EM	ANLS
Mixed-Mask	117	**70.68**	**68.90**	**75.77**
Dynamique Mask	133	68.16	66.98	74.70
Statique Mask	85	67.79	66.35	74.01

Table 3. Performance of impact type of tokenizer

Method	Pre-training duration (hours)	CHIC		
		F1	EM	ANLS
With Special token	117	**70.68**	**68.90**	**75.77**
Without Special Token	147	70.11	67.85	75.08

7 Conclusion and Future Work

In this paper, we propose the KAP pre-trained model on corporate documents. The KAP model allows taking into account the surrounding context of each word in order to get a local understanding. In addition to local understanding, we added the use of the global image to give each word a well understanding of whole the document. We have also made sure that the architecture is light enough to train very quickly. To do this we first reduced the size of the vocabulary by adding special characters to take into account only the most important words. Then we used a hidden size of 384 instead of 768. In future research, we will extend our ablation study to find the best hyperparameter. In particular by testing different vocabulary sizes (10k, 15k, 30k) and also increasing the size of the pre-training data. In this study, we will try to find a model to determine the subset that will bring the most impact to the pre-training and remove the data not contributing to performance in pre-training. In the second step, we will try to build a few-shot learning or zero-shot learning model for the corporate documents.

A Appendix

Model Configurations. We carried out all our experimentation on the same machine. This machine contains 4 Nvidia RTX A6000 GPUs 48GB and has 256 GB ram and 4 TV of SSD disk. To find out the environmental impact of all our experiments, we used [1]. Each experiment produced about 1.97 kg of CO2e and consumed 38.35 kWh. In total, to carry out all our experiments, we emitted 26.95 kg of CO2e and consumed 525.55 kWh. When we compare these consumptions to model consumptions like Bert, RoBERTa, and LayoutLMv3, KAP has less environmental impact while performing better on specific datasets. We have used the following hyperparameters. We used 110 batch size and trained our model 34 epochs. The average training time is 123 h on a single GPU. We used 12 layers and a hidden size of 384. We also kept the attention heads parameter at 12 and a dropout 0.1. Finally, we used the Adam optimizer with a learning rate of 1e−5.

Algorithm 1. Function to retrieve the surrounding context

1: **function** GET_CONTEXT($select\_index, bboxs, bb, direction$)
2: $near\_index \leftarrow \{\}$
3: $bboxs = yelo(near\_index)$ ▷ Normalisation of bouding boxes
4: **if** $direction$ = "$right$" **then**
5: $near\_index = np.where((bboxs[:, 1] > bb[1]-marge)and(bboxs[:, 1] < bb[1]+marge)and(bboxs[:, 0] < bb[0] - marge))[0]$
6: **else if** $direction$ = "$bottom$" **then**
7: $near\_index = np.where((bboxs[:, 0] > bb[0]-marge)and(bboxs[:, 0] < bb[0]+marge)and(bboxs[:, 1] > bb[1]))[0]$
8: **else if** $direction$ = "top" **then**
9: $near\_index = np.where((bboxs[:, 0] > bb[0]-marge)and(bboxs[:, 0] < bb[0]+marge)and(bboxs[:, 1] < bb[1]))[0]$
10: **end if**
11: $select\_index = keeps\_the\_nearest(near\_index)$ ▷ this function allows you to iteratively compare the distance between two words, if the distance with the last two words is much greater than that of the two previous words then you stop and keep the previous words.
12: **end function**

References

1. Lannelongue, L., Grealey, J., Inouye, M.: Green Algorithms: Quantifying the carbon footprint of computation (2020)
2. Xu, Y., et al.: XFUND: a benchmark dataset for multilingual visually rich form understanding (2022). https://doi.org/10.18653/v1/2022.findings-acl.253
3. Van Landeghem, J., et al.: Document understanding of everything
4. Sennrich, R., Haddow, B., Birch, A.: Neural machine translation of rare words with subword units (2016). arXiv:1508.0790
5. Mahamoud, I.S., Coustaty, M., Joseph, A., d'Andecy, V.P., Ogier, J.-M.: Corporate document for visual question answering (2022). https://arxiv.org/pdf/2305.01054.pdf
6. Mahamoud, I.S., Coustaty, M., Joseph, A., d'Andecy, V.P., Ogier, J.-M.: To reproduce our experiments we put in this git account, the analyses and all the hyperparameters. https://gitlab.com/contributions3/kap.git
7. Dosovitskiy, A., et al.: An image is worth 16×16 words: transformers for image recognition at scale (2020)
8. Park, S., et al.: Neural machine translation of rare words with subword units (2016). arXiv:1508.0790
9. https://www.iso.org/standard/81090.html
10. Zhang, H., et al.: Machine Learning (stat.ML), Machine Learning (cs.LG), FOS: Computer and information sciences, FOS: Computer and information sciences
11. Lu, J., Yang, J., Batra, D., Parikh, D.: Computer Vision and Pattern Recognition (cs.CV), Computation and Language (cs.CL), FOS: Computer and information sciences, FOS: Computer and information sciences. https://arxiv.org/abs/1606.00061
12. Xu, Y., et al.: Neural machine translation of rare words with subword units (2016). arXiv:1508.0790

13. Xu, Y., et al.: LayoutLMv3: pre-training for document AI with unified text and image masking (2022)
14. Xu, Y., et al.: LayoutLMv2: multi-modal pre-training for visually-rich document understanding (2020)
15. Mathew, M., Karatzas, D., Jawahar, C.V.: DocVQA: a dataset for VQA on document images (2021)
16. Devlin, J., Chang, M.-W., Lee, K., Toutanova, K.: BERT: pre-training of deep bidirectional transformers for language understanding (2018)
17. Liu, Y., et al.: RoBERTa: a robustly optimized BERT pretraining approach (2019)
18. Garncarek, Ł, et al.: LAMBERT: layout-aware language modeling for information extraction. In: Lladós, J., Lopresti, D., Uchida, S. (eds.) ICDAR 2021. LNCS, vol. 12821, pp. 532–547. Springer, Cham (2021). https://doi.org/10.1007/978-3-030-86549-8_34
19. Huang, Z., et al.: ICDAR2019 competition on scanned receipt OCR and information extraction (2019)
20. Harley, A.W., Ufkes, A., Derpanis, K.G.: Evaluation of deep convolutional nets for document image classification and retrieval (2015)
21. Majumder, B.P., Potti, N., Tata, S., Wendt, J.B., Zhao, Q., Najork, M.: Representation learning for information extraction from form-like documents (2020)

Transformer-Based Neural Machine Translation for Post-OCR Error Correction in Cursive Text

Nehal Yasin[1(✉)], Imran Siddiqi[2(✉)], Momina Moetesum[3(✉)], and Sadaf Abdul Rauf[4]

[1] Department of Computer Science, Bahria University, Islamabad, Pakistan
nehalyaseen22@gmail.com

[2] Xynoptik Pty Limited, Melbourne, VIC, Australia
imran.siddiqi@xynoptik.com.au

[3] School of Electrical Engineering and Computer Sciences, National University of Science and Technology, Islamabad, Pakistan
momina.moetesum@seecs.edu.pk

[4] Department of Software Engineering, Fatima Jinnah Women University, Rawalpindi, Pakistan

Abstract. This study investigates transformer-based Neural Machine Translation (NMT) for post-processing of noisy Optical Character Recognition (OCR) output. While recognition engines have matured significantly for most languages, the problem still remains challenging for text in cursive scripts reporting high character but relatively lower word recognition rates. This study targets post-processing of the noisy output of OCR for cursive text (using Urdu as a case study) and leverages a transformer-based NMT framework to correct the OCR errors. More specifically, we feed the noisy text as input and the correct transcription as target to the transformer. The model is trained and evaluated on such pairs of text collected from News tickers in videos and tweets from multiple News channels. A comprehensive experimental study shows significant performance improvement by introducing the proposed post-processing step.

Keywords: Post-OCR Error Correction · Cursive Text · Neural Machine Translation · Transformers

1 Introduction

Optical character recognition (OCR) is the set of processes that convert printed, handwritten, video, or scene text into a digital editable form. These processes include pre-processing of data, script-specific processing (recognition) of text, and post-processing. Despite the huge success of OCR in converting documents

S. A. Rauf—Contributing author.

M. Coustaty and A. Fornés (Eds.): ICDAR 2023 Workshops, LNCS 14194, pp. 80–93, 2023.
https://doi.org/10.1007/978-3-031-41501-2_6

into machine readable form, they are still behind humans because the output of OCR is noisy. The causes of these OCR errors include image degradation or noise, challenges in the segmentation of text, similarity in character shapes and insufficient knowledge of the text or language. It is also important to mention that OCR systems for many languages across the globe have matured significantly over the years and report close to 100% recognition performance. Research on Urdu OCR, on the other hand, is relatively recent [1] and consequently, most of the research attention has been on improving the Urdu OCR and post processing has remained less explored until now.

In the case of cursive text, like Urdu, most of the OCRs focus on recognizing characters rather than words. While recognizing characters is useful in the sense that it reduces total number of unique classes to be recognized, high character recognition rates do not map to high word recognition rates. Characters within the words can be misclassified and at times, several words may join into a single word or a single word may be split into multiple words. In other words, there can be situations where the high character recognition rates do not correspond to OCR output that is semantically meaningful [2].

A wide range of research endeavours exist on correcting human typing errors whether grammatical or spelling [3,4]. The OCR errors, however, are significantly different from human errors as the OCR system consists of different sets of processes sharing common characteristics causing the OCR system to make the same mistake every time the same word is encountered [5]. While for smaller texts the OCR errors can be corrected through human inspection, the process needs to be automated for large collections [6]. These include automatic post-processing techniques based on lexical and context aware error detection and correction. This study focuses on post processing the noisy OCR output of cursive text (using Urdu as a case study). The key highlights of this study are:

- Development of a transformer-based NMT technique to post-process the output of an OCR system
- Model training with textual transcriptions from video frames and twitter accounts of news channels to enhance its ability to handle noisy OCR output
- Comprehensive performance evaluation of the post-processing step validating significant enhancement in the recognition performance

In the next section we discuss the OCR errors in Urdu text and challenges that we encounter while post-processing of Urdu text. Section 3 presents work related to Urdu word segmentation techniques and error detection and correction in Urdu text. Section 4 describes our dataset and methodology. Section 5 gives details on experiments and result and in the last section we conclude our findings.

2 Urdu Optical Character Recognition

Optical character recognition systems find applications in a wide variety of domains ranging from document indexing and retrieval to assistive systems for the visually impaired [7]. Although the research interest in the development of

OCR systems for cursive scripts like Arabic and Urdu is relatively recent, thanks to the advancements in machine learning, a number of recognition systems have been researched and developed in this regard [8]. An important component of such systems is the post-processing of the noisy OCR output. Majority of the OCRs target high resolution scanned document images where the OCR performance is acceptable for most applications. For problems like handwriting recognition or video-OCRs, post processing is a much needed step as the OCR output in such scenarios is expected to be quite noisy.

Like many other domains, digitization of textual content in Urdu is also on the rise. However, there are several challenges associated with Urdu text which make its processing difficult as compared to text in other Western languages. The most notable of these challenges is the segmentation of text into words. While word segmentation is simple for many languages, in the case of Urdu text, the omission of space between two words and the insertion of space(s) within the same word makes it quite challenging. More specifically, if a word ends with a non-joiner character, the next word can start with or without a space [9]. Figure 1 shows an example of such a problem which shows the same text with and without spaces. In both cases, semantically, it stays the same text but without spaces, it becomes difficult to identify the word boundaries.

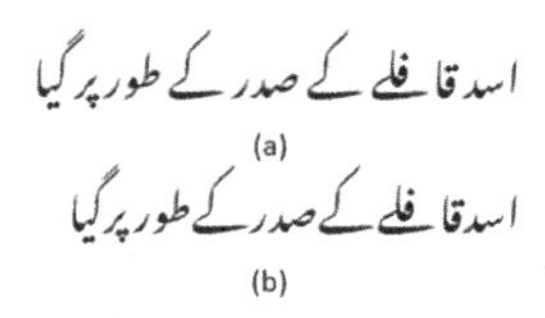

Fig. 1. (a) An Urdu sentence with spaces between words. (b) An Urdu sentence without spaces between words [10]

Another challenging task in Urdu is word variation as each word can be written in different forms and the meaning would still be the same. Examples of such words are shown in Table 1. Combining the words together not only changes the visual appearance, it also reduces the number of ligatures in the word. Another common scenario in Urdu is that several words contain multiple morphemes which combine together to create a single word. The problem occurs when one morpheme ends with a joiner character and it is required to insert an extra space within a word to produce a visually acceptable form. However, it also creates separate tokens for the same word. A few such examples are shown in Table 2. Such problems make post processing and classification of Urdu text a challenging problem.

Furthermore, the popular trend in recognition systems is the use of end-to-end trainable systems which take text line images as input along with the ground truth transcription and learn to recognize character shapes as well as

Table 1. Different Shapes of Same Words [11]

Combined	Separated
آپکا	آپ کا
اسوقت	اس وقت
محبوطن	محب وطن
محبوطن	ان جام
ابتک	اب تک

Table 2. Words With Multiple Tokens [11]

With Space	Without Space	Translation
احمقپن	احمق پن	Boobyism
مومبتی	موم بتی	Candle
خواہمخواہ	خواہ مخواہ	Unnecessary
احسانمند	احسان مند	Grateful

the segmentation points. The combination of convolutional and recurrent neural networks has been extensively employed for this problem [12]. Such systems report character recognition rates rather than word recognition rates. However, it is important to note that even very high character recognition rates may end up in relatively low word recognition performance as even a single character error in a word leads to the incorrect recognition of the complete word. Figure 2 illustrates this point with the OCR output for the video frame having character recognition errors resulting in high word recognition errors. In this example, 10 characters are incorrectly recognized from a total of 141 characters, resulting in a character recognition rate of 92%, which is quite an acceptable score. However, it is important to note that for most practical applications, word recognition rates are important as words represent the smallest semantically meaningful unit in the text. In the given example, 33 out of 42 words are correctly recognized resulting in a word recognition rate of 78%, a number which is fairly low for application development. It is also interesting that in many cases only 1–2 erroneous characters can lead to low word recognition rates. Correcting these errors using the knowledge of the language in a post processing step can significantly improve the word recognition rates.

(a) Video Frame

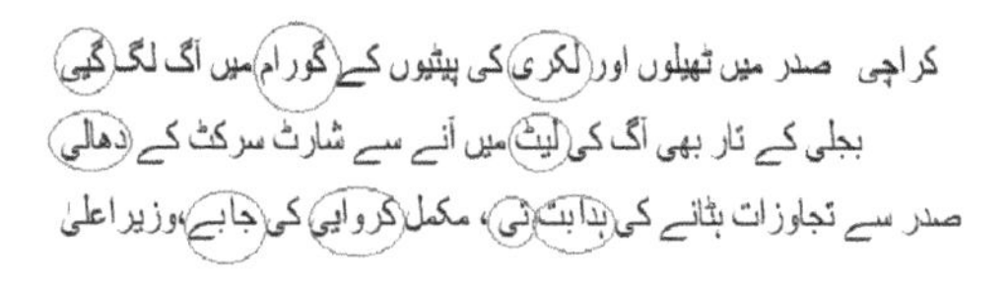

(b) Output of OCR

Fig. 2. News Channel Video Frame and Output of OCR

3 Related Work

This section presents a review of notable existing techniques for OCR post processing. For completeness, we first start with a discussion on the OCR systems followed by comprehensive review of techniques designed to post process the noisy OCR output. We next discuss the challenges and solutions to the segmentation of Urdu text.

Post-processing of text for error correction is an important step for any OCR output. While mature engines are available for most of the western languages; not much work has been done on cursive languages like Urdu and Arabic. Non-availability of benchmark datasets also hindered the progress of OCR systems for such languages. The development of datasets like UPTI [13] for printed text, UNHD [14] for handwriting and AcTiV [15] for video text contributed to this research area. While the performance of recognition systems for printed Urdu text is quite high [16], recognizing handwriting and video (caption) text is still challenging. On one hand, the recognition systems can be further investigated to improve the OCR performance while on the other hand, the noisy output of the OCR can be enhanced using effective post processing methods.

In practical applications where OCR generates transcription of huge collections, automatic post-processing is the only practical solution [17]. A common post processing method is the lexicon or dictionary-based error detection [18]. A major drawback of such method is that if the words are erroneous, the segmentation into words is not likely to be reliable. Furthermore, these methods do not take into account the context in which a word appears. To overcome these issues, context-based error detection and correction is commonly employed, which can effectively detect real-world errors according to the context of a word [19]. Aziz & Anwar [20], for instance, employ a dictionary to detect incorrect words, while candidate words are generated using Levenshtein Distance, Jaro Distance and Hamming Distance. The candidate words are ranked using three different methods namely Soundex, Shapex and the N-gram approach. They found Levenshtein distance and Jaro distance to produce better results as compared to the

Hamming distance. Doush et al. [21] proposed a hybrid system that uses both online and offline approaches for reducing OCR output errors. The offline part relies on language and error models, the language model being a table of words with frequencies while the error model exploits the edit distance to compute the similarity between two words. If the offline method fails to correct the error, the authors employ an online mode similar to what was proposed by Bassil et al. [22] where the tokens are fed to an online spell checker. The performance of the system was compared with that of a rule-based system using 1500 images containing Arabic text. Without post-processing, the error rate of the OCR was 20.99% which was reduced to 18.35% using the rule-based method while the proposed method reduced this number to 17.44%.

Nguyen et al. [23] used NMT with a pre-trained BERT model for post-ocr error correction. Their approach obtained results similar to the best-performing approaches on English datasets of ICDAR 2017/2019. In another work, Mukhtar et al. [24] employed an encoder-decoder framework of NMT with RNNs; sequence modelling was considered both at word and character levels. The experimental study was carried out on three different datasets and character-based sequence model ing was found to be better than word-level modeling.

The initial phase in any NLP task is typically word segmentation. In English and other western languages, words are separated through spaces but in case of Urdu text, space cannot be employed to tokenize the text in to words [25]. Consequently, intelligent techniques are required for segmenting words [26,27]. Such segmentation techniques are mainly divided into three categories:

- Dictionary or Lexicon-based segmentation techniques (dictionary matching, maximum matching and longest matching algorithms) [28–30].
- Linguistic or knowledge-based segmentation techniques [11].
- Machine learning-based segmentation techniques.
  - → 'Parts-of-Speech' (PoS) and 'Named Entity Recognition' (NER) [31].
  - → N-gram based language model [32].
  - → Neural network based models [33–35].

The output of word segmentation in Urdu is still problematic due to the cursive writing style, regardless of the method used. Table 3 provides a summary of techniques used for word segmentation and post-processing of cursive text.

Table 3. Literature Summary

Study	Dataset	Purpose	Method	Results
Akram et al. [25]	Urdu	Word segmentation	Statistical model using ligature and word statistics	96.10% word recognition rate
Manabu Sassano [28]	Japanese	Word segmentation	Maximum matching with fully lexicalized rules	91.7% words segmented correctly
Rabiya Rashid and Seemab Latif [30]	Urdu	Word segmentation	Dictionary based segmentation	97.2% words segmented correctly
Doush et al. [21]	Arabic	Error correction	Rule based system and hybrid system	Hybrid approach performed better than state-of-the-art rulebased approach [36]
Nguyen et al. [23]	English	Error detection and correction	NMT with pretrained BERT model	Their best model outperformed counterparts (ICDAR 2017, 2019 competition)
Mokhtar et al. [24]	Latin, German, English	Error correction	Word based NMT and character based NMT	Character based model performed better in all experiments

4 Proposed Methodology

The methodology for this study consists of three main steps: data collection and preparation, training a neural machine translation (NMT) transformer model and, evaluating the model using incorrect sentences. An overview of the entire methodology is presented in Fig. 3. First, a dataset from news sources (caption text and tweets) was collected. Caption text was the noisy output of a video OCR while errors we introduced in the tweets. The ground truth transcription was also available for performance evaluation. The prepared dataset was then used to train an NMT transformer model. Once the model was trained, it was fed with incorrect transcriptions and the recognition rates were computed with and without the proposed post-processing. Each of these steps is discussed in detail in the following.

4.1 Dataset

4.1.1 Data Collection

Data for this study was collected from two sources, caption text from local news channels videos as well as news tweets from the respective channels. A total of 4128 news tickers in Urdu were collected and to further enhance the dataset

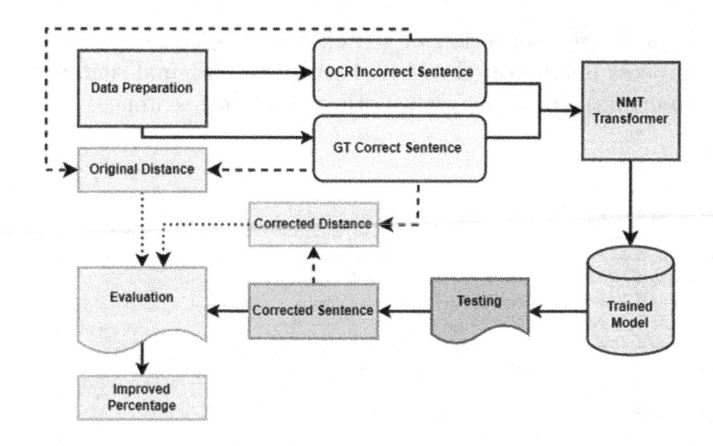

Fig. 3. An overview of the proposed method

size, tweets were also extracted from the official Twitter accounts of different Urdu news channels. Number of collected sentences from various sources are summarized in Table 4.

Table 4. Statistics of Data

Source	No. of Sentences
News channel videos	4,128
Samma Twitter	10,000
Geo Twitter	20,000
ARY Twitter	20,000
Dunya Twitter	10,000
Total	64,128

4.1.2 Data Cleaning

The data collected for this study was cleaned and processed to remove any extraneous or irrelevant information. This included removing special characters, emojis that were included in the text from tweets, and removal of redundant sentences.

4.1.3 Data Preparation

To create pairs of correct and incorrect sentences, for a given sentence, we subject it to an error introduction process multiple times. Errors were introduced randomly using insertion, deletion and substitution of characters. The number of errors to be introduced in a sentence was also varied and was a function of the length of the sentence. For each sentence, we repeated the process 9 times.

This resulted in a large collection of sentence pairs (64128×10) to train the model. The process is illustrated in Fig. 4 where the original sentence is paired with itself (no errors) as well as with 9 other erroneous sentences.

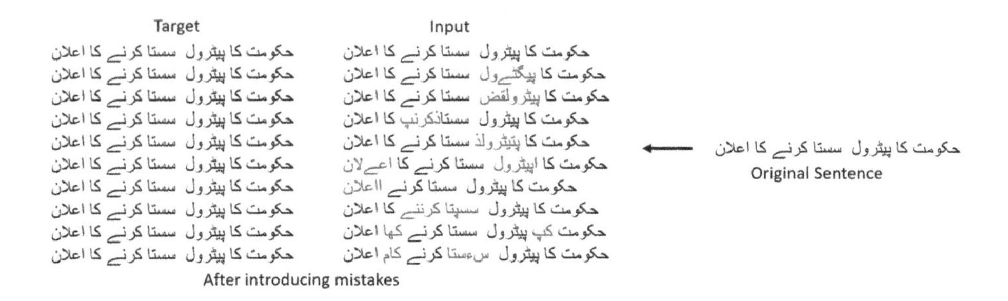

Fig. 4. Data preparation

Prior to feeding the sentence pairs to the model, we apply Byte Pair Encoding (BPE) with a character-level model. BPE is a data compression algorithm that employs a set of frequently occurring sub-strings, called **sentence** to represent the original data. By using a character-level model, we were able to encode the data at the level of individual characters as illustrated in Fig. 5. Such a representation is especially useful for handling languages like Urdu, which has a complex writing system and a large character set. It allowed us to effectively handle the inflections and variations present in the Urdu language.

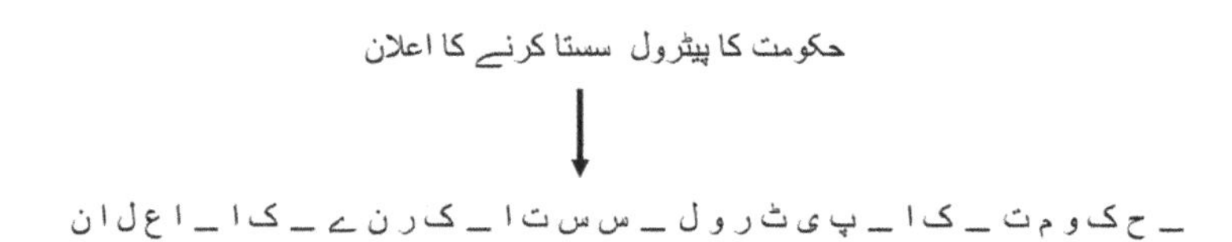

Fig. 5. Byte Pair Encoding of a sentence in Urdu

4.2 Model Architecture

We have chosen to employ a transformer-based neural machine translation (NMT) model to perform character-level error correction. The NMT system consists of an encoder module, which converts the source language (incorrect sentence) text into a continuous representation, and a decoder module, which generates the target language (correct sentence) translation using this representation.

A high-level model architecture is illustrated in Fig. 6 where the incorrect sentence is fed as input while the correct transcription is fed as the target. This type of neural network architecture uses self-attention mechanism [37] to capture dependencies between elements in the input and output sequences, allowing the model to effectively process long sequences and capture complex relationships within the data. The transformer model is known to have exceptional performance on machine translation task, and the current study explores its potential for error correction.

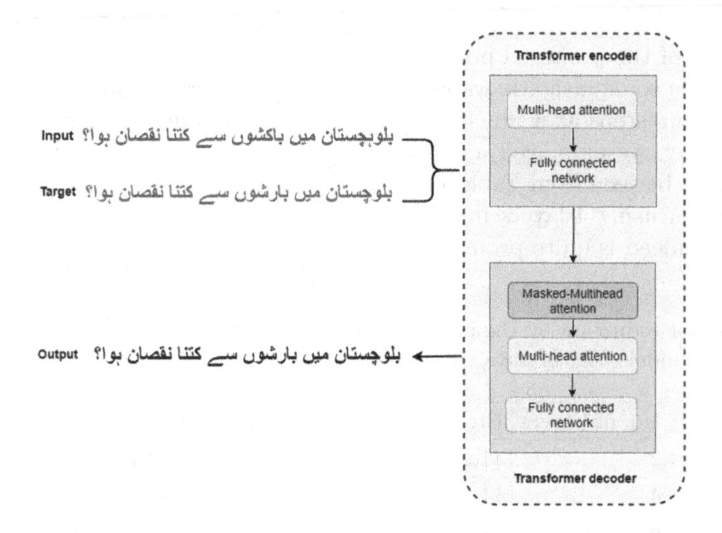

Fig. 6. High-level Model Architecture

During training, the model is fed incorrect and correct sentences in pairs to the encoder and the decoder respectively (with number of layers set to 4). Both encoder and decoder consist of multi-headed attention modules where the number of heads is set to 8. The purpose of multi-headed attention modules is to compute the attention weights for the input and producing an output vector with encoded information on how each word should attend to all other words in the sequence. Multi-headed attention modules are then connected to the fully connected network, which has a hidden dimension of 512, mapping the output of the self-attention layers to the final output predictions. The decoder consists of an additional module named masked multi-head attention, the purpose of this module in decoder is to perform self-attention on tokens up to the current position while preventing the model from accessing future information.

5 Experimentation and Results

For experimental study of our system, data was divided into the train (80%), validation (10%), and test (10%) sets. To evaluate the performance of our model, we calculated the improvement percentage (by employing Levenshtein distance) of the sentences corrected by the model. This was done by comparing the "original distance" between the ground truth correct sentence and the incorrect sentence, with the "corrected distance" between the ground truth correct sentence and the predicted corrected sentence produced by the model. The original distance represents the error present in the original, incorrect sentence, while the corrected distance reflects the error present in the model's corrected version of the sentence. Overall, the system reported an improvement of 56.92% with the introduction of the proposed post-processing step.

To provide some insights, we list (in Table 5) the errors before and after the post-processing steps as a function of number of layers in the model. Being an ablation study, we employed a subset (50% of data) of sentences to train the model. It can be seen from Table 5 that the number of errors reduce from 11,922 to 8,087 when using 4 layers in the model. This represents an improvement of 32% which indeed is quite promising.

Table 5. Error reduction by the proposed post-processing as a function of number of layers in the model (50% of data employed for training)

No. of Layers	Errors before	Errors After	Imp. perc.
2	11,922	9,234	21.79%
4	11,922	8,087	32.17%
6	11,922	8,432	29.27%

We also carried out a series of experiments to study the impact of the amount of training data on the overall performance. We varied the training data from 50% to 80% and the corresponding improvement in performance is illustrated in Fig. 7. It can be observed from the graph that the performance begins to stabilize with 70% of sentences in the training set.

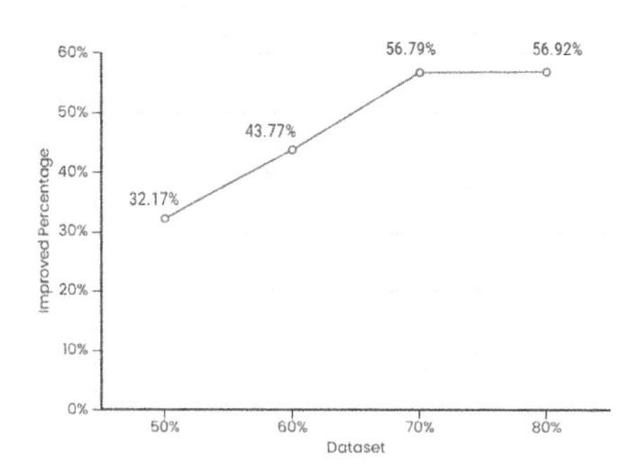

Fig. 7. Performance enhancement as a function of size of training data

6 Conclusion

In this study, we presented a transformer-based NMT model to correct errors in Urdu OCR output. Since word segmentation is known to be noisy for cursive text like Urdu, we employed a character-level Byte Pair Encoding (BPE) to represent the sentences. The NMT model is trained by feeding incorrect sentences to the encoder module and correct sentences to the decoder module as target. Consequently, the model learns to identify and correct the errors as a function of the context. Evaluation of the model on sentences collected from various news channels reported an improvement of nearly 57% by significantly reducing the errors. The findings of this pilot study are very encouraging and we intend to further enhance the system by employing a much larger dataset and more complex errors. We also plan to integrate this post-processing step with our existing caption detection and video-ocr system to provide a robust end-to-end solution.

References

1. Naz, S., Hayat, K., Razzak, M.I., Anwar, M.W., Madani, S.A., Khan, S.U.: The optical character recognition of Urdu-like cursive scripts. Pattern Recogn. **47**(3), 1229–1248 (2014)
2. Memon, J., Sami, M., Khan, R.A., Uddin, M.: Handwritten optical character recognition (OCR): a comprehensive systematic literature review (SLR). IEEE Access **8**, 142642–142668 (2020). https://doi.org/10.1109/ACCESS.2020.3012542
3. Naseem, T., Hussain, S.: A novel approach for ranking spelling error corrections for Urdu. Lang. Resour. Eval. **41**(2), 117–128 (2007)
4. Yuan, Z., Briscoe, T.: Grammatical error correction using neural machine translation. In: Proceedings of the 2016 Conference of the North American Chapter of the Association for Computational Linguistics: Human Language Technologies, pp. 380–386 (2016)
5. Hertel, M.: Neural language models for spelling correction. Methods **1**, 2 (2019)
6. El Atawy, S., Abd ElGhany, A.: Automatic spelling correction based on n-gram model, vol. 182, pp. 5–9 (2018)
7. Sharma, A., Srivastava, A., Vashishth, A.: An assistive reading system for visually impaired using OCR and TTS. Int. J. Comput. Appl. **95**(2) (2014)
8. Namysl, M., Konya, I.: Efficient, lexicon-free OCR using deep learning. In: 2019 International Conference on Document Analysis and Recognition (ICDAR), pp. 295–301. IEEE (2019)
9. Hassan, S., Irfan, A., Mirza, A., Siddiqi, I.: Cursive handwritten text recognition using bi-directional LSTMs: a case study on Urdu handwriting. In: 2019 International Conference on Deep Learning and Machine Learning in Emerging Applications (Deep-ML), pp. 67–72. IEEE (2019)
10. Zia, H.B., Raza, A.A., Athar, A.: Urdu word segmentation using conditional random fields (CRFs). arXiv preprint arXiv:1806.05432 (2018)
11. Durrani, N., Hussain, S.: Urdu word segmentation. In: Human Language Technologies: The 2010 Annual Conference of the North American Chapter of the Association for Computational Linguistics, pp. 528–536 (2010)

12. Shabbir, S., Siddiqi, I.: Optical character recognition system for Urdu words in Nastaliq font. Int. J. Adv. Comput. Sci. Appl. **7**(5) (2016)
13. Naz, S., et al.: Urdu Nastaliq recognition using convolutional-recursive deep learning. Neurocomputing **243**, 80–87 (2017)
14. Ahmed, S.B., Naz, S., Swati, S., Razzak, M.I.: Handwritten Urdu character recognition using one-dimensional BLSTM classifier. Neural Comput. Appl. **31**(4), 1143–1151 (2019)
15. Zayene, O., Hennebert, J., Touj, S.M., Ingold, R., Amara, N.E.B.: A dataset for Arabic text detection, tracking and recognition in news videos-AcTiV. In: 2015 13th International Conference on Document Analysis and Recognition (ICDAR), pp. 996–1000. IEEE (2015)
16. Sabbour, N., Shafait, F.: A segmentation-free approach to Arabic and Urdu OCR. In: Document Recognition and Retrieval XX, vol. 8658, pp. 215–226. SPIE (2013)
17. Hu, Y., Jing, X., Ko, Y., Rayz, J.T.: Misspelling correction with pre-trained contextual language model. In: 2020 IEEE 19th International Conference on Cognitive Informatics & Cognitive Computing (ICCI* CC), pp. 144–149. IEEE (2020)
18. Naseem, T.: A hybrid approach for Urdu spell checking. Master of Science (Computer Science) thesis at the National University of Computer & Emerging Sciences (2004)
19. Chen, S., Misra, D., Thoma, G.R.: Efficient automatic OCR word validation using word partial format derivation and language model. In: Document Recognition and Retrieval XVII, vol. 7534, pp. 217–224. SPIE (2010)
20. Aziz, R., Anwar, M.W.: Urdu spell checker: a scarce resource language. In: Bajwa, I.S., Sibalija, T., Jawawi, D.N.A. (eds.) INTAP 2019. CCIS, vol. 1198, pp. 471–483. Springer, Singapore (2020). https://doi.org/10.1007/978-981-15-5232-8_40
21. Doush, I.A., Alkhateeb, F., Gharaibeh, A.H.: A novel Arabic OCR post-processing using rule-based and word context techniques. Int. J. Doc. Anal. Recognit. (IJDAR) **21**(1), 77–89 (2018)
22. Bassil, Y., Alwani, M.: OCR post-processing error correction algorithm using google online spelling suggestion. arXiv preprint arXiv:1204.0191 (2012)
23. Nguyen, T.T.H., Jatowt, A., Nguyen, N.-V., Coustaty, M., Doucet, A.: Neural machine translation with BERT for post-OCR error detection and correction. In: Proceedings of the ACM/IEEE Joint Conference on Digital Libraries in 2020, pp. 333–336 (2020)
24. Mokhtar, K., Bukhari, S.S., Dengel, A.: OCR error correction: state-of-the-art vs an NMT-based approach. In: 2018 13th IAPR International Workshop on Document Analysis Systems (DAS), pp. 429–434. IEEE (2018)
25. Akram, M., Hussain, S.: Word segmentation for Urdu OCR system. In: Proceedings of the Eighth Workshop on Asian Language Resources, pp. 88–94 (2010)
26. Mahmood, A.: Arabic & Urdu text segmentation challenges & techniques. Int. J. Comput. Sci. Technol. **4**, 32–34 (2013)
27. Daud, A., Khan, W., Che, D.: Urdu language processing: a survey. Artif. Intell. Rev. **47**(3), 279–311 (2017)
28. Sassano, M.: Deterministic word segmentation using maximum matching with fully lexicalized rules. In: Proceedings of the 14th Conference of the European Chapter of the Association for Computational Linguistics, Volume 2: Short Papers, pp. 79–83 (2014)
29. Sproat, R., Shih, C., Gale, W., Chang, N.: A stochastic finite-state word-segmentation algorithm for Chinese. arXiv preprint cmp-lg/9405008 (1994)

30. Rashid, R., Latif, S.: A dictionary based Urdu word segmentation using maximum matching algorithm for space omission problem. In: 2012 International Conference on Asian Language Processing, pp. 101–104. IEEE (2012)
31. Khan, N.H., Adnan, A.: Urdu optical character recognition systems: present contributions and future directions. IEEE Access **6**, 46019–46046 (2018)
32. Singh, U.: Word segmentation problem in Urdu text and language model based solution (2016)
33. Cai, D., Zhao, H.: Neural word segmentation learning for Chinese. arXiv preprint arXiv:1606.04300 (2016)
34. Cai, D., Zhao, H., Zhang, Z., Xin, Y., Wu, Y., Huang, F.: Fast and accurate neural word segmentation for Chinese. arXiv preprint arXiv:1704.07047 (2017)
35. Raj, S., Rehman, Z., Rauf, S., Siddique, R., Anwar, W.: An artificial neural network approach for sentence boundary disambiguation in Urdu language text. Int. Arab J. Inf. Technol. (IAJIT) **12**(4) (2015)
36. Al Azawi, M., Breuel, T.M.: Context-dependent confusions rules for building error model using weighted finite state transducers for OCR post-processing. In: 2014 11th IAPR International Workshop on Document Analysis Systems, pp. 116–120. IEEE (2014)
37. Vaswani, A., et al.: Attention is all you need. CoRR abs/1706.03762 (2017). arXiv:1706.03762

Arxiv Tables: Document Understanding Challenge Linking Texts and Tables

Karolina Konopka[1], Michał Turski[1,2](✉), and Filip Graliński[1,2]

[1] Adam Mickiewicz University, Poznań, Poland
karkon13@st.amu.edu.pl, {mturski,filipg}@amu.edu.pl
[2] Applica.ai, Warsaw, Poland
{michal.turski,filip.gralinski}@applica.ai

Abstract. We introduce Arxiv Tables, a novel challenge for Document Understanding focused on tables, but in relation to text passages. In order to build the data set, we leverage arXiv, a large open-access archive of scholarly papers. We use both LaTeX source codes and graphical renderings of papers and combine tables with their references in the main text to create a quasi-Question Answering dataset by masking selected fragments available in the table. What distinguishes the dataset is that (1) the domain is science, (2) the input texts are longer than in typical Document Understanding Question Answering tasks, and (3) both the input and output contain non-standard characters used in scientific notation. For easier comparison for future research using this dataset, strong baselines are also given.

Keywords: Document understanding · OCR · Question Answering · table processing

1 Introduction

Recently, there has been a lot of progress in the relatively new domain of Document Understanding encompassing tasks such as classification, information extraction, question answering done for documents of rich layout and graphical structures, including elements such as tables, graphs, listings, formulae. A number of Transformer-based models were proposed for processing documents, such as LayoutLM [30], LAMBERT [9], TILT [23], also including end-to-end models, i.e. working without assuming an external OCR module, such as Donut [16], Dessurt [7], or Pix2Struct [17].

What is even more fundamental is that a number of Document Understanding datasets, challenges, and benchmarks have been proposed, examples include SROIE [13], Kleister [26], DUE [2]. Question Answering challenges are of special interest as they fit the generative nature of modern language models, see e.g. the DocVQA challenge [19]. It still seems that opportunities offered by some raw

K. Konopka—Work done as master thesis.

M. Coustaty and A. Fornés (Eds.): ICDAR 2023 Workshops, LNCS 14194, pp. 94–107, 2023.
https://doi.org/10.1007/978-3-031-41501-2_7

data sets have not been fully exploited. In this paper, we are using data available at arxiv.org, open-access archive for over 2.1M scholarly articles to create a novel Document Understanding task related to tables.

The advantage of arXiv is that not only final documents (mostly in the PDF format), but also their sources (mostly in LaTeX) are available. LaTeX is a rich language that can express not only text formatting but also tables and references between them and the main text (the `\label` and `\ref` commands). We leveraged this data to create a quasi-QA task related to tables, called Arxiv Tables.

The main contributions of this paper are as follows:

- we prepared and publicly shared a new large Document Understanding task[1],
- the evaluations on a number of non-trivial baselines were carried out[2].

2 Related Work

A number of challenges or even benchmarks (see e.g. the DUE benchmark [2]) has been proposed for Document Understanding. Some of them are in the Question Answering (QA) setup and one of the most popular is DocVQA [19], in which the documents were sourced from UCSF Industry Documents Library, "a digital archive of documents created by industries that influence public health", in general, is a popular source of documents for Document Understanding challenges. The questions for DocVQA were crowd-sourced, the average length of a question is only 8.12 words and about 1/4 of the questions are of 'table/list' type.

2.1 Tabular Question Answering

There are also a few QA datasets with tabular input. BioTABQA [18] is a Question Answering dataset in a biomedical domain. The authors used templates to create questions and answers to tables from a medical textbook. As all the tables have the same format, a template approach was applicable. Unfortunately, the dataset diversity is small: there are only 22 question-and-answer templates and only one table format. WikiTableQuestions [22] is a dataset of crowd-sourced question-and-answer pairs to randomly selected 2,108 Wikipedia tables. Another work in the domain of QA using tables from Wikipedia is HybridQA [5]. The authors created the dataset, where a table is contextualized by text, namely they used text passages linked by hyperlinks from the table cells. Question and answer pairs to these table-and-text examples were obtained by crowd-sourcing.

Arxiv Table dataset differs from the one mentioned about by domain – there has not been any QA or quasi-QA dataset using tables for the scientific domain.

1 The dataset is available at Gonito.net platform: https://gonito.net/challenge/arxiv-tables.

2 Scripts and reproduction instructions are available at https://github.com/applicaai/arxiv-tables-baselines.

This particular domain is challenging because of specific terminology and non-standard characters present in scientific texts, especially in tables with measurements. Our dataset is also bigger, both in terms of tables and number of examples (see Table 1).

Table 1. Comparison of different Question Answering and quasi-Question Answering tasks in domain of table understanding.

Dataset	Domain	Tables #	Examples #
BioTABQA	Biomedical	513	31k
WikiTableQuestions	Wikipedia	2k	22k
HybridQA	Wikipedia	13k	70k
Arxiv Table (ours)	Scientific	96k	127k

2.2 Table to Text

There is another line of work in Document Understanding called Table to Text. In this paradigm, the goal is to generate a textual description of a table. One of the most popular datasets of this kind is ROTOWIRE [29], where a model is required to generate a summary of an NBA match using statistics about it presented in a form of a table. Another dataset is LogicNLG [4], where the inputs are tables from Wikipedia and the goal is to "generate natural language statements that can be logically entailed by the facts in the table". The statements to be entailed were written by crowd-workers. NummericNLG [27] is a dataset from a scientific domain, where the source is articles from ACL Anthology website[3]. The tables are automatically extracted from the articles and the expected texts are obtained by a heuristic, which matches the table to its description in the text using the table's reference number. SciXGen [3] applies a very similar idea to scientific articles from arXiv, but model input is contextualized by a text before the reference. SciGen [20] is also a dataset of arXiv tables, but the expected output texts are table descriptions, which were manually written by experts in a particular scientific domain. There are also two datasets, where the expected output is a summary only of a particular part of a table, and the remaining content of the table serves as a context. The first of them is WikiTableText [1], in which the goal is to generate a summary of a highlighted row from a table. The second one is ToTTo [21], in which a model is required to generate a summary of a few highlighted cells. Both datasets use Wikipedia as a table source and the expected outputs were manually written.

The dataset presented in this paper differs from the one described in this section because of its character. While Table to Texts challenges are focused on generating text, the goal in our dataset is to infer one value. Because of that different metrics are used: Table to Text challenges are evaluated using text

[3] https://aclanthology.org/.

generation metrics like perplexity or BLEU, while Arxiv Tables uses accuracy. Text generation metric put more focus on generating coherent, grammatically correct text, while accuracy score (in the way we use it) promotes solution which is the best at extracting data from a table.

2.3 Table Extraction

Table Extraction is a problem where the goal is to extract either table structure, contents, or both from an image of a table. This task has been addressed in a number of datasets, see, for instance, TabLex [8] and PubTables-1M [25] as recent examples of large datasets of this type. Tables generated from LaTeX source codes were also included in these datasets. AxCell [15] (also known as PWC) is a dataset closely related to the table extraction problem, where the goal is to extract a machine learning leaderboard from a scientific paper. A model is required to present a leaderboard in a standardized manner, hence this task requires a model to perform some reasoning and normalization.

By comparison, Arxiv Tables is *not* a table extraction challenge, it is concerned more with understanding tables in the context of a text fragment.

3 Dataset

The Arxiv Tables dataset is based on articles published on the arXiv.org[4] website, which provides publicly available archives of scientific documents. It collects articles in mathematics, physics, astronomy, electrical engineering, computer science, quantitative biology, statistics, mathematical finance, and economics, and many of these include tables (examples are given in Fig. 1).

Each instance in the Arxiv Tables dataset consists of an image of a table, the quasi-question (referring to a given table) with masked information, and the expected answer (see an example in Fig. 2).

3.1 Preparation

The source data of the articles, which consist of compressed packages with LaTeX files, were collected in bulk from Amazon S3. Then the documents were automatically searched for the existence of tables and paragraphs that include a LaTeX reference (`\ref{}` command) to a table.

In the last stage, each paragraph with reference had key information masked, and the tables were saved as images. Training, test, and validation sets were created with a structure suitable for the Gonito.net[5] challenge [10].

Most of the packages are *TAR* archives with *GNU zip* the format or single files compressed to *GNU zip* format. All of them include LaTeX files, which were submitted to arXiv.org to produce PDF files after automatic processing by

[4] https://arxiv.org/.
[5] https://gonito.net/.

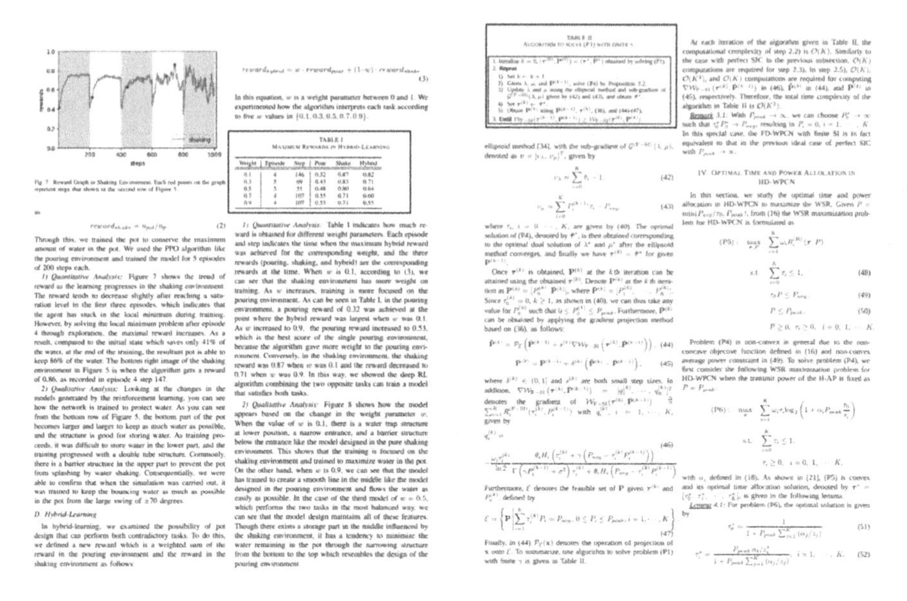

Fig. 1. Two different examples of pages from documents published on arXiv.org with tables marked. The table on the left contains numeric data [6] and the table on the right has one row that contains a description of an algorithm [14].

AutoTeX software. It should be noted that the Amazon S3 storage sometimes includes PDF files instead of source files and we filtered them out since they could not be automatically processed.

Each compressed file or archive was automatically checked for the presence of tables and LaTeX references. It is worth noting that only those paragraphs that referred to exactly one table were used for the final set to avoid ambiguity (the same value or name may appear in multiple tables, it's hard to say which table is the right context for a particular value). However, one table may have multiple LaTeX references that are used in the dataset. After extraction of every paragraph that includes LaTeX reference to only one table and creating images of the tables, the paragraph was automatically converted into a quasi-question that has one key information masked and the masked key information becomes an expected answer.

Expected answer in our dataset is a specific piece of information from a table, either:

- a text containing between 3 and 15 characters present explicitly in the table (an 'extractive' example),
- or one of the eight listed adjectives: `largest`, `smallest`, `better`, `worse`, `best`, `worst`, `more`, `less` ('Comparative' examples[6]; they do *not* have to be present in the table, the assumption is that we can check deeper reasoning capabilities of a model, not just ability to carry out information extraction, this way).

[6] On the evaluation platform they are denoted as 'degree'.

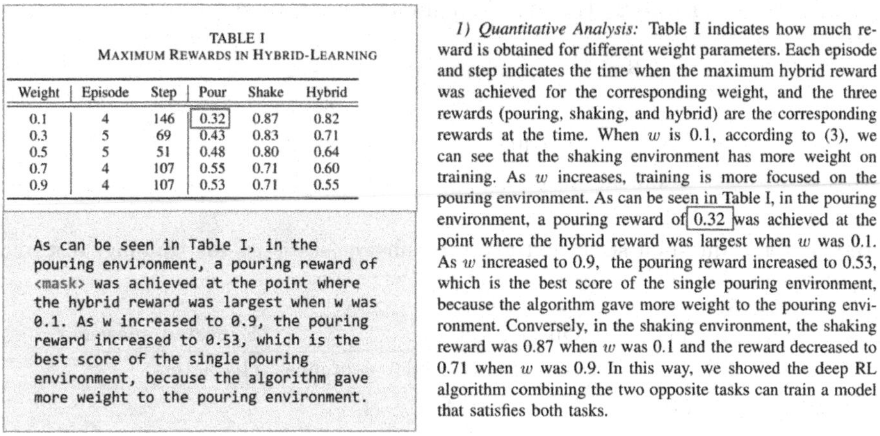

TABLE I
MAXIMUM REWARDS IN HYBRID-LEARNING

Weight	Episode	Step	Pour	Shake	Hybrid
0.1	4	146	0.32	0.87	0.82
0.3	5	69	0.43	0.83	0.71
0.5	5	51	0.48	0.80	0.64
0.7	4	107	0.55	0.71	0.60
0.9	4	107	0.53	0.71	0.55

```
As can be seen in Table I, in the
pouring environment, a pouring reward of
<mask> was achieved at the point where
the hybrid reward was largest when w was
0.1. As w increased to 0.9, the pouring
reward increased to 0.53, which is the
best score of the single pouring
environment, because the algorithm gave
more weight to the pouring environment.
```

1) Quantitative Analysis: Table I indicates how much reward is obtained for different weight parameters. Each episode and step indicates the time when the maximum hybrid reward was achieved for the corresponding weight, and the three rewards (pouring, shaking, and hybrid) are the corresponding rewards at the time. When w is 0.1, according to (3), we can see that the shaking environment has more weight on training. As w increases, training is more focused on the pouring environment. As can be seen in Table I, in the pouring environment, a pouring reward of 0.32 was achieved at the point where the hybrid reward was largest when w was 0.1. As w increased to 0.9, the pouring reward increased to 0.53, which is the best score of the single pouring environment, because the algorithm gave more weight to the pouring environment. Conversely, in the shaking environment, the shaking reward was 0.87 when w was 0.1 and the reward decreased to 0.71 when w was 0.9. In this way, we showed the deep RL algorithm combining the two opposite tasks can train a model that satisfies both tasks.

Fig. 2. An example of an image of a table [6] with text referring to it, along with a selected piece of information that can be masked because it is present in the table. In this case value 0.32 can be masked because it is referenced in the table.

However words such as `and`, `or`, `from`, `of`, `the`, `model`, `for`, `table`, were explicitly excluded as answers because of a lack of specificity.

3.2 Dataset Statistics

The total number of downloaded and correctly processed documents is 52883.

The resulting corpus, divided into a test, training, and validation sets, has the sizes presented in Table [2]. Expected answers in each subset contain around 25% of adjectives and 11% of numbers from the table, the remaining 64% are other kinds of entities. See Table 5 for some examples of expected answers.

The training subset has 3698 unique expected answers and the most frequent phrases are `more`(478), `better`(433), `best` (341), `less` (131), `largest` (82), `worse` (37), `smallest` (26), `worst` (25), `100` (22), `200` (17), `AUC` (12), `mean` (12), `0.5` (10), `300` (10), `1000` (10), `LSTM` (10), `time` (9), `RMSE` (9), `models` (9), `CNN` (9). 3035 values occur only once.

Table 3 presents statistics related to the length of quasi-questions. From these data, it can be concluded that most quasi-questions contain up to 500 words, on average they include almost 7 sentences, and quasi-questions that have around 13,000 words are an exception that is only present in the training set.

Table 2. Training, validation and test set sizes.

Subset	Quasi-questions #	Images #
train	114157	86978
test	6403	4932
validation	6039	4612

Table 3. Basic quasi-question and expected answer statistics for training, test and validation sets.

	Subset	Quasi-question			Expected answer
		characters	words	sentences	characters
MIN	train	8	4	1	3
	test	36	12	1	3
	validation	32	7	1	3
MAX	train	256382	12950	122	15
	test	18337	1250	50	15
	validation	11233	2234	84	15
MEAN	train	874.88	177.34	6.94	7.39
	test	864.74	175.16	6.85	7.06
	validation	868.69	176.32	6.94	7.11
MEDIAN	train	763	154	6	7
	test	763	154	6	6
	validation	761	154	6	6

3.3 Evaluation Procedure

The solutions are evaluated using GEval evaluation tool [11] using Accuracy evaluation metric. Values of Accuracy for the two subsets of items (adjectives and strings present in the tables) are also calculated as supplementary metrics.

4 Baseline Models

4.1 LayoutLMv3

The first of our baselines is LayoutLMv3 [12]. It is a Transformer-based [28] model for document understanding (mostly information extraction) tasks. As input to the model, we need an image of a document (in our case it is a table image), document tokens with bounding boxes, and a prompt. To produce tokens and bounding boxes of a table we used Microsoft Read v.3.2. The maximum length of a model input equals 512 tokens (for a quasi-question and document tokens together). Because of that, we decided to truncate a quasi-question to 25 words only, we use 17 words of context before `[Mask]`, the `[Mask]` token, and 7 words of context after the [Mask] . This set-up was motivated by the intuition that the context before the [Mask] is more important, as it is the natural order of reading. We feed the model with all the tokens from the truncated quasi-question and as many tokens from the table, as it fits in a vector of size 512.

LayoutLMv3 uses a sequence-labeling approach, hence to calculate loss we need annotations at the token level. Because of that the proposed baseline can only work with 'extractive' examples. To provide labels of 'extractive' datapoints at a token level, we proposed a heuristic for token labeling. It finds the expected answer in the table using fuzzy matching. The whole pipeline of training pipeline is presented in Fig. 3.

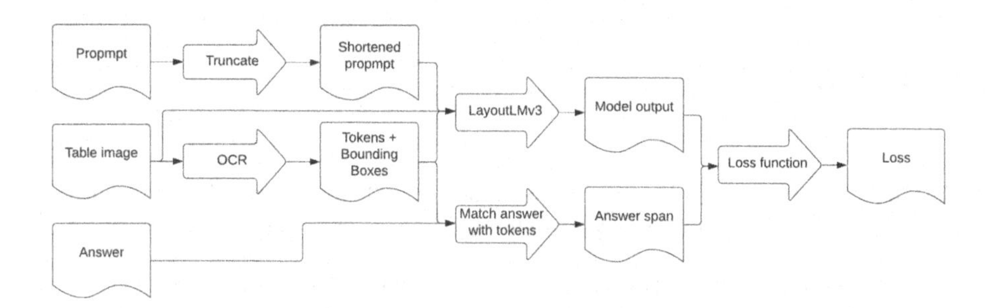

Fig. 3. Pipeline for training LayoutLMv3 baseline.

4.2 T5

We also provide some baselines in the sequence-to-sequence paradigm, namely T5 [24] and its adaptation for document understanding T5+2D [2]. Input to both models is a quasi-question together with tokens from a document. T5+2D uses also information about the bounding boxes of the tokens. Once again to provide tokens and bounding boxes we used Microsoft OCR v.3.2. The T5 family of models has no predefined maximum length of the input, hence the inputs we provided to the models were only limited by GPU memory (40GB), namely not more than 1450 tokens. Because of that, we did not truncate the quasi-question, an input was built out of the whole quasi-question and as many tokens from the table as possible. In more than 90% of cases, 1450 tokens input vector was long enough to contain the whole quasi-question and all the tokens from the table. As the model produces sequence as an output, we do not need any knowledge about exact answer span to calculate loss, we can just compare generated tokens with the expected answer. The whole pipeline is presented in Fig. 4.

Following the authors of the document understanding benchmark DUE [2], we trained 4 models using such a pipeline: the T5 model, the T5+2D model, which uses positions of tokens on a page, and their pretrained versions. T5 and T5+2D models are publicly available and pretrained ones were shared with us thanks to the courtesy of authors of the DUE benchmark.

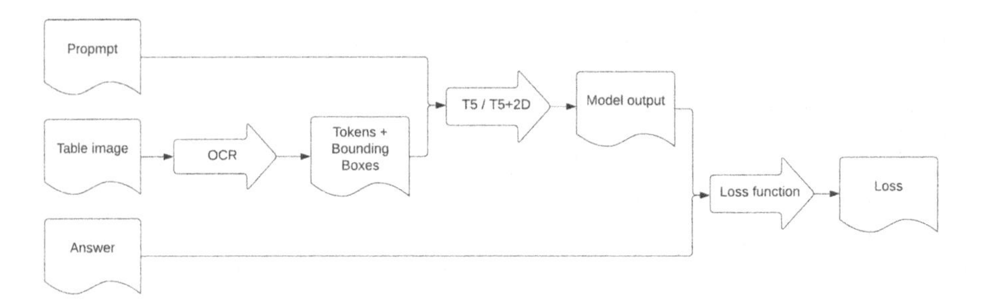

Fig. 4. Pipeline for T5 and T5+2D baselines. Bounding boxes are used only by the T5+2D model, plain T5 ignores them.

5 Results

Table 4. The detailed results (average accuracy over 3 runs) of our baselines on the test set. U stands for unsupervised pretraining. Value after ± is standard deviation.

Datapoints	LayoutLMv3	T5	T5+2D	T5+U	T5+2D+U
Extractive	$49.5_{\pm 1.1}$	$60.7_{\pm 0.4}$	$59.1_{\pm 1.1}$	$62.6_{\pm 0.8}$	$63.5_{\pm 0.3}$
Comparative	—	$86.4_{\pm 0.5}$	$86.0_{\pm 0.3}$	$86.0_{\pm 0.4}$	$85.1_{\pm 0.2}$
All	—	$66.9_{\pm 0.4}$	$65.6_{\pm 0.9}$	$68.3_{\pm 0.7}$	$68.7_{\pm 0.3}$

The detailed results are presented in Table 4. The weakest of the presented models is LayoutLMv3. Its performance was probably limited by three factors: length of the input sequence, OCR mistakes, and quality of token labeling. Firstly, we were not able to feed the whole quasi-question and all the tokens from a table to the model, for some tables expected answers were not even present in the tokens provided. Secondly, the OCR engine used works well with Latin characters, but when the table contains other symbols, such as Greek letters or special characters denoting units of measurement, it fails. The sequence labeling model does not have any mechanism for correcting OCR mistakes: it may only compose an answer from the input tokens. Finally, the heuristic for token labeling introduces its own mistakes, especially when applied to noisy OCR output. The sequence labeling approach cannot be applied to datapoints other than 'extractive' one, hence 'comparative' and 'all' scores were not presented.

Models from the T5 family were also affected by OCR mistakes, but theoretically, sequence-to-sequence models can generate answers not present in the input and, thanks to that, correct OCR (even if the OCR engine returned wrong predictions). Unfortunately, the T5 dictionary is also limited in terms of special characters, hence sometimes the models have no capability of generating a correct answer. They cannot even read these kinds of characters and they internally

Table 5. Sample of 40 errors for random tables from the validation set for the T5+2D+U model.

Expected answer	Actual answer
SFR	log [LIR,SF]
R[O III]	D600
W80,max	W80
D600	max
(3)	(1)
(5)	v2
(6)	(1)
64 × 4	64 4
128 × 8	128 8
r0/a	ZP
largest	smallest
0.07	0.02
100	NS2
less	more
changes	users
ICD-11	iCAT
ϵ/σ	/0
7 M$\odot$	= 0.5
best	bSNR
better	more
MNRM	coarser mesh
ML3	PGM3
const $- \sqrt{\epsilon}$	1/
best	smallest
s)	k + n2k
6.3	236.5
i = E*	i =
i=1 τ *	i=1
on-the-loop	mix
largest	smallest
slope	mass
SM(f1, f2, f3)	f' 3
OtWΦ	OtB
1998	1986
Ca i	FeI
Fe i	FeI
1988	X6708
Nijm I	AV18
more	less
0 1	1

represent them as special tokens [unknown]. As one may expect, pretrained models perform significantly better than their not pretrained counterparts. The not pretrained T5+2D model performed worse than the base T5 model. We hypothesize that it is caused by randomly initialized weights in layout related part of the model: they probably introduce more noise than useful information for the rest of the model. As far as the comparison between pre-trained T5 and T5+2D is concerned, there is not any significant difference in their performance. We think both models reached their glass ceiling caused by the level of OCR mistakes and limited dictionary, hence layout information does not help the T5+2D+U model to overcome these obstacles.

Note that the information about item types ('extractive' vs 'comparative') was not used during model training and testing, in particular models were not explicitly constrained to the eight adjectives/adverbs.

6 Error Analysis

A random sample of 40 errors for the current best (T5+2D+U) model is presented in Table 5. Six errors (15%) were directly caused by an OCR error (e.g. `64 4` instead of `64 × 4`), four errors (10%) are the wrong forms of an adjective/adverb ('comparative' type, e.g. `smallest` instead of `largest`), in one case of the 'comparative' type a non-adjective/adverb were returned (`bSNR` instead of `best`).

7 Conclusions

By leveraging the availability of a large number of scientific papers and their source codes, we introduced a new, relatively large challenge for the domain of document understanding. The results of strong baselines still show that reasoning about a text in the context of tables requires new ideas for document understanding models, as even layout-aware models brought no improvement over their non-layout-aware counterparts.

In the future, a similar approach can be applied to figures/graphs (`figure` environment in LaTeX) leading to a the quasi-Question-Answering challenge with even more pronounced visual aspects.

Acknowledgments. The Smart Growth Operational Programme partially supported this research under projects no. POIR.01.01.01-00-0144/17-00 (*Robotic processes automation based on Artificial Intelligence and deep neural networks*) and POIR.01.01.01-00-1624/20 (*Hiper-OCR - an innovative solution for information extraction from scanned documents*).

References

1. Bao, J., et al.: Table-to-text: describing table region with natural language. In: Proceedings of the AAAI Conference on Artificial Intelligence, vol. 32, no. 1 (2018). https://doi.org/10.1609/aaai.v32i1.11944. https://ojs.aaai.org/index.php/AAAI/article/view/11944

2. Borchmann, Ł., et al.: DUE: End-to-end document understanding benchmark. In: Thirty-Fifth Conference on Neural Information Processing Systems Datasets and Benchmarks Track (Round 2) (2021)
3. Chen, H., Takamura, H., Nakayama, H.: SciXGen: a scientific paper dataset for context-aware text generation. In: Findings of the Association for Computational Linguistics: EMNLP 2021, pp. 1483–1492. Association for Computational Linguistics, Punta Cana, Dominican Republic (2021). https://doi.org/10.18653/v1/2021.findings-emnlp.128. https://aclanthology.org/2021.findings-emnlp.128
4. Chen, W., Chen, J., Su, Y., Chen, Z., Wang, W.Y.: Logical natural language generation from open-domain tables. In: Proceedings of the 58th Annual Meeting of the Association for Computational Linguistics, pp. 7929–7942. Association for Computational Linguistics, Online (2020). https://doi.org/10.18653/v1/2020.acl-main.708. https://aclanthology.org/2020.acl-main.708
5. Chen, W., Zha, H., Chen, Z., Xiong, W., Wang, H., Wang, W.: HybridQA: a dataset of multi-hop question answering over tabular and textual data. In: Findings of EMNLP 2020 (2020)
6. Choi, J., Hyun, M., Kwak, N.: Task-oriented design through deep reinforcement learning (2019). https://doi.org/10.48550/ARXIV.1903.05271. https://arxiv.org/abs/1903.05271
7. Davis, B., Morse, B., Price, B., Tensmeyer, C., Wigington, C., Morariu, V.: End-to-end document recognition and understanding with Dessurt. arXiv e-prints, pp. arXiv-2203 (2022)
8. Desai, H., Kayal, P., Singh, M.: TabLeX: a benchmark dataset for structure and content information extraction from scientific tables. In: Lladós, J., Lopresti, D., Uchida, S. (eds.) ICDAR 2021. LNCS, vol. 12822, pp. 554–569. Springer, Cham (2021). https://doi.org/10.1007/978-3-030-86331-9_36
9. Garncarek, Ł, et al.: LAMBERT: layout-aware language modeling for information extraction. In: Lladós, J., Lopresti, D., Uchida, S. (eds.) ICDAR 2021. LNCS, vol. 12821, pp. 532–547. Springer, Cham (2021). https://doi.org/10.1007/978-3-030-86549-8_34
10. Gralinski, F., Jaworski, R., Borchmann, Ł., Wierzchon, P.: Gonito. net-open platform for research competition, cooperation and reproducibility. In: Proceedings of the 4REAL Workshop: Workshop on Research Results Reproducibility and Resources Citation in Science and Technology of Language, pp. 13–20 (2016)
11. Graliński, F., Wróblewska, A., Stanisławek, T., Grabowski, K., Górecki, T.: GEval: tool for debugging NLP datasets and models. In: Proceedings of the 2019 ACL Workshop BlackboxNLP: Analyzing and Interpreting Neural Networks for NLP, pp. 254–262. Association for Computational Linguistics, Florence, Italy (2019). https://www.aclweb.org/anthology/W19-4826
12. Huang, Y., Lv, T., Cui, L., Lu, Y., Wei, F.: LayoutLMV3: pre-training for document AI with unified text and image masking. In: Proceedings of the 30th ACM International Conference on Multimedia (2022)
13. ICDAR: competition on scanned receipts OCR and information extraction (2019). https://rrc.cvc.uab.es/?ch=13. Accessed 01 Feb 2023
14. Ju, H., Zhang, R.: Optimal resource allocation in full-duplex wireless-powered communication network (2014). https://doi.org/10.48550/ARXIV.1403.2580. https://arxiv.org/abs/1403.2580
15. Kardas, M., et al.: Axcell: Automatic extraction of results from machine learning papers. In: Proceedings of the 2020 Conference on Empirical Methods in Natural Language Processing (EMNLP), pp. 8580–8594 (2020)

16. Kim, G., et al.: OCR-free document understanding transformer. In: Avidan, S., Brostow, G., Cissé, M., Farinella, G.M., Hassner, T. (eds.) ECCV 2022. LNCS, vol. 13688, pp. 498–517. Springer, Cham (2022). https://doi.org/10.1007/978-3-031-19815-1_29
17. Lee, K., et al.: Pix2Struct: screenshot parsing as pretraining for visual language understanding. arXiv preprint arXiv:2210.03347 (2022)
18. Luo, M., Saxena, S., Mishra, S., Parmar, M., Baral, C.: Biotabqa: Instruction learning for biomedical table question answering. In: CEUR Workshop Proceedings, vol. 3180, pp. 291–304 (2022). Publisher Copyright: 2022 Copyright for this paper by its authors.; 2022 Conference and Labs of the Evaluation Forum, CLEF 2022; Conference date: 05–09-2022 Through 08–09-2022
19. Mathew, M., Karatzas, D., Jawahar, C.: DocVQA: a dataset for VQA on document images. In: Proceedings of the IEEE/CVF Winter Conference on Applications of Computer Vision (WACV), pp. 2200–2209 (2021)
20. Moosavi, N.S., Ruckl'e, A., Roth, D., Gurevych, I.: Learning to reason for text generation from scientific tables. ArXiv abs/2104.08296 (2021)
21. Parikh, A., et al.: ToTTo: a controlled table-to-text generation dataset. In: Proceedings of the 2020 Conference on Empirical Methods in Natural Language Processing (EMNLP), pp. 1173–1186. Association for Computational Linguistics (2020). https://doi.org/10.18653/v1/2020.emnlp-main.89. https://aclanthology.org/2020.emnlp-main.89
22. Pasupat, P., Liang, P.: Compositional semantic parsing on semi-structured tables. In: Proceedings of the 53rd Annual Meeting of the Association for Computational Linguistics and the 7th International Joint Conference on Natural Language Processing, vol. 1 (Long Papers), pp. 1470–1480. Association for Computational Linguistics, Beijing, China (2015). https://doi.org/10.3115/v1/P15-1142. https://aclanthology.org/P15-1142
23. Powalski, R., Borchmann, Ł, Jurkiewicz, D., Dwojak, T., Pietruszka, M., Pałka, G.: Going full-TILT boogie on document understanding with text-image-layout transformer. In: Lladós, J., Lopresti, D., Uchida, S. (eds.) ICDAR 2021. LNCS, vol. 12822, pp. 732–747. Springer, Cham (2021). https://doi.org/10.1007/978-3-030-86331-9_47
24. Raffel, C., ET AL.: Exploring the limits of transfer learning with a unified text-to-text transformer. J. Mach. Learn. Res. **21**(140), 1–67 (2020). http://jmlr.org/papers/v21/20-074.html
25. Smock, B., Pesala, R., Abraham, R.: Pubtables-1m: towards comprehensive table extraction from unstructured documents. In: Proceedings of the IEEE/CVF Conference on Computer Vision and Pattern Recognition (CVPR), pp. 4634–4642 (2022)
26. Stanisławek, Tomasz: Kleister: key information extraction datasets involving long documents with complex layouts. In: Lladós, Josep, Lopresti, Daniel, Uchida, Seiichi (eds.) ICDAR 2021. LNCS, vol. 12821, pp. 564–579. Springer, Cham (2021). https://doi.org/10.1007/978-3-030-86549-8_36
27. Suadaa, L.H., Kamigaito, H., Funakoshi, K., Okumura, M., Takamura, H.: Towards table-to-text generation with numerical reasoning. In: Proceedings of the 59th Annual Meeting of the Association for Computational Linguistics and the 11th International Joint Conference on Natural Language Processing, vol. 1 (Long Papers), pp. 1451–1465. Association for Computational Linguistics (2021). https://doi.org/10.18653/v1/2021.acl-long.115. https://aclanthology.org/2021.acl-long.115

28. Vaswani, A., et al.: Attention is all you need. In: Proceedings of the 31st International Conference on Neural Information Processing Systems, pp. 6000–6010. NIPS 2017, Curran Associates Inc., Red Hook, NY, USA (2017)
29. Wiseman, S., Shieber, S., Rush, A.: Challenges in data-to-document generation. In: Proceedings of the 2017 Conference on Empirical Methods in Natural Language Processing, pp. 2253–2263. Association for Computational Linguistics, Copenhagen, Denmark (2017). https://doi.org/10.18653/v1/D17-1239. https://aclanthology.org/D17-1239
30. Xu, Y., Li, M., Cui, L., Huang, S., Wei, F., Zhou, M.: LayoutLM: pre-training of text and layout for document image understanding. In: Gupta, R., Liu, Y., Tang, J., Prakash, B.A. (eds.) KDD 2020: The 26th ACM SIGKDD Conference on Knowledge Discovery and Data Mining, Virtual Event, CA, USA, 23–27 August 2020, pp. 1192–1200. ACM (2020). https://dl.acm.org/doi/10.1145/3394486.3403172

Subgraph-Induced Extraction Technique for Information (SETI) from Administrative Documents

Dipendra Sharma Kafle[1(✉)], Eliott Thomas[1], Mickael Coustaty[1], Aurélie Joseph[2], Antoine Doucet[1], and Vincent Poulain d'Andecy[2]

[1] Université de La Rochelle, L3i Avenue Michel Crépeau, 17042 La Rochelle, France
{dipendra.sharma_kafle,eliott.thomas,mickael.coustaty, antoine.doucet}@univ-lr.fr
[2] Yooz, Aimargues, France
{aurelie.joseph,Vincent.PoulaindAndecy}@getyooz.com

Abstract. Information Extraction plays a key role in the automation of auditing processes in administrative documents. However, variety in layout and language is always a challenging task. On the other hand, large volumes of public training datasets related to administrative documents such as invoices are rare to find. In this work, we use Graph Attention Network model for information extraction. This type of model makes it easier to understand the mechanism as compared to classical neural networks due to the visualization of link between entities in the graph. Moreover, it maximizes the layout and structure retrieval which is a crucial advantage in administrative documents. From the same graph, our model learns at different graph levels to encapsulate dynamic and more enriched knowledge in each batch, thus maximizing the generalization on smaller dataset. We present how the model learns in each graph level and compare the results with baselines on private as well as public datasets. Our model succeeds in improving recall and precision scores for some classes in our private dataset and produces comparable results for public datasets designed for Form Understanding and Information Extraction.

Keywords: Information Extraction · Invoice · Graph Neural Network

1 Introduction

Documents are an integral part of our society for recording and retrieving information. Especially, their wide use for our daily administrative tasks makes them highly relevant to automation. Since digitization and automation in the processing of business documents such as invoices, orders, payslips, etc. can save a lot of time and resource, it is necessary to make use of information extraction.

Extracting information from semi-structured documents like invoices has always been a challenge. Indeed, the quality of extraction tends to degrade when

M. Coustaty and A. Fornés (Eds.): ICDAR 2023 Workshops, LNCS 14194, pp. 108–122, 2023.
https://doi.org/10.1007/978-3-031-41501-2_8

introduced with challenging factors such as new layout for the document. Therefore, there is a crucial need of a unified approach that can generalize all sorts of document layouts while retaining the quality of extraction.

Apart from variety in layout and language, another issue includes information tags with high similarity to each other in terms of syntax, lexical semantics and position in the document. Based on this similarity, we distinguish them into two classes: major class and minor class. For our dataset with eight tags, we have five distinct tags like Document Type, Document Date, Document Number, Amount and Currency. Out of these five tags, two of them can be further divided into sub-tags. For example, Date can be divided into Document Date and Due Date. Similarly, Amount can be divided into Net Amount, Tax Amount and Total Amount. Hence, we consider the five distinct tags as major classes and their sub-divisions as minor classes.

It is difficult for the learning model to differentiate between minor classes as compared to major classes because minor classes have high similarity score for their respective text embeddings and they tend to be found near to each other in the documents. So, based on both the embeddings and the layout information, minor classes confuse the model during inference.

Nevertheless, we consider that along with text embeddings and layout information, the additional information of relation or links between words in the documents helps the model to infer better. Therefore, in this paper, we deal with these issues with the aid of graph structures with links.

In the recent past, there have been several works based on Transformers, Layout-methods as well as Graph-based methods which managed to achieve decent extraction results. However, not all of them focus on the minor classes too [2,7,11,18]. Contrary to those earlier works, we deal with both major and minor classes. Yet, we consider the fact that different classes might need different type of knowledge to learn meaningful patterns. Therefore, we train our model with different layout settings to examine the difference in learning for each individual class. In other words, the model learns at different graph-levels in order to capture more enriched knowledge. We test our method in Sect. 5 and compare our result with the baseline.

Overall, we aim to tackle the following two research questions:

1. How does the learning at different graph levels impact the model?
2. What are the advantages of Graph-based method and Transformers-based method over each other?

We use a dataset composed of English and French invoices to train and test our model. Furthermore, we compare our work with some related works which have used different approaches for the same task. Finally, we conclude with advantages and limitations of the approaches mentioned.

2 Related Work

To utilize Graph Neural Networks (GNN), we need to extract certain attributes or features from the graph and feed them into the model for learning. In this case, the features are text and layout information, which we extract embeddings and coordinates from respectively. Previous studies have used BERT, a state-of-the-art language model, to obtain text embeddings. As BERT [4] is based on attention mechanism, we choose to use a neural network with attention, such as GAT [18], to combine the strengths of both transformers and graphs.

2.1 BERT

BERT is a neural network that uses multi-head and multi-layer self-attention to summarize a sequence. Self-attention relates different positions in the sequence to create a summary representation. BERT is used in some papers for extracting information from invoice-like documents, often through sequence labeling tasks that identify tags like date, amount and address in the document.

In another study, BERT model was used for sequence labeling with Named Entity Reocgnition (NER) [6]. Bert architecture was combined with Conditional Random Field (CRF) in the final layer to extract major as well as minor classes. Layout embeddings were added to text embeddings from BERT and fed to the CRF layer. CRF has shown good results in the area of information extraction [5]. In fact, CRF used to be one of the most popular method for information extraction before the rise of neural networks.

Although BERT is effective at 1D information processing, it struggles with capturing multi-dimensional information. One approach to address this limitation was tackled via the use of 2D grids in the document as word-piece embedding vectors [3], but this method did not focus on minor classes. To capture subtle knowledge and relationships between words, lines, and blocks, Graph Neural Networks (GNNs) are needed to compute both local and global information.

2.2 GAT

Graph structure when combined with node level features produces better results in comparison to their individual performance [10]. Such mechanism of graph neural networks is called Graph Convolutional Network(GCN). Yet, GCN is a lesser suitable option for unseen data of different graph structure [10] as its aggregation function is dependent on the structure of the graph.

Since we aim to improve the generalization of our model, we use GAT for its special type of aggregation function. In contrast to GCN, GAT uses weights defined by attention mechanism for the node features instead of convolution. This technique is feature-dependent and includes structure-free normalization in the case of inference. Moreover, attention in GAT Version 2 is more dynamic than GAT, which yields better results [1]. The main difference between GAT and GATv2 is the linear combination of different attention mechanisms with varying number of heads, while GAT uses a single attention mechanism with a fixed number of heads.

Even if the Neural Network used for the embedding step is important, another main feature is related to the graph itself and its structure. Star graph is one of the popular layout as it is built by linking the surrounding nodes in all the directions, thus giving it the shape of a star. Usually, it is combined with GCN to extract information from invoices [13]. Conversely, we aim at minimizing the number of almost irrelevant links between nodes while retaining the quality of extraction. The irrelevant nodes can be defined as the words which are out of context and provide no meaningful information for a particular word.

Previous works have utilized image embedding for table and field representation in graph-based approaches. A recent study proposed a two-stage approach for invoice table extraction, with each stage dedicated to image embedding and graph representation, respectively [17]. However, our focus is on building a computation efficient model with less training data, so we only use text embeddings.

Apart from transformers and graph, Hamdi et al. [6] used a traditional class-based classification method for information extraction, achieving good results despite language and layout variations. The approach aimed to predict each word's class within its own context, using feature vectors like textual, spatial, structural, and logical to represent the document's words.

Since our work is based on the same training and testing data as that of Hamdi et al. [6], we consider their work as our baseline. In addition to that, we also compare our work with Cloudscan [15], which is referred as the state of the art in [6]. Cloudscan was trained on a recurrent neural network (RNN) model over 300k invoices to recognize eight tags. Lastly, we test our method with additional datasets such as FUNSD: Form Understanding in Noisy Scanned Documents [9] for form understanding and SROIE:Scanned Receipt OCR and Information Extraction [8] for invoice-like documents. These datasets are public but they are small and lack adequate amount of examples to train a good deep neural network.

In the case of SROIE, the baseline scores for information extraction task provided by the top 10 teams are recorded from a range of 77.38% to 90.49%. On the other hand, FUNSD provides a baseline of 57% for entity labeling task. Besides the original paper, FUNSD also have been used by some graph-based methods. One of them is Reading Order Equivariant Positional Encoding (ROPE) [11], a positional encoding technique that is designed to aid the sequential presentation of words in documents with the help of GCN. It defines unique reading order codes for neighbourhood words relative to the target word given a word-level graph connectivity. Their score on FUNSD for word labeling task is 57.22%. Another paper based on GAT architecture improved this score to 64% and claimed to be the state of the art [2]. However, in a recent paper, the tasks of entity labeling and relation extraction were combined together and the model identified relation between group of words with layout information [19]. The model was based on GCN and LSTM. Their score on FUNSD is 65.96%. On the other hand, layout method LayoutLMv3 [7] scored F1 accuracy of 92.08% and stands as the state of the art.

3 Methodology

3.1 Features

In order to capture the text syntax and lexical semantics, we introduce Boolean features that specify the content of the word. Apart from learning from the text embeddings obtained from the pre-trained BERT model, these Boolean features help the model distinguish between words related to different classes. For instance, amounts are almost always in numbers. These features are described as follows:

- isAlphabets: This indicates if the word is plain text with only alphabets.
- isPunctuation: This tells if the word-string is just a punctuation mark, for instance; full-colon, hyphen and so on.
- isNumeric: This indicates whether the word is numeric with digits.
- isMixed: This specifies if the word is mixed with two or more of the aforementioned features.

3.2 Model

Our model is based on GAT (Version 2) [1] works on self-attention mechanism. It has 5 layers, 16 attention-heads and 512 hidden dimensions with a Leaky Relu activation function. It computes individual weights for every node (word) of the graph (document), given the node features. We use Focal Loss for our loss function. Besides that, we select a learning rate of 1e−5 with Adam optimizer. All the hyper-parameters were selected on the basis of hyper-parameter tuning experiments.

3.3 Method

Our approach begins with the conversion of input document into graph representation. We consider every word in the document as nodes in the graph. As for edges, we create a local neighbourhood for nodes with K-Nearest Neighbours [16] algorithm and connect those nodes which are horizontally or vertically aligned with each other.

Since BERT sequence is one dimensional, we bring two dimensional knowledge with the help of graph. To achieve this, the coordinates of tokens are used as positions for nodes. Text embeddings from BERT combined with our Boolean features are encoded in the node features. For text embeddings, we use Camembert [14], which is a French language pre-trained model (to better fit to the dataset content).

However, we do not consider all the text embeddings in the document for training the model. In order to target only the relevant information, we implement selective-context approach. In this approach, we keep a distance threshold and neglect those embeddings that do not have one of our eight tags nearby. Hence, the number of nodes are decreased in the graph yet the relevant information about the context of the words is still preserved.

Since our data is highly skewed, general loss functions such as Cross Entropy Loss (shown below in Eq. 1) did not produce good prediction results. In order to preserve the invaluable node-to-node information, we avoid sampling techniques. Instead, we implement loss function such as Focal Loss. On top of that, we introduce mask for words that are not one of the 8 tags and ultimately ignore them in the loss function. Focal Loss [12] (shown in Eq. 2) is an extension of Cross Entropy Loss for class imbalance problem. It assigns higher weights to scarce examples and down-weights easily classified labels. Focal Loss penalizes even harder with (1- pt)**gamma to the cross-entropy loss, where gamma is a focusing tunable parameter.

$$\mathcal{L}_{CE}(y, \hat{y}) = -\sum_{i=1}^{n} y_i \log(\hat{y_i}) \tag{1}$$

where y represents the true label distribution and $\hat{y}$ represents the predicted probability distribution over C classes.

$$\mathcal{L}_{FL}(p_t) = -\alpha_t(1 - p_t)^{\gamma} \log(p_t) \tag{2}$$

where p_t is the predicted probability of the positive class, α_t is a balancing parameter, and γ is a focusing parameter.

Capturing more knowledge from the document leads to better learning for the model. Learning at graph level is decent, but going deeper and learning dynamic patterns improves it even further. This is why our model learns from three levels of graph. They are described as follows:

Graph Level. Graph level in our case corresponds to document level. In other words, model learns at document level. As we already have the coordinates of each word in the dataset, creating nodes at their specific locations is an easy task. The difficult task is to create edges between nodes in the next step. The edges must be optimized in such a way that a minimal number of edges should capture maximum neighbourhood information in order to save the computation cost while improving inference. One way of creating edges between relevant nodes is to look for the alignment of the nodes. As the key-value pair are usually located horizontally or vertically to each other, we create edges in the same fashion. However, the dataset provided by OCR might not contain exactly correct coordinates and so the nodes might deviate from straight angle by certain value. For that reason, we consider a threshold of N degree to create edges. In order to save high computation cost, we apply another threshold to create edges between closest K nodes. We use K-Nearest Neighbours [16] for this purpose. In this way, we filter redundant links between nodes in the graph. We take 5 documents/graphs per batch to train the model. An illustration of this is shown in Fig. 1.

Local Sub-graph Level. Contrary to Graph level, here we select only 1 document at once and extract sub-graphs from the document to feed them as batch for the model. In each batch, 5 nodes are chosen at random. For each of those

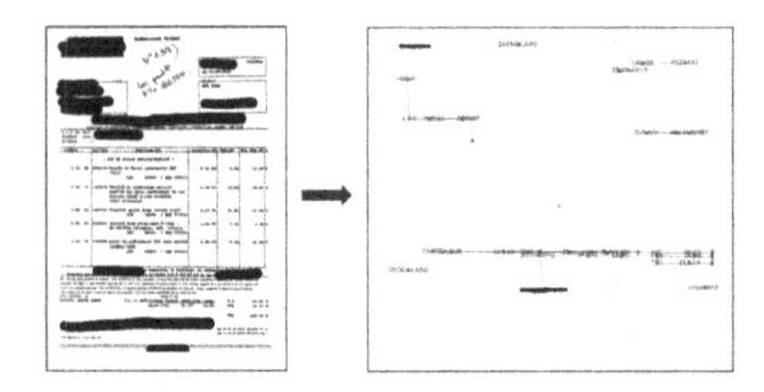

Fig. 1. Illustration of conversion of document into graph with selective context.

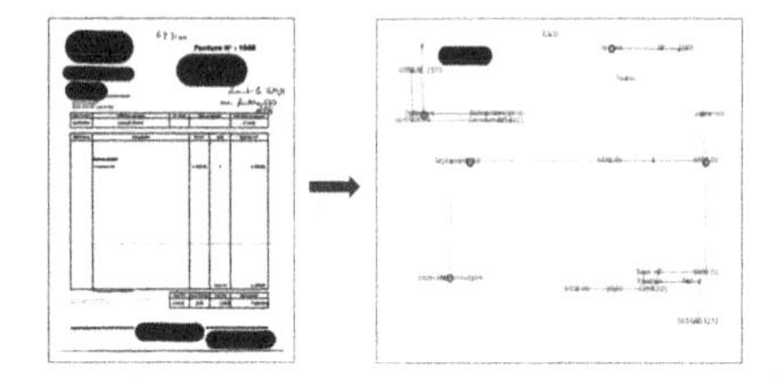

Fig. 2. Illustration of batch selection at local sub-graph level. From each document, the five nodes are selected for each batch (shown in yellow) at random. (Color figure online)

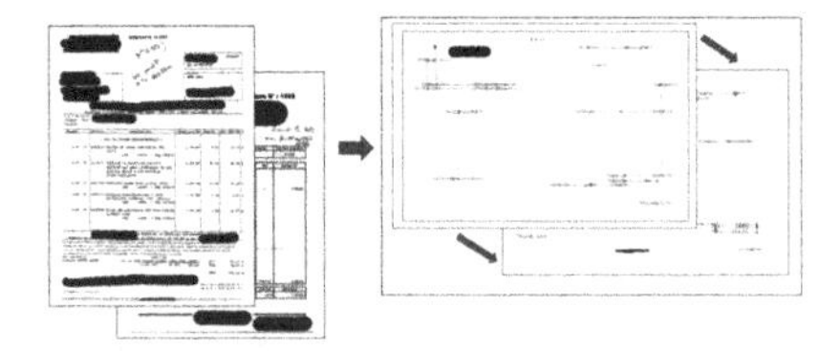

Fig. 3. Illustration of batch selection at global sub-graph level. Five nodes are selected for each batch (shown in yellow) at random from any document in the dataset. (Color figure online)

5 nodes, the neighbourhood will be considered for up to 5 hops from that particular node. In other words, in each batch 5 sub-graphs are taken. After all the nodes have been visited for the batch, we move on to the next document and follow the same procedure once again. Figure 2 describes how local sub-graphs are created and used as batch for the model.

Global Sub-graph Level. Lastly, we take all the documents in the training set and create one big global graph. A key note here is that though it is one big graph, the nodes from different documents are not connected to each other. Therefore, we can describe it as graphs stacked on top of each other, making it a heap of graphs. Next, we again select 5 nodes at random with upto 5 hops. However, this time the nodes can be selected from any document. Hence, in each batch 5 sub-graphs will be selected from any document in the training set.

4 Data

We use 20K dataset [6] of 19,775 (20k) documents in French and English languages for training. The test dataset [6] is a separate dataset of 4K documents in French and English with similar class distributions as the training dataset. The dataset contains 3 major classes: Document Type, Document Number and Currency, and 5 minor classes: Document Date, Due Date, Net Amount, Tax Amount and Total Amount. The class distribution of each tag in the training dataset are stated in the Table 1.

Table 1. Distribution of tags in 20k training dataset

Tag	Distribution
DOC_TYPE	14,301
DOC_NBR	18,131
DOC_DATE	17,739
DUE_DATE	9,554
NET_AMT	7,323
TAX_AMT	12,509
TOT_AMT	15,505
CURRENCY	12,111

As already mentioned above in Sect. 2, we also use FUNSD and SROIE to compare our model for the tasks of information extraction as well as form understanding tasks. FUNSD is designed for Text Detection, Optical Character Recognition, Spatial Layout Analysis as well as Form Understanding. It comprises 199 fully annotated forms. The train and test sets are split into 75% and 25% respectively. SROIE includes 1000 scanned receipt images and annotations for the competition on scanned receipts OCR and key information extraction (SROIE). There are 653 documents in the training set, and 357 documents in the test set.

5 Experiments and Results

In this section, we perform a series of experiments on our private dataset. In addition to that, we also experiment with public datasets such as SROIE and FUNSD to test its application to form understanding tasks.

5.1 Extraction at Different Graph Levels

In the first set of experiments, we show the performance of the model at different graph levels. The results of the experiments show that the model learns some tags in the graph better than the other tags at particular graph level.

Table 2 indicates that the best recall score for Document Number and Total Amount is achieved at the graph level classification, whereas the best recall score for document date and Net Amount is obtained at the local sub-graph level. Global sub-graph level is found to be the best for Document Type, Due Date, Tax amount, and Currency. For precision, local sub-graph level works best for document Total Amount and Currency, while global sub-graph level has the highest precision scores for Document Type, Document Number, Due Date, Net Amount, and Tax Amount. Graph level scored the best precision for Document Date.

Table 2. Comparison of our Graph-based model at different graph levels. Graph denotes Graph-level, Local means Local Sub-graph level and Global represents Global Sub-graph level. Highest Scores are shown in **bold**. All the models are trained with 20K dataset.

Tags	Recall			Precision		
	Graph	Local	Global	Graph	Local	Global
DOC_TYPE	86	**97**	**97**	94	89	**97**
DOC_NBR	**93**	77	91	60	85	**86**
DOC_DATE	47	**88**	36	**75**	66	72
DUE_DATE	41	20	**59**	51	67	**73**
NET_AMT	45	**85**	62	55	48	**57**
TAX_AMT	49	55	**69**	58	66	**68**
TOT_AMT	**65**	29	58	46	**85**	68
CURRENCY	80	90	**93**	84	**96**	92

We can infer that the prediction quality improved while going from graph level to sub-graph level. Nevertheless, we still need graph-level information because some best scores are associated with it. The level of graph hints at the type of knowledge the model needs to learn in the best way possible. The tags which need more specific and enriched knowledge of its local surrounding can be tackled with local sub-graph method. On top of that, the tags which also require more dynamic knowledge for generalization amongst various layouts works well at global sub-graph level. Hence, the extraction results answer our first research question by showing that the extraction quality especially for minor classes can be improved by using sub-graph level information in each batch.

5.2 Extraction by Our Method Vs Baseline

We compare the performance of our Graph method (SETI) with NER-based, Class-based, and Cloudscan approaches. We also compare BERT vs. SETI to answer our second research question. While the NER-based and Class-based methods were trained on 20k as well as 100k documents separately, Cloudscan was trained on 300k documents. On the other hand, our method was trained on the same 20K documents. Table 3 and Table 4 show the results for our method vs. baseline trained on 20k. Table 5 and Table 6 show results for our method (trained on 20K) and baseline (trained on 100k) as well as Cloudscan (trained on 300k).

In Table 3, we can observe that our method improved recall scores for Document Type, Document Number, Document date and Net Amount compared to the baselines. 3 out of 4 tags are improved at sub-graph level. Plus, our method performs better than BERT-based NER method for all the tags except for Due Date.

Table 3. Comparison of recall between Named Entity Recognition, Class-based and our Graph-based model at different graph levels. Best scores are highlighted in **bold**. Best scores for NER (BERT) vs Graph (ours) are associated with asterisk (*). All the models are trained with 20K dataset.

Recall					
Tags	NER-Based	Class-Based	Ours (Graph)	Ours (Local)	Ours (Global)
DOC_TYPE	81	85	86	97	**97***
DOC_NBR	67	86	**93***	77	91
DOC_DATE	78	33	47	**88***	36
DUE_DATE	**74***	70	41	20	59
NET_AMT	47	78	45	**85***	62
TAX_AMT	49	**78**	49	55	69*
TOT_AMT	49	**87**	65*	29	58
CURRENCY	82	**96**	80	90	93*

Table 4. Comparison of precision between Named Entity Recognition, Class-based and our Graph-based model at different graph levels. All the models are trained with 20K dataset. Best scores are represented by **bold**. Best scores for NER (BERT) vs Graph (ours) are associated with asterisk(*).

Precision					
Tags	NER-Based	Class-Based	Ours (Graph)	Ours (Local)	Ours (Global)
DOC_TYPE	**98***	97	94	89	97
DOC_NBR	74	**86**	60	85	**86***
DOC_DATE	**95***	92	75	66	72
DUE_DATE	**93***	91	51	67	33
NET_AMT	58*	**82**	55	48	57
TAX_AMT	66	**87**	58	66	68*
TOT_AMT	61	**89**	46	85*	68
CURRENCY	83	83	84	**96***	92

On the other side, our method improved precision scores for tags such as Document Number and Currency. For BERT vs. Graph, our method increased precision scores for 4 tags; Document Number, Tax Amount, Total Amount and Currency. Out of them, 2 are minor classes. Thus, in summary, our Graph-based method (SETI) has the potential to perform better than Transformers-based methods such as BERT for majority of tags.

In Tables 5 and 6, we compare our results with other baseline models trained with bigger datasets (20k vs 100k and 300k). In short, our model still managed to perform best for Document Type and Document Number in recall. In case of precision, NER-based and Class-based performed better than our method. For BERT vs Graph, our method still got better recall scores for all the tags except Due Date. Whereas, our Graph method could only improve precision scores for Total Amount and Currency.

Table 5. Comparison of recall between Named Entity Recognition, Class-based and our Graph-based models at different graph levels. NER-based and Class-based models are trained with 100K dataset. Cloudscan is trained on 300k dataset. Whereas, our Graph-based method is trained on 20k dataset. Best scores are shown in **bold**. Best scores for NER (BERT) vs Graph (ours) are associated with asterisk (*). (-) denotes tag is not available in Cloudscan.

Recall						
Tags	Cloudscan	NER-Based	Class-Based	Ours (Graph)	Ours (Local)	Ours (Global)
DOC_TYPE	-	79	90	86	**97***	**97***
DOC_NBR	84	69	89	**93***	77	91
DOC_DATE	77	78	**94**	47	88*	36
DUE_DATE	-	74*	**90**	41	20	59
NET_AMT	**93**	47	81	45	85*	62
TAX_AMT	**94**	49	79	49	55	69*
TOT_AMT	**92**	44	87	65*	29	58
CURRENCY	78	76	**98**	80	90	93*

5.3 Extraction by BERT vs SETI Analysis

Based on the results mentioned in Table 3 and Table 4, we present our findings as the answer to our second research question. We describe the advantages and limitations of both methods on a sample document from our dataset in Figs. 4 and 5.

The findings are described based on the following two factors:

1. **Structure**: We can find two types of information in a business document: structured and unstructured information. Structured information is presented in a consistent format across documents.Conversely, unstructured information is not organized in a specific format and does not follow a consistent pattern or arrangement. For our dataset, Due Date, Net Amount and Currency are unstructured information. Whereas, remaining five of them are structured (as shown in Fig. 6). From Tables 3 and 4, it can be observed that BERT mostly performed better for unstructured information. This is because BERT can leverage its pre-training on large amounts of text data to understand the contextual meaning of words and phrases to accurately extract the information. However, graph-based SETI performed better for structured information as it can leverage the structured nature of the information in documents and capture the relationships between different entities.
2. **Sequence**: Though BERT is bi-directional, it is still limited to one-dimensional sequence representation. This can be an issue because it will depend on how the OCR reads the document. As shown in Fig. 4, the OCR can read and store the n-grams in blue sequence or red sequence. This means BERT will process the n-grams in the same order. Thus, whenever the order

Table 6. Comparison of precision between Named Entity Recognition, Class-based and our Graph-based models at different graph levels. NER-based and Class-based models are trained with 100K dataset. Cloudscan is trained on 300k dataset. Whereas, our Graph-based method is trained on 20k dataset. Best scores are shown in **bold**. Best scores for NER (BERT) vs Graph (ours) are associated with asterisk (*). (-) denotes tag is not available in Cloudscan.

Precision						
Tags	Cloudscan	NER-Based	Class-Based	Ours (Graph)	Ours (Local)	Ours (Global)
DOC_TYPE	-	**99***	97	94	89	97
DOC_NBR	88	85*	**89**	60	85*	76
DOC_DATE	89	**96***	**96**	75	66	72
DUE_DATE	-	**96***	93	51	67	33
NET_AMT	**95**	62*	86	55	48	57
TAX_AMT	**94**	70*	90	58	66	68
TOT_AMT	94	63	**95**	46	85*	68
CURRENCY	**99**	90	84	84	96*	92

is wrong, BERT is vulnerable to learn with wrong contextual information. On the other hand, with SETI, it does not matter which words/n-grams the OCR reads at first. As long as their coordinates are preserved, SETI always builds sub-graphs based on proximity and alignment (which is the relevant contextual information). Unlike BERT, SETI makes it possible to collect relevant information even from distant line of the document with the aid of multiple hops between nodes (as shown in Fig. 5).

5.4 Extraction on FUNSD and SROIE

From Table 7, we can observe that our method SETI performs better than any other graph-based approaches. However, it does not provide excellent prediction results as LayoutLMv3 (based on multimodal transformers). However, we should also note the fact that LayoutLMv3 is based on text as well as image embeddings. It is evident from many works that multimodal approach [7] which also includes image embeddings performs better than a single text model. In contrast to that, we do not use image encoding in our model as our methodology is to experiment only with the text. Table 6 shows individual F1 accuracy scores for each model. For the SROIE dataset, the SETI approach outperforms nine other teams and reaches second place with a 0.20% difference with the top scoring team [8]. The results are presented in Table 8.

5.5 Advantages of SETI over Baseline Methods

Some of the main advantages of SETI over other methods such as BERT, Cloudscan, etc. are:

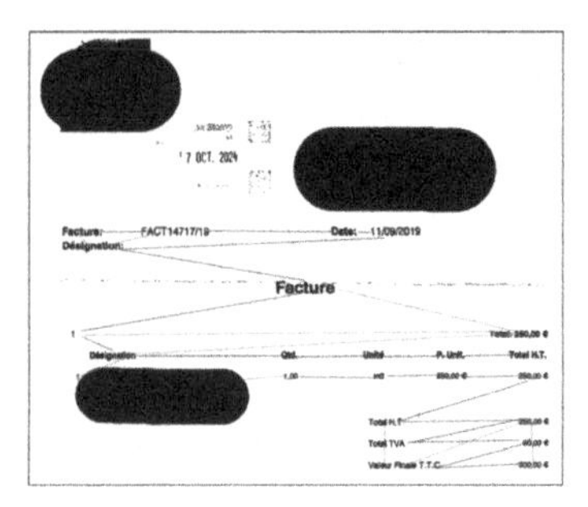

Fig. 4. Illustration of sequence followed by BERT to process the n-grams in the document. Red and blue links are alternate sequences which depend on the OCR algorithm. (Color figure online)

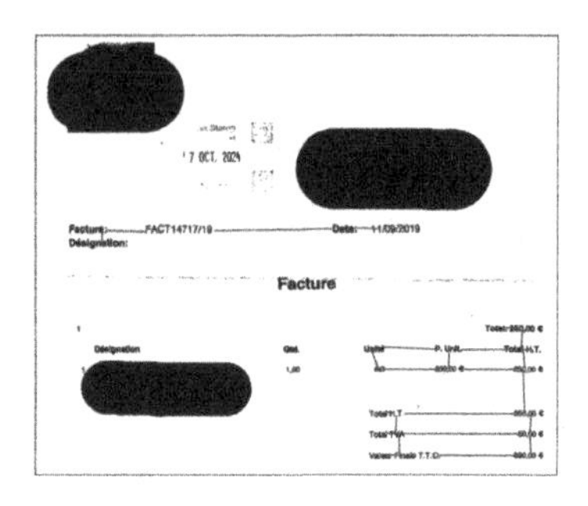

Fig. 5. Illustration of the formation of sub-graphs in SETI. The green links between n-grams represent edges formed based on proximity and alignment of n-grams (nodes). (Color figure online)

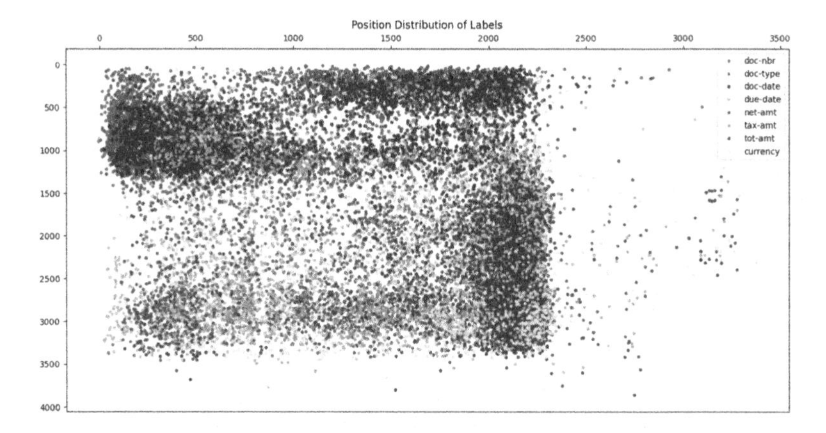

Fig. 6. Demonstration of the distribution of tags based on relative position of the tags in the documents in 20k dataset.

1. SETI represents information in a structured form, which makes it easier to analyze and interpret the information. It identifies patterns in the data by capturing entity relationships. It is more transparent than other methods such as BERT. Thus, it is easier to interpret, notably compared to BERT, which tends to be opaque.
2. It requires a lesser amount of data to produce results comparable to the baselines. Since our model learns at different graph (document) levels, it is able to grasp more enriched information with the same or an even lesser amount of data.

Table 7. Comparison of f1-scores between different related works based on graph, BERT and layout that used FUNSD as their dataset. Asterisk(*) represents best score out of graph-based approaches.

Model	F1-micro
FUNSD [9]	57.00
ROPE [11]	57.22
NER with Graph [2]	64.00
Dependency Parsing [19]	65.96
LayoutLMv3 [7]	**92.08**
SETI [Ours]	66.00*

Table 8. Comparison of F1-scores between top scoring team mentioned in SROIE paper and SETI.

Model	F1-micro
Top Socring Model in [8]	**90.49**
SETI [Ours]	90.29

6 Conclusion

In this work, we have presented different ways of targeting dynamic layout information for graph-based text models. We have shown how different tags require different type of neighbourhood information. With our more enriched knowledge from sub-graphs, we have succeeded to improve recall and precision scores for some of the tags in our private dataset. For public dataset, results from our method are comparable to the related works which are based on text models. With these findings, for future work, we aim to improve our selective-context approach and implement a cascading pipeline for better extraction since it aids in capturing a wider range of patterns and relationships in the data.

Acknowledgement. This work was supported by the French government in the framework of the France Relance program and by the YOOZ company. We would also like to thank Jérôme Lacour, Jonathan Ouellet and Mohamed Saadi from YOOZ for their support.

References

1. Brody, S., Alon, U., Yahav, E.: How attentive are graph attention networks? arXiv preprint arXiv:2105.14491 (2021)
2. Carbonell, M., Riba, P., Villegas, M., Fornés, A., Lladós, J.: Named entity recognition and relation extraction with graph neural networks in semi structured documents. In: 2020 25th International Conference on Pattern Recognition (ICPR), pp. 9622–9627. IEEE (2021)

3. Denk, T.I., Reisswig, C.: BERTgrid: contextualized embedding for 2D document representation and understanding. arXiv preprint arXiv:1909.04948 (2019)
4. Devlin, J., Chang, M.W., Lee, K., Toutanova, K.: BERT: pre-training of deep bidirectional transformers for language understanding. arXiv preprint arXiv:1810.04805 (2018)
5. Finkel, J.R., Grenager, T., Manning, C.D.: Incorporating non-local information into information extraction systems by Gibbs sampling. In: Proceedings of the 43rd Annual Meeting of the Association for Computational Linguistics (ACL 2005), pp. 363–370 (2005)
6. Hamdi, A., Carel, E., Joseph, A., Coustaty, M., Doucet, A.: Information extraction from invoices. In: Lladós, J., Lopresti, D., Uchida, S. (eds.) ICDAR 2021. LNCS, vol. 12822, pp. 699–714. Springer, Cham (2021). https://doi.org/10.1007/978-3-030-86331-9_45
7. Huang, Y., Lv, T., Cui, L., Lu, Y., Wei, F.: LayoutLMv3: pre-training for document AI with unified text and image masking. In: Proceedings of the 30th ACM International Conference on Multimedia, pp. 4083–4091 (2022)
8. Huang, Z., et al.: ICDAR 2019 competition on scanned receipt OCR and information extraction. In: 2019 International Conference on Document Analysis and Recognition (ICDAR), pp. 1516–1520. IEEE (2019)
9. Jaume, G., Ekenel, H.K., Thiran, J.P.: FUNSD: a dataset for form understanding in noisy scanned documents. In: 2019 International Conference on Document Analysis and Recognition Workshops (ICDARW), vol. 2, pp. 1–6. IEEE (2019)
10. Kipf, T.N., Welling, M.: Semi-supervised classification with graph convolutional networks. arXiv preprint arXiv:1609.02907 (2016)
11. Lee, C.Y., et al.: ROPE: reading order equivariant positional encoding for graph-based document information extraction. arXiv preprint arXiv:2106.10786 (2021)
12. Lin, T.Y., Goyal, P., Girshick, R., He, K., Dollár, P.: Focal loss for dense object detection. In: Proceedings of the IEEE International Conference on Computer Vision, pp. 2980–2988 (2017)
13. Lohani, D., Belaïd, A., Belaïd, Y.: An invoice reading system using a graph convolutional network. In: Carneiro, G., You, S. (eds.) ACCV 2018. LNCS, vol. 11367, pp. 144–158. Springer, Cham (2019). https://doi.org/10.1007/978-3-030-21074-8_12
14. Martin, L., et al.: Camembert: a tasty French language model. arXiv preprint arXiv:1911.03894 (2019)
15. Palm, R.B., Winther, O., Laws, F.: CloudScan-a configuration-free invoice analysis system using recurrent neural networks. In: 2017 14th IAPR International Conference on Document Analysis and Recognition (ICDAR), vol. 1, pp. 406–413. IEEE (2017)
16. Peterson, L.E.: K-nearest neighbor. Scholarpedia **4**(2), 1883 (2009)
17. Saout, T., Lardeux, F., Saubion, F.: A two-stage approach for table extraction in invoices. arXiv preprint arXiv:2210.04716 (2022)
18. Veličković, P., Cucurull, G., Casanova, A., Romero, A., Lio, P., Bengio, Y.: Graph attention networks. arXiv preprint arXiv:1710.10903 (2017)
19. Zhang, Y., Zhang, B., Wang, R., Cao, J., Li, C., Bao, Z.: Entity relation extraction as dependency parsing in visually rich documents. arXiv preprint arXiv:2110.09915 (2021)

Document Layout Annotation: Database and Benchmark in the Domain of Public Affairs

Alejandro Peña[1(✉)], Aythami Morales[1], Julian Fierrez[1], Javier Ortega-Garcia[1], Marcos Grande[1], Íñigo Puente[2], Jorge Córdova[2], and Gonzalo Córdova[2]

[1] BiDA - Lab, Universidad Autónoma de Madrid (UAM), 28049 Madrid, Spain
alejandro.penna@uam.es
[2] VINCES Consulting, 28010 Madrid, Spain

Abstract. Every day, thousands of digital documents are generated with useful information for companies, public organizations, and citizens. Given the impossibility of processing them manually, the automatic processing of these documents is becoming increasingly necessary in certain sectors. However, this task remains challenging, since in most cases a text-only based parsing is not enough to fully understand the information presented through different components of varying significance. In this regard, Document Layout Analysis (DLA) has been an interesting research field for many years, which aims to detect and classify the basic components of a document. In this work, we used a procedure to semi-automatically annotate digital documents with different layout labels, including 4 basic layout blocks and 4 text categories. We apply this procedure to collect a novel database for DLA in the public affairs domain, using a set of 24 data sources from the Spanish Administration. The database comprises 37.9K documents with more than 441.3K document pages, and more than 8M labels associated to 8 layout block units. The results of our experiments validate the proposed text labeling procedure with accuracy up to 99%.

Keywords: Document Layout Analysis · Legal domain · Data Curation · Natural Language Processing

1 Introduction

Nowadays, the Portable Document Format (PDF), originally developed by Adobe and standardized [11] in 2008, has become one of the most important file formats for digital document storing and sharing. The reason behind this success is the possibility to present documents including a variety of components (e.g. text, multimedia content, hyperlinks, etc.) in a format independent from the software, hardware and operating system. Furthermore, this file format allows encryption, compression, digital signature, and even interactive editing (e.g. form filling).

M. Coustaty and A. Fornés (Eds.): ICDAR 2023 Workshops, LNCS 14194, pp. 123–138, 2023.
https://doi.org/10.1007/978-3-031-41501-2_9

The advantages of the PDF format have converted it in a basic document tool for governments, administrations or enterprises. However, despite its usefulness, automatic processing of digital PDF documents remains as a difficult task. To correctly process and extract information from a document, it is required first to understand how the different components of the document are structured and how they interact with each other. For instance, processing information contained in a table usually requires to previously detect its basic structure. Even when it comes to text processing, text blocks in documents can be grouped into a variety of semantic levels (e.g. body text, titles, captions, etc.), which have different relevance and presentation formats. The way in which basic elements are presented in a document to effectively transmit its message is known as document layout. Once the document layout is clear, then modern Natural Language Processing (NLP) technologies (e.g., transformers [3,12] with attention mechanisms [21]) can be applied for generating useful outputs from segmented text blocks.

Document Layout Analysis (DLA) is a task that aims to detect and classify the basic components of a document. As we previously introduced, this task is a crucial component within the automatic document processing pipeline. Nevertheless, its usefulness is proportional to its difficulty. The main reason behind this fact is the large variability inherent in the problem. In this work, we propose a method to semi-automatically annotate a large number of digital PDF documents with their basic layout components. Our method combines a document collection procedure, the use of PDF miners to extract layout information, as well as a human-assisted process of data curation. We use this pipeline to generate a corpus of official documents for DLA in the legislative domain, which we call Public Affairs Layout (PAL) database. The source of the documents in this work are official gazettes from different institutions of the Spanish Administration. Official gazettes are periodical publications,[1] in which administrations include legislative/judicial information and announcements. Despite the fact that they originate from different administrations and countries, these documents usually present common features related with the spatial location and visual characteristics of the different text blocks. Take for example the document page images presented in Fig. 1. While Fig. 1.a is an example of a spanish official gazette, its layout it's similar to that of Fig. 1.b and Fig. 1.c, where page images from the french and the EU official gazettes are respectively depicted. Independently of the language of the document, a reader can easily identify the different text blocks (e.g., titles, body, summary).

In this context, the main contributions of this work are the following:

- We present Public Affairs Layout (PAL) database,[2] a new publicly available dataset for DLA, collected from a set of 24 different legislative sources from official organisms. The database comprises nearly 37.9K documents, 441.3K pages, and more than 8M layout labels.

[1] https://op.europa.eu/en/web/forum/european-union.
[2] https://github.com/BiDAlab/PALdb.

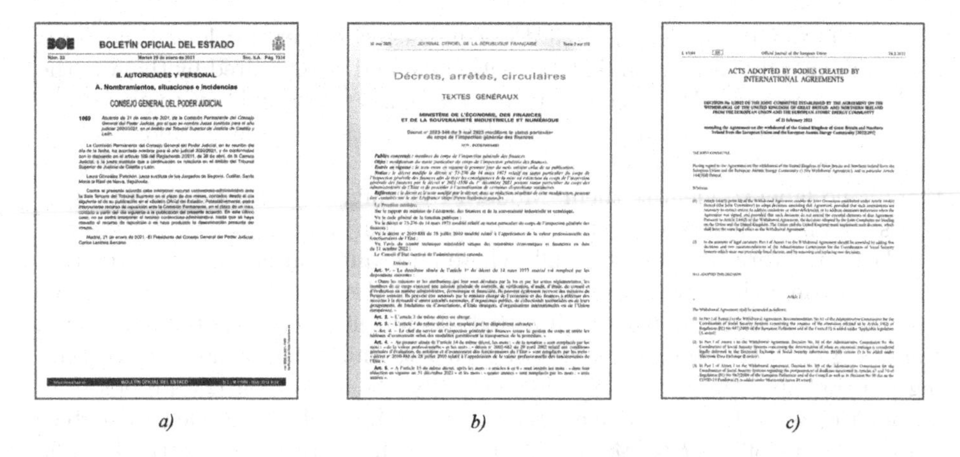

Fig. 1. Visual examples of page document images from different official gazettes: a) spanish gazette (i.e., BOE); b) french gazette (i.e., Légifrance); and c) Official Journal of the European Union.

- We provide layout information extracted from the documents, including the pre-processed cleaned text from the text blocks detected. Thus, in addition to the DLA dataset, a large corpus of public affairs text in spanish, and other 4 co-official languages, was collected for NLP pre-training and domain adaptation.
- We assess our text labeling strategy with different experiments, in which we prove the usefulness of the information extracted to classify text blocks into different semantic classes defined after an empirical analysis of the data sources.

The rest of the paper is structured as follows: Sect. 2 provides a review of different works concerning document layout analysis and automatic digital document processing. In Sect. 3 we present our semiautomatic procedure to collect and annotate our legislative document database, as well as its details. Section 4 presents the experiments and results of this work. Finally, Sect. 5 summarizes the main conclusions.

2 Related Work

The literature on Document Layout Analysis (DLA) differences between two types of PDF documents: 1) native or digital-born documents and 2) document image. The former are those originally created from a digital version of the documents, while the latter are scanned images captured from a physical document, or digital-born documents which were converted to images. This distinction leads to different approaches on how to extract their main layout components.

With regard to PDF native documents, the availability of all document information within the internal PDF structure makes the use of PDF miner tools the most common approach to extract layout information. A large number of tools exist for layout extraction in digital documents. However, for long time these tools were mainly focused on text extraction [2], and were limited by the way the PDF format processed their components (it specifies where and how to place individual components, without using high level semantic information about them). This behavior also makes difficult to detect layout elements such as tables, as there is no label to identify them. Modern PDF miners had learn to work with this structure, but elements such as tables remain difficult to detect. We can cite here the work of Bast *et al.* [2], where an evaluation of 14 text extraction tools, including their own, was conducted. They also proposed a benchmark for text extraction methods from digital PDF documents, and collected a database consisting on 12K scientific articles from *arXiv*, which they annotated by parsing the corresponding TeX files. More recently, the authors of [23] proposed an automatic method to annotate a large corpus of digital PDF articles, by matching the output of the *PDFMiner*[3] library with the XML representations of the articles. They released a page image database, known as PubLayNet, and later made available the original native PDF documents used to create it.

On the other hand, DLA on document images has been addressed as an image processing task with the use of Computer Vision techniques. By not having access to internal information of the documents, especially when working with scanned documents, databases in this domain were mostly annotated at hand, which ultimately limited their size. For instance, we can cite here the datasets collected for the ICDAR document processing challenges [1,5,6], which included complex documents with realistic layouts, the ICDAR 2013 Table Recognition Challenge dataset [9], or the UW III and UNLV datasets. Other works present their own manually annotated databases based on scientific articles [14,20], with the previously mentioned PubLayNet [23] being the larger database (i.e. nearly 350K page images) thanks to an automatic annotation method.

Early approaches for conducting DLA on document images included text segmentation techniques [4,8] or the use of HoG features [13,18] to perform the task. More recently, deep learning-based methods have been applied, specially the use of R-CNN object detection models [10,17]. In [20] a combination of F-RCNN [17] with contextual information was proposed to perform this task. The authors of PubLayNet [23] used both F-RCNN and M-RCNN [10] in their experiments on the novel database. Oliveria *et al.* proposed [14] the use of 1-D CNNs as both an efficient and fast solution for DLA. They used the running length algorithm [22] to detect regions of information in grayscale images, and detect blocks as regions connected after a 3×3 dilatation operation. Then, the network classifies the blocks using both vertical and horizontal projections of these blocks.

3 Semiautomatic Document Layout Annotation

In this Section, we will present the Public Affairs Layout (PAL) Database, a new database for DLA in the legal domain, with special focus in official documents.

[3] https://github.com/euske/pdfminer.

More specifically, Sect. 3.1 presents the data sources, our data collection method, and the details of the final database. Section 3.2 describes the tools used to extract layout information from the documents, the features extracted, and the different layout components of the database. Finally, in Sect. 3.3 we explain our semiautomatic method to classify the text blocks extracted into different semantic categories, and the following data curation procedure.

3.1 Document Collection

Spain has a wide variety of sources of public affairs documents. All the judicial, royal and governmental decrees, as well as the laws approved by the congress, have to be published in the daily official gazette. Our data collection involves 24 public sources including 3 national administrations, 19 regional administrations, and 2 local administrations in charge of the different territories in which Spain is politically/administrative divided (i.e. autonomous communities/cities). There are 5 co-official languages in the spanish territory, and each region has the freedom to publish official documents in their own format. All of this generates a great variety of styles and formats.

We included in our database a total 24 different sources of information from official organisms. The main characteristics of these data sources are summarized as follows:

1. The availability of historical repositories of PDF files with more than ten years of almost daily publications (i.e. there are usually no publications on Sundays).
2. The diversity of document layouts, which is different for each source and it has been changing over the years for each one.

Table 4 in the Appendix presents the full list of data sources used in this work. Since we had access to historical repositories of all the publications, we used an automatic web scrapping method to download all the documents. We then filtered documents published before 2014, and used the most current ones in our work. This allowed us to discard scanned-image PDF files corresponding to old publications, which were left out of the scope of this work. We use as web scraping backbone the Python library Scrapy[4], concretely the Spider class, where we defined how the website would be parsed.

The PAL database comprises 37,910 documents, in which we have 441.3K document pages and 138.1M tokens. Attending to the layout labels, we can find 1M images, 118.7K tables, 14.4K links, and 7.1M text blocks. The database is divided into a train set, and a validation set, where text labels were validated by a human supervisor, as we will comment later in Sect. 3.3. The train set is composed of 36,466 documents, 422K document pages and 130.5M tokens, with 1.1M images, 145.2K tables, 16.3K links and 8.8 M text blocks. The list of sources we used to collect our data, as well as the number of documents, pages, tokens and examples from each layout category included in our validation set

[4] https://scrapy.org/.

Table 1. Description of the PAL database (validation set). We provide statistics on the number of documents, pages and tokens for each source, along with the number of examples of each layout category.

Source ID	#Doc.	#Pages	#Tokens	Layout Components						
				#Images	#Tables	#Links	#ID	#Title	#Summary	#Body
1	193	602	182.5K	1206	179	1	2296	2463	192	5540
2	16	231	143.3K	0	0	0	715	1040	11	2138
3	14	224	48.9K	19	150	0	743	1369	62	2562
4	28	857	329.5K	80	141	0	2735	6287	199	9518
5	50	403	176.6K	845	106	2	810	1647	114	5293
6	30	884	189K	65	449	105	3034	3108	144	6718
7	122	393	205.9K	2	93	1	786	1695	128	7166
8	44	649	299.1K	5593	176	3	793	1702	283	7052
9	96	570	139.1K	6	346	103	1709	1189	102	5114
10	13	1046	736.8K	1023	279	268	275	5394	214	13.3K
11	75	476	170K	102	141	3	928	1496	76	4277
2	41	742	367.3K	725	128	145	2232	5146	251	15.7K
13	43	310	118K	600	17	27	930	727	55	3742
14	43	281	131.2K	865	98	275	774	1164	42	3670
15	50	225	69.5K	8	38	0	659	852	50	2737
16	13	383	204.4K	1064	140	22	382	1209	73	4100
17	142	315	143.9K	7	72	372	941	1283	157	4899
18	35	1064	297.2K	12304	273	5	2127	1410	224	9766
19	10	1302	532.5K	1511	4384	0	2592	1962	470	15.4K
20	40	1049	348.3K	1049	63	0	2098	1951	44	32.1K
21	32	887	323.8K	862	114	268	1779	2320	293	15.8K
22	57	549	453.3K	626	309	0	547	1941	210	4534
23	61	4771	1.45M	15468	1919	413	9608	12563	882	48.9K
24	197	1064	454.8K	1071	17	23	1049	2006	282	14.4K
Total	**1444**	**19276**	**7.52M**	**45.1K**	**9632**	**2036**	**40.5K**	**61.9K**	**4558**	**244.4K**

are presented in Table 1. We set a minimum of 10 documents and 200 pages for each data source in the validation set. Note that some sources present significant differences in the pages per documents relation (e.g. for Source 1 the relation is roughly 3 pages per document, while for Source 19 is 130). This is due to the nature of the documents we had access to. While all the documents we downloaded are official gazettes, these gazettes were available in two different formats: 1) full gazette contained in one document, and 2) the gazette divided in different, individual documents, each one containing a section or announcement. Most gazettes are available in both formats, but the access we had to these using the Spiders varies between publications. Another interesting relation is the number of tokens vs the number of pages, which is significantly higher for two sources, namely Source 22 (i.e. 825.7 tokens per page) and Source 10 (i.e. 704.4 tokens per page). These publications are the only ones among the 24 to present a two-columns format, so the number of tokens per page is naturally higher.

All the documents we collected are PDF native format, that is, they are not scanned images of a document, rather they were originally created from a digital version of the document. This fact allowed us to use PDF miner tools to identify the main document layout components, and extract information related to these elements. In the following Section, we will introduce the different layout components extracted in this work, as well as our semiautomatic annotation algorithm.

3.2 Layout Components Extraction

We considered 4 main layout categories or *blocks* in this work: 1) image; 2) table; 3) link (i.e. a region in the document associated to an external URL), and 4) text blocks. We used 2 different PDF miner libraries to extract these components from the documents:

- **PyMuPDF.**[5] This is a Python binding for MuPDF, a powerful PDF viewer and toolkit. We use this tool to extract image, link, and text blocks from each page of the documents. PyMUPDF not only allows us to detect these blocks (i.e. returning their position as a 4-tuple bounding box (x_0, y_0, x_1, y_1)), but also returns information about them (e.g. the raw text of the blocks, font type, size, etc.)
- **Camelot.**[6] A Python library to extract tables from PDFs. This library detects the position of the tables in each page by getting both the vertical and horizontal lines composing the table, and computing their boundaries (again as a 4-tuple bounding box). Then, it extracts their information in a *pandas dataframe* that preserves their structure, which can be exported in different formats.

It's important to note that these libraries work only for PDF native documents, therefore any document image scanned included in them was treated as a simple image. For each input PDF file, we extract the layout information and the information from all the tables detected. We also generate a version of the input PDF file annotated with layout information, which allowed us to visually assess the results of the extraction.

When extracting tables with Camelot, we adapted the bounding boxes to PyMuPDF's coordinate system, which considers the origin (0, 0) in the top left corner (see Fig. 2). We also had to consider the case of rotated tables (i.e. wide tables that appear rotated to fit in a whole page), which presented another coordinate system different from normal tables.

We extracted the features presented in Table 2 for the 4 layout blocks studied in this work. As we commented before, thanks to PyMUPDF's tools we had access to different font characteristics from the text blocks, including size, font type, bold or italic information. This allowed us to define several features describing the text blocks. Note here that our approach is limited to the data contained in the PDF structure of the document. Hence, the feature extraction depends on this information, and ultimately on the editor used to create the files. We

[5] https://pymupdf.readthedocs.io/en/latest/.
[6] https://pypi.org/project/camelot-py/.

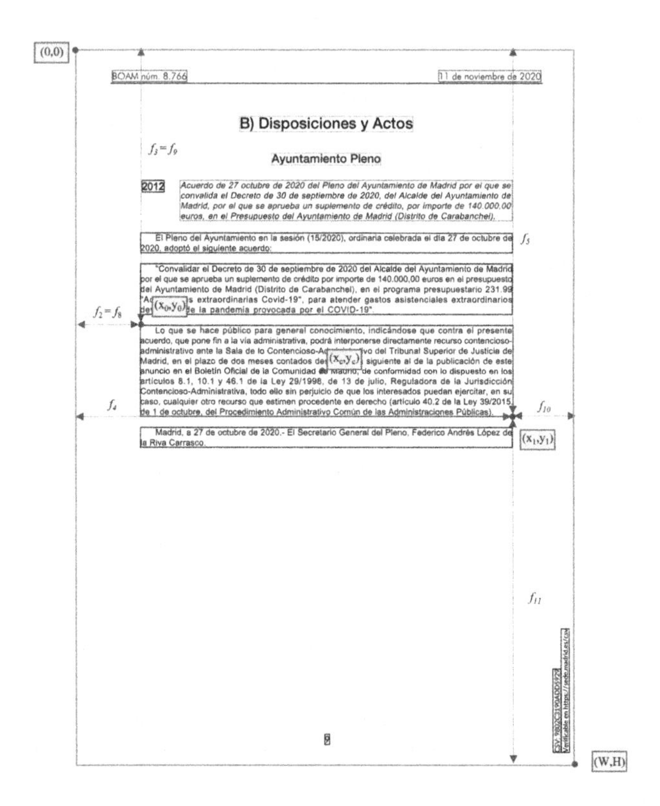

Fig. 2. Document page image with Layout component annotations (L in Table 2). The color codification is green for *Identifier*, pink for *Title*, cyan for *Summary*, and black for *Body*. We also illustrate the coordinate system and the different positional features for an example text block. (Color figure online)

extracted text blocks in reading order (i.e. following a top-left to bottom-right schema) at line level, merging close lines with similar font features (except for size, which we averaged using the number of tokens of each size in the resulting blocks). In this step, we pre-processed the raw text to remove line breaks, excessive white spaces, and \uFFFF Unicode characters, which were found in substitution of white spaces in the raw text extracted from Source 5 documents. It's worth to mention here the case of Source 4 documents, where the raw text appeared without white spaces when trying to extract it. This was probably due to the original PDF editor, which instead of using white spaces just put each word in its corresponding place. We could extract each word individually with PyMuPDF, and reconstruct the original text using their block and line references. Finally, we removed any text blocks with an overlap over 70% with a table detected.

Note that not all the text blocks have the same semantic role within a document. Some document components, such as tables and images, usually have a clearly defined purpose. However, a text block may be a paragraph inside the

Table 2. Layout features extracted for each layout block detected by our algorithm.

Feature	Description	Layout Blocks			
		Image	Table	Link	Text
f_1	Page number in which the block was detected, starting from 0	✓	✓	✓	✓
f_{2-5}	A 4-tuple (x_0, y_0, x_1, y_1) bounding box that defines the block's region in the page (see the coordinate system in Fig. 2)	✓	✓	✓	✓
f_{6-7}	A 2-tuple (x_c, y_c), where $x_c = (x_0 + x_1)/2$, $y_c = (y_0 + y_1)/2$	✓	✓	✓	✓
f_{8-11}	A 4-tuple with the block distance to each limit of the page (see the coordinate system in Fig. 2)	✓	✓	✓	✓
f_{12}	Important data about the block (output CSV file path for tables, URL for links, and the pre-processed text for text blocks)	✗	✓	✓	✓
f_{13}	Proportion of bold tokens in the text block	✗	✗	✗	✓
f_{14}	Proportion of italic tokens in the text block	✗	✗	✗	✓
f_{15}	Average font size of the text block	✗	✗	✗	✓
f_{16}	A tuple with the different font types in the text block	✗	✗	✗	✓
f_{17}	Proportion of capital letters in the text block	✗	✗	✗	✓
f_{18}	Number of elements separated by simple space in the text block	✗	✗	✗	✓
B	Type of block detected (0 for image, 1 for table, 2 for link, and 3 for text)	✓	✓	✓	✓
L	Layout component label (0 for image, 1 for table, 2 for link, 3 for identifier, 4 for title, 5 for summary, and 6 for main text)	✓	✓	✓	✓

body text, a title, or a caption, among other options. Furthermore, these semantic roles of the text are usually denoted in a document by using different layout features (e.g. the use of bold or italics, variety of fonts, different sizes, specific positions, etc.). For this reason, we inspected the documents from each source, and defined 4 different text categories within them:

1. **Identifier**. A text block that identifies the document. Here we can find the date, the number of the publication, or even source-specific identifiers.
2. **Summary**. A text block that can be found at the beginning of some announcements, and summarizes their content.
3. **Title**. A text block which identifies different sections within the document, or has a significant higher importance than the body text. They usually present different font characteristics than regular text blocks.
4. **Body**. The text blocks composing the main content of the document.

Some visual examples of these text categories can be found in Fig. 2. Considering the text categories, layout components in our documents are labeled into 1 among 7 possible categories. As these text categories have a semantic meaning in our documents, but they might not have it for some other applications, we made an explicit distinction between *Block* (B) and *Label* (L) annotations in our dataset (see Table 2). Note that for tables, links and images, the value of both annotations are the same.

3.3 Text Block Labeling and Data Curation

As we introduced in the previous Section, we defined 4 different text categories for the text blocks of our documents. After extracting the layout components into layout files, we had a set of 18 different features describing each text block. We proceeded to annotate each text block based on the features in a two-step process, which is illustrated in Fig. 3.

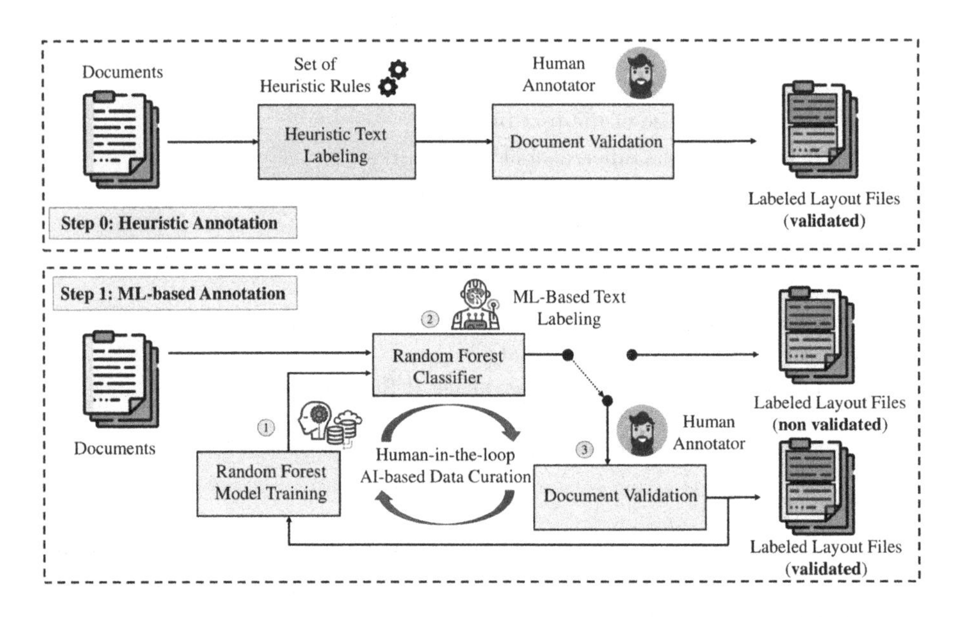

Fig. 3. Text block label's data curation process. During Step 0 we assign labels to text blocks using a set of heuristic rules, which are validated and corrected by a human supervisor. Then, we use the validated documents to train a text labeling classifier per source, hence reducing the number of errors and accelerating the labeling process.

In the first step, or initialization step (Step 0 in Fig. 3), we defined a set of heuristic rules after an initial inspection of the feature values for each source and text category (e.g. text blocks with a proportion of bold tokens over 0.5

are labeled as *Title*). The goal of the heuristic rules wasn't to perform a perfect labeling of the text blocks, but to have an initial set of documents with noisy labels. We defined these rules based only in the font text features (f_{13}-f_{18} in Table 2), except the font types (f_{16}) and the presence of some key words in the text (f_{12}) to detect the identifiers (e.g. the Spanish words for the days of the week). We developed an application to help a human supervisor validate the resulting noisy-labeled documents, and correct wrong labels. For an input file, this application displayed each page annotated with the current bounding boxes and labels. By clicking inside a bounding box the supervisor switch to the block label to the next value. This allowed us to obtain an initial set of correctly labeled documents for each document source in a "quick" way. All the documents were validated by a unique human supervisor, with the aim of preventing subjective biases of different supervisors from creating a disparity in the labeling strategy (specially when it came to title labeling). However, the resulting validated documents were assessed by the different authors, who agreed with the labeling criteria.

Once we had the initial set of validated documents for a specific source, we moved to the next step (Step 1 in Fig. 3). We considered a threshold of 50 pages as the minimum set to proceed. In this step, we started by training a classifier for each source with their validated documents. We decided to use a Random Forest classifier for this task, as the nature of our rule-based decision making labeling was close to the hierarchical logic behind such classifier.

We qualitatively assessed that, by using the trained models to validate new documents, the number of errors significantly dropped. Our experiments in Sect. 4.1 will demonstrate quantitative results supporting this fact. Hence, from this point we continue with a Human-in-the-loop AI-based Data Curation process. The use of the AI-based labeling speeded up the validation process significantly, as many errors made by the rule-based labeling were corrected. This allowed us to increase the validated sets, and retrain the models with more data, which ultimately ended up reducing further the errors. We repeated periodically this iterative process, and end up with models whose outputs were satisfactory. At this point, we use the AI-labeling process to create a large set of unvalidated data whose labels were clean enough for a training set. A visual inspection of an arbitrary selection of documents assessed this hypothesis.

4 Experiments

In this Section, we present the main experimental setup and results of this work. Our experiments aim to validate the usefulness of the layout features extracted from the text blocks to train text labeling classifiers.

4.1 Automatic Text Blocks Labeling

As we previously introduced in Sect. 3.3, we applied a data curation method based on the use of a text labeling classifier to annotate the text blocks in our database. In this Section, we report an experiment to quantitative assess our strategy, and the usefulness of the text features extracted (see Table 2) to discriminate between different semantic text categories. Recalling Sect. 3.2, we defined 4 different text categories in our work after an initial inspection of the data sources: 1) *Identifier*; 2) *Title*; 3) *Summary*; and 4) *Body Text*. Among these, the most common one is the *Body Text* category, with *Summary* being the least frequent (see Table 1).

During our text labeling procedure (presented in previous Sect. 3.3), an individual classifier was trained for each document source using the validated documents. Here, we will evaluate the performance of these individual classifiers. We chose a Random Forest (RF) model as classifier with a maximum depth of 1000, inspired by the low number of features and the hierarchical decision logic of the problem, for which we consider the RF model to be the perfect suit. We use as input for the classifiers all the features depicted in Table 2, except for feature f_{12} (i.e. the raw text data itself). We normalized features f_2 - f_{11} with respect to the dimensions of the document. For feature f_{16}, we created a dictionary to assign a value to each unique configuration of text fonts (remember that we can find different font types within the same text block, which we stored as a tuple).

We use 80% of the documents from the validation set (see Table 1) for training, and 20% for testing. Note that we decided to make the train/test splitting at document level, instead of page level, so we could take into account potential intra-document biases (i.e. the existence of significant layout differences within a document with respect to the classic source's template, where annexes, for instance, play an important role). For each source, we repeated the experiment 10 times with arbitrary train/test splits, and report as result the mean and the standard deviation of both overall and per-class accuracies.

The results are presented in Table 3. We report both the overall accuracy and the accuracy per class. We can observe that the best results per class are those obtained for *Identifier*, with a mean accuracy over 99% in all cases except for two sources, Source 6 and Source 22. These two cases also show a standard deviation higher than 3, which further highlights the increased difficulty in detecting identifiers in these sources. The good results obtained for the *Identifier* class could be expected, as these text blocks present low variability in the documents (i.e. they usually have the same position in the documents, number of tokens, and font characteristics). After them, the *Body* class presents the best results, with all the mean accuracies over 95% and low standard deviations in general. On the other hand, the "worst" performance is obtained for the *Title* category. We consider this class as the less objective during labeling, as different considerations of what is a title can be correct, especially when encountered within the body text of the announces, or in the annexes. This subjective nature makes titles harder to classify. Another reason behind this performance is the potential mistakes between *Title* and *Body text* classes, as the difference in the presence of these classes penalizes errors more for the former. Nevertheless, the mean performance in the *Title* class is over 84% in all cases, with 19 sources reaching a performance over 90%. Attending to the overall accuracies, all the sources obtain

Table 3. Overall and per class accuracies of the text block classifiers for each data source. We report the accuracy in terms of $mean_{std}$ over a 10 Folds Cross Validation protocol.

Source ID	Accuracy (%)				
	Overall	ID	Title	Summary	Body
1	$98.08_{1.17}$	$99.54_{0.41}$	$96.20_{4.35}$	$99.47_{1.05}$	$98.48_{1.18}$
2	$96.59_{3.61}$	$100.0_{0.00}$	$91.57_{6.10}$	100_{0}	$97.99_{2.36}$
3	$98.26_{1.66}$	100_{0}	$96.72_{3.53}$	$97.75_{4.53}$	$98.91_{0.08}$
4	$99.03_{1.00}$	$99.99_{0.02}$	$98.33_{1.81}$	$96.55_{4.68}$	$99.24_{0.78}$
5	$96.26_{2.67}$	100_{0}	$94.03_{4.95}$	100_{0}	$96.22_{3.73}$
6	$95.93_{2.61}$	$98.67_{3.51}$	$88.48_{7.07}$	$90.98_{6.44}$	$98.92_{0.80}$
7	$97.90_{1.02}$	100_{0}	$96.56_{1.76}$	$99.20_{1.60}$	$97.87_{1.61}$
8	$94.81_{2.03}$	100_{0}	$86.39_{9.85}$	$89.08_{4.67}$	$97.62_{1.55}$
9	$97.19_{1.61}$	100_{0}	$89.10_{6.56}$	$98.52_{2.26}$	$98.66_{1.48}$
10	$98.70_{0.95}$	$99.85_{0.30}$	$97.60_{2.47}$	$96.84_{4.48}$	$99.26_{0.46}$
11	$96.54_{2.89}$	$99.96_{0.13}$	$96.02_{4.06}$	$98.75_{2.50}$	$95.45_{4.21}$
12	$99.35_{0.55}$	100_{0}	$98.03_{1.95}$	$98.16_{2.66}$	$99.74_{0.42}$
13	$98.25_{1.91}$	100_{0}	$93.90_{6.84}$	$97.98_{4.07}$	$98.74_{1.52}$
14	$96.52_{1.95}$	$99.69_{0.63}$	$94.65_{5.36}$	100_{0}	$96.27_{3.17}$
15	$94.64_{3.27}$	$99.33_{1.18}$	$89.13_{7.40}$	$93.09_{6.37}$	$95.54_{3.56}$
16	$93.37_{3.20}$	$99.89_{0.34}$	$84.35_{10.72}$	$61.08_{12.35}$	$96.67_{2.30}$
17	$97.16_{0.83}$	100_{0}	$93.25_{2.99}$	$97.37_{2.00}$	$97.68_{0.84}$
18	$98.94_{0.91}$	$99.93_{0.09}$	$94.95_{4.23}$	$97.96_{3.49}$	$99.35_{0.79}$
19	$98.00_{1.21}$	$99.91_{0.19}$	$90.50_{3.53}$	$99.58_{0.65}$	$98.89_{1.09}$
20	$99.93_{0.07}$	100_{0}	$99.44_{1.47}$	$99.09_{2.73}$	$99.95_{0.05}$
21	$99.38_{0.52}$	100_{0}	$96.61_{3.14}$	$93.96_{5.88}$	$99.76_{0.45}$
22	$98.22_{1.28}$	$98.87_{3.38}$	$96.87_{2.39}$	$95.03_{5.55}$	$98.90_{0.84}$
23	$97.36_{1.01}$	$99.96_{0.04}$	$91.08_{3.65}$	$99.55_{0.47}$	$98.25_{1.62}$
24	$99.27_{0.44}$	100_{0}	$95.25_{3.97}$	$99.63_{0.75}$	$99.79_{0.06}$

good results, which surpass 96% for 20 different sources. The lowest performance is obtained in the documents from Source 16, which also shows an outlier-like performance in the summaries (i.e. 61.08%) and the lowest performance in titles (i.e. 84.35%).

5 Conclusions

In this work, we developed a new procedure to semi-automatically extract layout information from a set of digital documents, and provide annotations about the main layout components. Our methods are based on the use of web scraping tools to collect documents from different pre-defined sources, and extract layout information with classic PDF miners. The miners not only detect tables, links, images, and text blocks in the documents, but provide us with different information about these blocks, including font characteristics of the text blocks. We then defined a set of text features, which are useful to describe the text blocks and discriminate between semantic categories.

We applied our procedure to generate a new DLA database composed of public affairs documents from spanish administrations sources. After an initial inspection of the documents from each source, we defined 4 different text categories, and classify the text blocks in these categories using a Human-in-the-loop AI-based Data Curation process. Our data curation process trains text labeling models using human-validated documents in an iterative way (i.e. the output of the models are validated and corrected by a human supervisor, leading to new validated documents to train more accurate models).

Our experiments assessed the usefulness of the text features to discriminate between the previously defined classes, thus validating our data curation procedure. We then explored the use of Random Forests to train a classifier per source, whose results validated the proposed strategy. As future work, we suggest exploring other text labeling models, such as recurrent models, which could allow the flow of information between blocks in the page. Other text semantic classes could be studied, depending on the nature of the source documents, or potential applications after the DLA such as topic classification [15]. Finally, the scanned images left out in this work could be processed using Computer Vision models trained with the PAL database, and added in future versions of the database. Multimodal methods for integrating native digital and image-based information for improved DLA will be also investigated [16], in addition to analysis [7] and compensation [19] of possible biases in the machine learning processes involved in our developments.

Acknowledgments. Support by VINCES Consulting under the project VINCESAI-ARGOS and BBforTAI (PID2021-127641OB-I00 MICINN/FEDER). The work of A. Peña is supported by a FPU Fellowship (FPU21/00535) by the Spanish MIU.

Annex

Table 4. List of data sources used to collect the PAL Database.

Data Source	ID	Language	Access
Boletín Oficial del Estado	1	Spanish	boe.es/diario_boe
Boletín del Congreso de los Diputados	2	Spanish	congreso.es/indice-de-publicaciones
Boletín del Senado	3	Spanish	senado.es/web/actividadparlamentaria/ publicacionesoficiales/senado/boletinesoficiales
Boletín de la Comunidad de Madrid	4	Spanish	bocm.es
Boletín de la Rioja	5	Spanish	web.larioja.org/bor-portada
Boletín de la Región de Murcia	6	Spanish	borm.es
Boletín del Principado de Asturias	7	Spanish	sede.asturias.es/ast/servicios-del-bopa
Boletín de Cantabria	8	Spanish	boc.cantabria.es/boces/
Boletín Oficial del País Vasco	9	Spanish, Basque	euskadi.eus/y22-bopv/es/bopv2/datos/Ultimo.shtml
Boletín de Navarra	10	Spanish, Basque	bon.navarra.es/es
Boletín de la Junta de Andalucía	11	Spanish	juntadeandalucia.es/eboja.html
Boletín de Aragón	12	Spanish	boa.aragon.es
Boletín de Islas Canarias	13	Spanish	gobiernodecanarias.org/boc
Boletín de Islas Baleares	14	Spanish, Catalan	caib.es/eboibfront/
Boletín de Castilla y León	15	Spanish	bocyl.jcyl.es
Boletín de la Ciudad de Ceuta	16	Spanish	ceuta.es/ceuta/bocce
Boletín de Melilla	17	Spanish	bomemelilla.es/bomes/2022
Diario de Extremadura	18	Spanish	doe.juntaex.es/
Diario de Castilla–La Mancha	19	Spanish	docm.jccm.es/docm/
Diario de Galicia	20	Galician	xunta.gal/diario-oficial-galicia/
Diari de la Generalitat Valenciana	21	Spanish, Valencian	dogv.gva.es/es
Diari de la Generalitat Catalana	22	Spanish, Catalan	dogc.gencat.cat/es/inici/
Boletín del Ayuntamiento de Madrid	23	Spanish	sede.madrid.es/portal/site/tramites/ menuitem.944fd80592a1301b7ce0ccf4a8a409a0
Boletín del Ayuntamiento de Barcelona	24	Spanish, Catalan	w123.bcn.cat/APPS/egaseta/ home.do?reqCode=init

References

1. Antonacopoulos, A., et al.: A realistic dataset for performance evaluation of document layout analysis. In: ICDAR, pp. 296–300 (2009)
2. Bast, H., Korzen, C.: A benchmark and evaluation for text extraction from PDF. In: 2017 ACM/IEEE Joint Conference on Digital Libraries (2017)
3. Brown, T., et al.: Language models are few-shot learners. In: Advances in Neural Information Processing Systems, vol. 33, pp. 1877–1901 (2020)
4. Bukhari, S., et al.: Improved document image segmentation algorithm using multiresolution morphology. In: Document Recognition and Retrieval XVIII, vol. 7874, pp. 109–116 (2011)
5. Clausner, C., et al.: The ENP image and ground truth dataset of historical newspapers. In: ICDAR, pp. 931–935 (2015)
6. Clausner, C., et al.: ICDAR2017 competition on recognition of documents with complex layouts-RDCL2017. In: ICDAR, vol. 1, pp. 1404–1410 (2017)

7. DeAlcala, D., Serna, I., Morales, A., Fierrez, J., et al.: Measuring bias in AI models: an statistical approach introducing N-Sigma. In: COMPSAC (2023)
8. Eskenazi, S., et al.: A comprehensive survey of mostly textual document segmentation algorithms since 2008. Pattern Recogn. **64**, 1–14 (2017)
9. Göbel, M., et al.: ICDAR 2013 table competition. In: Proceedings of the International Conference on Document Analysis and Recognition, pp. 1449–1453 (2013)
10. He, K., et al.: Mask R-CNN. In: Proceedings of the IEEE International Conference on Computer Vision and Pattern Recognition, pp. 2961–2969 (2017)
11. Document management - Portable document format - Part 1: PDF 1.7. Standard, International Organization for Standardization (ISO) (2008)
12. Kenton, J., et al.: BERT: pre-training of deep bidirectional transformers for language understanding. In: Proceedings of NAACL-HLT, pp. 4171–4186 (2019)
13. Lang, T., et al.: Physical layout analysis of partly annotated newspaper images. In: Proceedings of the 23rd Computer Vision Winter Workshop, pp. 63–70 (2018)
14. Oliveira, D., Viana, M.: Fast CNN-based document layout analysis. In: Proceedings of the IEEE International Conference on Computer Vision, pp. 1173–1180 (2017)
15. Peña, A., Morales, A., Fierrez, J., et al.: Leveraging large language models for topic classification in the domain of public affairs. In: ICDAR (2023)
16. Peña, A., Serna, I., et al.: Human-centric multimodal machine learning: recent advances and testbed on AI-based recruitment. SN Comput. Sci. **4**, 434 (2023)
17. Ren, S., et al.: Faster R-CNN: towards real-time object detection with region proposal networks. In: Advances in Neural Information Processing Systems, vol. 28 (2015)
18. Sah, A., et al.: Text and non-text recognition using modified HOG descriptor. In: Proceedings of the IEEE Calcutta Conference, pp. 64–68 (2017)
19. Serna, I., et al.: Sensitive loss: improving accuracy and fairness of face representations with discrimination-aware deep learning. Artif. Intell. **305**, 103682 (2022)
20. Soto, C., Yoo, S.: Visual detection with context for document layout analysis. In: EMNLP-IJCNLP, pp. 3464–3470 (2019)
21. Vaswani, A., et al.: Attention is all you need. In: Advances in Neural Information Processing Systems, vol. 30 (2017)
22. Wong, K., et al.: Document analysis system. IBM J. Res. Dev. **26**(6), 647–656 (1982)
23. Zhong, X., et al.: PubLayNet: largest dataset ever for document layout analysis. In: ICDAR, pp. 1015–1022 (2019)

A Clustering Approach Combining Lines and Text Detection for Table Extraction

Karima Boutalbi[1,2](✉), Visar Sylejmani[2], Pierre Dardouillet[1,2], Olivier Le Van[2], Kave Salamatian[1], Hervé Verjus[1], Faiza Loukil[1], and David Telisson[1]

[1] Savoie-Mont-Blanc University, Annecy, France
{karima.boutalbi,pierre.dardouillet,kave.salamatian, herve.verjus,faiza.loukil,david.telisson}@univ-smb.fr
[2] Cegedim-SRH, Lyon, France
{karima.boutalbi,visar.sylejmani,pierre.dardouillet, olivier.levan}@cegedim-srh.com

Abstract. Table detection is a crucial step in several document analysis applications as tables are used to present essential information to the reader in a structured manner. In companies that deal with a large amount of data, administrative documents must be processed with reasonable accuracy, and the detection and interpretation of tables are crucial. Table recognition has gained interest in document image analysis, particularly in unconstrained formats (absence of rule lines, unknown information of rows and columns). This problem is challenging due to the variety of table layouts, encoding techniques, and the similarity of tabular regions with non-tabular document elements. In this, paper, we make use of the location, context, and content type, thus it is purely a structure perception approach, not dependent on the language and the quality of the text reading. We evaluate our model on invoice-like documents and the proposed method showed good results for the task of table extraction.

Keywords: Table detection · Table extraction · Document analysis · Table recognition · PDF accessibility

1 Introduction

Many non-editable documents are shared in PDF (Portable Document Format). They are typically not accompanied by tags for annotating the page layout, including the position of the table. One of the major challenges for the analysis and understanding of such documents is how to extract tables from them. Table extraction as a part of table understanding includes two stages: table detection, i.e. recovering the bounding box of a table in a document, and table data extraction by structure recognition, i.e. recovering its rows, columns, and cells.

In our work, we deal with payslip documents, which consist of a header, the principal table (body of the document) that we want to extract, and a footer.

M. Coustaty and A. Fornés (Eds.): ICDAR 2023 Workshops, LNCS 14194, pp. 139–145, 2023.
https://doi.org/10.1007/978-3-031-41501-2_10

Figure 1 represents three templates of payslips with different characteristics. We observe that all tables are separated by columns but there are no horizontal row lines; in addition, the header and footer may also contain tables or have a similarity of tabular regions with non-tabular document elements.

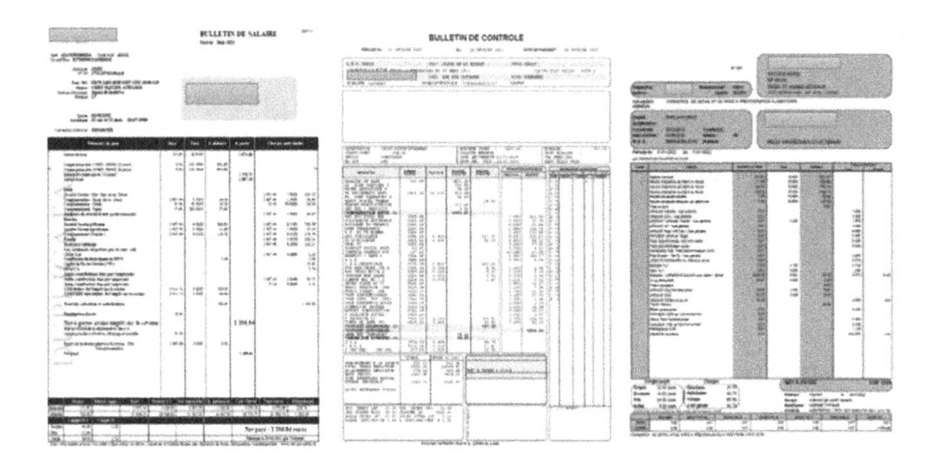

Fig. 1. Examples of payslip templates.

We first detect all the content of the document including the table content, and then we define the boundary of the table as a filter that separates the table content from the header and the footer of the document, assuming that a table and its content have certain properties related to vertical lines and text position.

Thus, rather than converting PDFs to images or HTML and then processing with other methods (e.g., OCR), we propose a method that combines lines and text detection for table extraction. Our approach is purely a perception approach, we make use of the location of the text, not its context and its signification. The preprocessing stage is crucial to the effectiveness of our method; we then perform a clustering algorithm for column classification. In the final stage, the distinctive size and succession of rectangles around the items allow us to separate the table from the rest of the document.

Considering the above contributions, this paper is organized as follows. Section 2 discusses existing papers that have dealt with table detection. Section 3 describes the steps of the adopted method. Section 4 presents the study results. Finally, Sect. 5 concludes the paper and presents some enhancements.

2 Related Work

Many works have been achieved in document table detection. Mikhailov et al. [2] proposed a novel approach for table detection in untagged PDF documents. They first use deep neural networks to predict some table candidates, then they select probable tables from the candidates by verifying their graph representation. Thus, they build a weighted directed graph from text blocks inside a predicted area of a table. Riba et al. [5] proposed a graph-based approach for detecting tables in document images by using Graph Neural Networks (GNNs) to describe the local repetitive structural information of tables in invoice documents. Authors in [1] demonstrate the performance improvement of the proposed approach for detecting tabular regions from document images. They use Deep learning-based Faster R-CNN, assuming that tables tend to contain more numerical data and hence it applies color coding/coloration as a signal for separate apart numerical and textual data. In [4], the authors tackle the problem of table detection by proposing a bi-modular approach based on the structural information of tables. This structural information includes bounding lines, row/column separators, and space between columns.

Another popular method for extracting tables is proposed in [3] where authors focus on extracting information from tables with various layouts, maintaining their structure, and processing them into textual data from images.

3 Methodology

We have collected 16 payslips templates, with a variety of table structures. We focus on the detection and extraction of data from the table. We use text boxes and columns to define the table.

Definition 1 (text box): A text box is a rectangle surrounding a string, and the position of the box is defined by four points (x_1, y_1, x_2, y_2).

Definition 2 (column): a column is a vertical line surrounding the content of a table and it's defined by four points (x_1,y_1, x_2, y_2) where $x_1 = x_2$.

We can further suppose that every item i is defined with a pair $< b,e>$of a start point b of the item and an end point$e$ of the item.

Definition 3 (cluster): Let's S be a set of tokens (text boxes or columns). S is a cluster if $\forall t_k \in S\ \exists t'_k \in S$ such that: $\forall t''_k \notin S |center(t_k) - center(t'_k)| \leq |center(t_k - t''_k)|$.

As clustering algorithms, we use K-means and DBSCAN (Density-Based Spatial Clustering of Applications with Noise) algorithms which both are based on Euclidean distance.

K-means algorithm is an iterative algorithm that splits a dataset into K predefined number of distinct non-overlapping subgroups (clusters) where each data point belongs to only one group. DBSCAN Finds core samples of high density and expands clusters from them. It is useful for data that contains clusters of similar density. Also, it is able to find arbitrary-shaped clusters and clusters with noise.

3.1 Data Processing

We start by extracting all vertical lines from the document. Due to the encoding techniques, some lines appear to be continuous, but there is a serie of overlapping and non-continuous lines. Also, some lines that are very close to each other appear visually to belong to the same line as shown in Fig. 2a. We group together all those lines that visually appear to belong to the same vertical line by using the rules we have defined; Fig. 2b illustrates the vertical lines after data preprocessing. Another way to regroup the series of non-continuous lines is vertical line clustering but we rather use a rule-based method since the vertical lines will be our input data to the proposed approach.

3.2 Line Detection-Based Approach

A simple first approach for extracting tables is to find fixed columns surrounding text. We assume that each column belongs to a cluster. Let S be a set of columns' table that we want to extract. We wonder if we can extract tables only by using vertical lines. Since we assume that each column should be separated enough to belong to a different cluster. According to Definition 3, we can partition the entire set of vertical lines into clusters as shown in Fig. 2b. The clustering was performed with K-means and DBSCAN, as presented in Definition 2. A column is defined by a start point $y - 1$ and an endpoint y_2. A vertical line is a line that their $x_1 = x_2$. Thus, the input of the clustering algorithm will be a one-dimensional vector where each instance is x_1.

Fig. 2. Vetical lines processing.

3.3 Proposed Approach: Clustering Using Lines and Text Box Detection

The first step of our method consists in detecting a unique position of the text box. Since a text box surrounds a string of characters or numerical values, separated by more than one space. So we did a preprocessing to define all page boxes. Each item is surrounded by a rectangular box that will be our reference for the position of textual data. We extend all boxes according to the right and left surroundings' vertical lines as illustrated in Fig. 3, thus all boxes belonging to the same columns will have the same size. Figure 4 shows an example of a payslip after box extension.

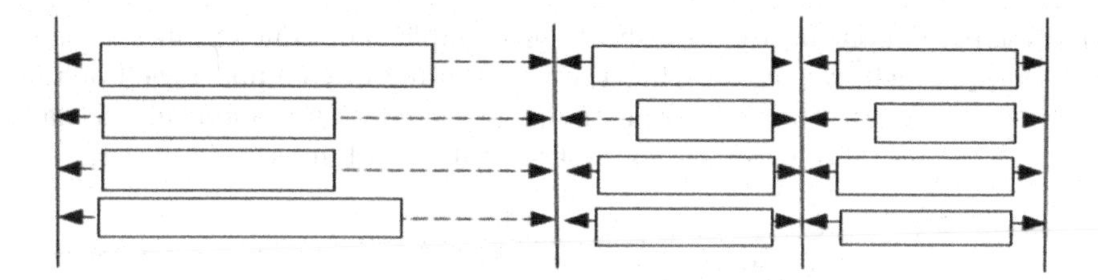

Fig. 3. An illustration of text boxes extension.

After extending boxes to vertical lines each box belonging to the same column will have the same size; the goal of this step is to obtain vertically aligned boxes. Thus, we perform clustering on all aligned boxes. The input of the clustering algorithm is 2-dimensional data (x_1 and x_2) corresponding to the beginning and the end of each box.

RUBRIQUES	BASE	TAUX SALARIAL	PART SALARIALE (EUR)		PART PATRONALE (EUR)	
			A PAYER	A DEDUIRE	TAUX	MONTANT
A02 SALAIRE BRUT MENS			9000,00			
U88 Tot.Cot.soc.sal				2118,87		
U89 Tot.Cot.fisc.sal				662,58		
V87 Remuneration nette			6218,55			
V96 Prime qualite vie			1218,75			
VP7 Prime expatriation			975,00			

Fig. 4. Example of payslip after text boxes extension.

4 Results

We compare the two approaches, the clustering of vertical lines and the second one combining lines and text position. To evaluate our method we used two measures, namely *WRCTC* (well-recognized columns over the total number of columns) and *WPTT* (well-partitioned tables over the total number of tables).

WRCTC: represents the rate of the well-recognized columns over the total number of columns per table.

WPTT: represents the rate of well-partitioned tables over the total number of tables.

We observe that K-means is more efficient with the first approach even if is more time-consuming due to the number of cluster research in terms of the number of columns recognition with 93% and this method aims to recognize 75% of the tables in the collected documents. In the second approach, we observe that we obtain almost the same result using K-means, but the results are considered improved in terms of the number of column recognition with 97% and 81% of well table extraction. In addition, with the proposed approach, the time consumption is very low compared with our other experiments and it takes 0.0032 s for table extraction per document. Figure 5 shows an example of obtained results using the proposed approach, we can see that the partitioning of columns is well defined, thus we can separate the table from the page's header and footer on one hand, on the other hand, it allows us to extract the table's columns and cells (Table 1).

Table 1. Obtained results comparison.

Method	Algorithm	WRCTC	WPTT	Time(S)
Line	K-means	0.93	0.75	0.4313
Line	DBSCAN	0.54	0.25	0.0030
Proposed method	K-means	0.94	0.75	1.1688
Proposed method	DBSCAN	0.97	0.81	0.0032

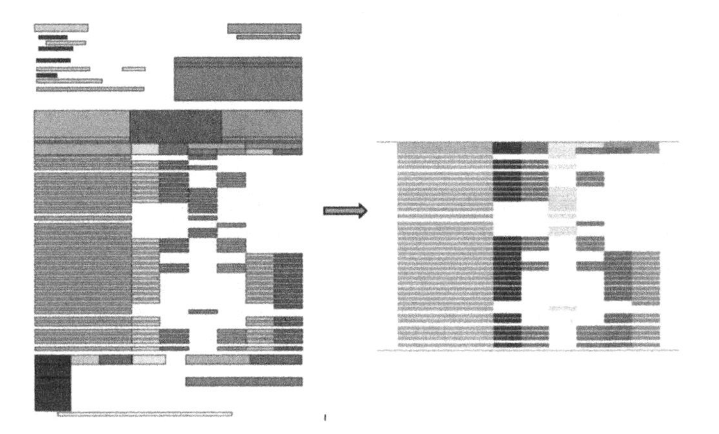

Fig. 5. Result example of the proposed table extraction approach.

5 Conclusion

We have proposed a method for table extraction combining vertical lines extracted from PDF documents and text box location. Our method does not use the context or the text content of the document, we only make use of the location. Our approach has been tested on real data of our company that need to extract information from payslips' tables. This will be useful for a lot of applications, such as deducing the calculation modes, triggers remuneration, frequencies, etc. Our problem is difficult due to the similarity of tabular regions with non-tabular document elements. However, we have obtained an accuracy of 81% over 16 extremely different payslip templates.

References

1. Arif, S., Shafait, F.: Table detection in document images using foreground and background features. In: 2018 Digital Image Computing: Techniques and Applications (DICTA), pp. 1–8 (2018). https://doi.org/10.1109/DICTA.2018.8615795
2. Mikhailov, A., Shigarov, A., Rozhkov, E., Cherepanov, I.: On graph-based verification for PDF table detection. In: 2020 Ivannikov Ispras Open Conference (ISPRAS), pp. 91–95 (2020). https://doi.org/10.1109/ISPRAS51486.2020.00020
3. Nidhi, Saluja, K., Mahajan, A., Jadhav, A., Aggarwal, N., Chaurasia, D., Ghosh, D.: Table detection and extraction using opencv and novel optimization methods. In: 2021 International Conference on Computational Performance Evaluation (ComPE), pp. 755–760 (2021). https://doi.org/10.1109/ComPE53109.2021.9752204
4. Ranka, V., Patil, S., Patni, S., Raut, T., Mehrotra, K., Gupta, M.K.: Automatic table detection and retention from scanned document images via analysis of structural information. In: 2017 Fourth International Conference on Image Information Processing (ICIIP), pp. 1–6 (2017). https://doi.org/10.1109/ICIIP.2017.8313719
5. Riba, P., Dutta, A., Goldmann, L., Fornés, A., Ramos, O., Lladós, J.: Table detection in invoice documents by graph neural networks. In: 2019 International Conference on Document Analysis and Recognition (ICDAR), pp. 122–127 (2019). https://doi.org/10.1109/ICDAR.2019.00028

WML

WML 2023 Preface

Our heartiest welcome to the proceedings of the ICDAR 2023 Workshop on Machine Learning (WML 2023, 4th edition) which was organized in conjunction with the 17th International Conference on Document Analysis and Recognition (ICDAR) held in San José, California, USA during August 21–26, 2023.

In the present age, machine learning, in particular deep learning, is an incredibly powerful way to make predictions based on large amounts of available data. There are many applications of machine learning in computer vision and pattern recognition, including document analysis, medical image analysis, etc. We organized this workshop for August 26, 2023 in order to facilitate innovative collaboration and engagement between the document analysis community and other research communities such as computer vision and images analysis, etc. The workshop provides an excellent opportunity for researchers and practitioners at all levels of experience to meet colleagues and to share new ideas and knowledge about machine learning and its applications in document analysis and recognition areas. The workshop enjoys strong participation from researchers in both industry and academia.

In this 4th edition of ICDAR WML we received 15 submissions. Each submission was reviewed by at least two expert reviewers. The Program Committee of the workshop comprised 24 members from different countries. Taking into account the recommendations of the Program Committee members, we selected 11 papers for presentation in the workshop, resulting in an acceptance rate of 73.3%.

The workshop was a one-day program and included oral presentations from 11 papers with two keynote talks from well-known researchers. The first keynote talk was delivered by Cheng-Lin Liu of NLPR, CAS, China, and the second keynote was given by Lianwen Jin, South China University of Technology, China. Our sincere thanks to Cheng-Lin and Lianwen for accepting our invitations to deliver the keynotes.

We wish to thank all the researchers who showed interest in this workshop by sending contributed papers. We also thank our program committee members for their time and effort in reviewing submissions. We would also like to thank the ICDAR 2023 organizing committee for supporting our workshop. Finally, we wish to thank the other members of the workshop team.

June 2023

Umapada Pal
Dacheng Tao
Xiao-jun Wu
Eric Granger
Lianwen Jin
Saumik Bhattacharya

Organization

Workshop Chairs

Umapada Pal	Indian Statistical Institute, Kolkata, India
Dacheng Tao	University of Sydney, Australia
Xiao-jun Wu	Jiangnan University, China

Program Chairs

Eric Granger	ETS, Montreal, Canada
Lianwen Jin	South China University of Technology, China
Saumik Bhattacharya	Indian Institute of Technology Kharagpur, India

Organizing Chairs

Shivakumara Palaiahnakote	University of Malaya, Malaysia
Ashish Ghosh	Indian Statistical Institute, India
Michael Blumenstein	University of Technology Sydney, Australia

Program Committee

Alaei, Alireza	Southern Cross University, Australia
Bhattacharya, Saumik	Indian Institute of Technology Kharagpur, India
Britto, Alceu	PUCPR, Brazil
Chan, Chee Seng	University of Malaya, Malaysia
Chanda, Sukalpa	Østfold University College, Norway
Chen, Shanxiong	Southwest University, China
De, Ishita	Barrackpore Surendranath College, India
Gao, Liangcai	Peking University, China
Harit, Gaurav	IIT Rajasthan, India
Impedovo, Donato	Dipartimento di Informatica - UNIBA, Italy
Iwana, Brian Kenji	Kyushu University, Japan
Lian, Zhouhui	Peking University, China
Luqman, M. Muzzamil	L3i Laboratory, University of La Rochelle, France
Pal, Umapada	Indian Statistical Institute, Kolkata, India
Pal, Srikanta	Maynooth University, Ireland
Palaiahnakote, Shivakumara	University of Malaya, Malaysia
Raghavendra, R.	Norwegian Biometrics Laboratory, Norway

Roy, Partha Pratim	Indian Institute of Technology, India
Roy, Kaushik	West Bengal State University, India
Roy, Swalpa Kumar	North Bengal Engineering College, India
Saini, Rajkumar	LTU, Sweden
Sun, Jun	Jiangnan University, China
Sundaram, Suresh	Indian Institute of Technology Guwahati, India
Zhang, Heng	Institute of Automation, CAS, China

Absformer: Transformer-Based Model for Unsupervised Multi-Document Abstractive Summarization

Mohamed Trabelsi(✉) and Huseyin Uzunalioglu

Nokia Bell Labs, Murray Hill, NJ, USA
{mohamed.trabelsi,huseyin.uzunalioglu}@nokia-bell-labs.com

Abstract. Multi-document summarization (MDS) refers to the task of summarizing the text in multiple documents into a concise summary. Abstractive MDS aims to generate a coherent and fluent summary for multiple documents using natural language generation techniques. In this paper, we consider the unsupervised abstractive MDS setting where there are only documents with no ground truth summaries provided, and we propose Absformer, a new Transformer-based method for unsupervised abstractive summary generation. Our method consists of a first step where we pretrain a Transformer-based encoder using the masked language modeling (MLM) objective as the pretraining task in order to cluster the documents into groups with semantically similar documents; and a second step where we train a Transformer-based decoder to generate abstractive summaries for the clusters of documents. To our knowledge, we are the first to successfully incorporate a Transformer-based model to solve the unsupervised abstractive MDS task. We evaluate our approach using three real-world datasets, and we demonstrate substantial improvements in terms of evaluation metrics over state-of-the-art abstractive-based unsupervised methods.

Keywords: unsupervised multi-document abstractive summarization · pretrained language model

1 Introduction

In this era of big data, the fast increase of the availability of text data makes understanding and mining this data a very time-consuming task. One of the most important text mining tasks is text summarization which helps to analyze the text data efficiently and effectively. Important information can be summarized from multiple documents which refers to the case of multi-document summarization (MDS). Multiple deep learning-based methods for MDS rely on the presence of a large amount of annotated data in terms of document-summary pairs for the supervised training. However, in many domains, such supervised data is not available and expensive to produce. Therefore, researchers have focused on the unsupervised learning setting where there are only documents available

M. Coustaty and A. Fornés (Eds.): ICDAR 2023 Workshops, LNCS 14194, pp. 151–166, 2023.
https://doi.org/10.1007/978-3-031-41501-2_11

without gold summaries. Existing methods [5,6,13] in MDS incorporate recurrent architectures, such as LSTM and GRU, in autoencoder-based models. In recent years, deep contextualized language models, such as BERT [7] and DistilBERT [18], have been proposed to solve multiple NLP tasks by taking advantage of the Transformer [19] which captures long-range dependencies better than the recurrent architectures. In particular, the Transformer-based seq2seq models [3,12,14,20] have led to state-of-the-art results in the supervised MDS. The Transformer-based models are not explored in the unsupervised MDS setting. For instance, incorporating Transformers in an autoencoder-based model for the unsupervised MDS remains a challenging task because gold summaries are not available for training an end-to-end model. Therefore, the state-of-the-art end-to-end Transformer-based encoder-decoder models, such as PEGASUS [20], cannot be directly applied to generate summaries in the unsupervised MDS.

We present ***Absformer***, a novel *Trans**former***-based model for unsupervised multi-document ***Abs**tractive* summarization. Our objective consists of generating a summary, that is composed of a set of sentences, for each group of semantically similar documents. Therefore, Absformer is a two-phased method for the unsupervised MDS. In phase I, we divide the set of documents into groups of similar documents using either an unsupervised clustering algorithm when no criteria of partition is given, or a supervised classification when labeling information is available. We refer to this phase as the embedding-based document clustering which requires a pretrained or fine-tuned encoder in order to produce the embeddings of documents. We show that pretraining the encoder on a standard Masked Language Modeling (MLM) objective leads to clusters with semantically similar documents. In addition, these clusters can be enhanced when labeling information is available for fine-tuning of the encoder. In phase II, we generate a summary for every cluster using our new Transformer-based decoder. In the training phase, the decoder model generates the documents from the embeddings that are obtained from the frozen encoder. Then, the trained decoder is used to generate summaries from the embeddings of the centers of clusters as in previous works [4–6]. An important design choice in Absformer is that the encoder should be kept frozen during the summary generation training in order to preserve the document clustering information in terms of document embeddings and cluster center embeddings. Therefore, compared to the traditional text-to-text encoder-decoder models, we decouple the encoder and decoder, so that in phase II, Absformer takes directly an embedding-level information as input instead of the token-level information, and it generates as output a token-level sequence. We show that initializing the embedding layer, the decoder blocks, and the language model head of the Absformer's decoder using the corresponding weights of the frozen encoder leads to better results in terms of generated summaries and faster convergence in the training phase. We note that both the encoder and decoder have the same number of layers as DistilBERT which has a reduced size and comparable performance to BERT, because our objective is to obtain both accurate document clusters and concise summaries while keeping the time and memory complexity reasonable.

In summary, we make the following contributions: (1) We propose a new Transformer-based method, called Absformer, for unsupervised MDS. Absformer is a two-phased method. In phase I, we propose to cluster the documents into groups of semantically similar documents by pretraining the encoder on a MLM objective. The clusters can be enhanced when labeling information is available to fine-tune the encoder. (2) In phase II, we train a Transformer-based decoder model that generates the documents from the embeddings that are obtained from the frozen encoder. Then, the trained decoder is used to generate summaries from the embeddings of the centers of clusters. (3) To improve the time and memory complexity of Absformer, we use the same architecture of DistilBERT for the encoder and decoder. To achieve faster convergence, we initialize the embedding layer, the decoder blocks, and the language model head of the Absformer's decoder using the corresponding weights of the frozen encoder.

2 Related Work

2.1 Multi-document Summarization

Multi-document summarization (MDS) [15] refers to the task of summarizing the text in multiple documents into a concise summary. Based on the availability of ground truth summaries, MDS can be categorized into two groups: the first group is denoted as supervised MDS [3,12,14] when ground truth summaries are available in the training phase, and unsupervised MDS [4–6,16] when there is no available ground truth summaries in the training. In the past few years, deep learning (DL) has led to a significant improvement in multiple tasks. Following the success of DL models, researchers have focused on exploring DL in MDS. Different neural architectures are used for MDS. For instance, recurrent architectures, including Recurrent Neural Networks (RNNs), Long Short-Term Memory (LSTM), and Gated Recurrent Unit (GRU), have been proposed as the main component in multiple methods [5,6,13]. In recent years, deep contextualized language models, such as BERT [7], and DistilBERT [18] have been proposed to solve multiple NLP tasks. Building on the Transformer, multiple methods [3,12,14] achieved state-of-the-art results in the supervised MDS.

2.2 Unsupervised Multi-document Summarization

Researchers have focused on developing abstractive methods for unsupervised MDS. LSTM is used as the main component in multiple methods for the unsupervised MDS [1,5,6]. Chu et al. [5] proposed an autoencoder-based method, called Meansum, where LSTM is incorporated in both the encoder and decoder. Meansum generates summaries for a fixed number of documents (8 documents). LSTM was also used as a main component in the method proposed by Coavoux et al. [6]. This method uses an aspect-based clustering technique that groups reviews that are about the same aspect. Then, an LSTM-based decoder generates a summary for each aspect. Amplayo et al. [1] proposed another LSTM-based method, called DenoiseSum, where denoising autoencoders is used for unsupervised summaries. Multiple unsupervised MDS methods [4,10,11,16] incorporate the variational autoencoder (VAE) model to generate summaries. Brazinskas et al. [4] proposed a VAE-based model, called CopyCat, for opinion summarization. Iso et al. [10] showed that the averaging aggregation can cause the summary degeneration problem that leads to generic summaries. The authors proposed a method, called convex aggregation for opinion summarization, that computes a weighted average of the latent vectors of reviews as the summary representation. Nguyen et al. [16] proposed a class-specific VAE-based model for the unsupervised review summarization. The model uses an independent classifier that assigns each review to the predefined classes. Isonuma et al. [11] uses a recursive Gaussian mixture to model the sentence granularity in a latent space, and generate summaries with multiple granularities.

3 Problem Statement

In the unsupervised MDS, multiple documents $D = \{d_1, d_2, \ldots, d_n\}$ are given, where n is the total number of documents. The task consists of predicting a summary, that is composed of a set of sentences, for each group of similar documents. This means that the set of documents D should be divided into groups of similar documents using either an unsupervised clustering algorithm when no criteria of partition is given, or a supervised classification when labeling information is available. For the rest of the paper, we refer to the step of grouping similar documents as the document clustering regardless of the technique that is used for this purpose. The document clustering step results in K groups of documents $D = \{g_1, g_2, \ldots, g_K\}$, where every group g_i contains similar documents. The objective is to learn two models F and N, where (1) F is used for document clustering in order to partition the input space, and (2) N is used to generate a coherent summary for every cluster obtained in the document clustering step.

4 Absformer: Document Clustering

We introduce our proposed method Absformer which is a two-phased Transformer-based method for abstractive MDS. We first describe phase I which consists of the document clustering step, and then present phase II which consists of the summary generation step. The overview of our method is shown in Fig. 1.

Existing methods in the unsupervised MDS [1,4–6] use LSTM as the main component to obtain either document- or sentence-level representations. In our proposed model, we use the Transformer as the main component in both phases. In particular, we use DistilBERT, which is formed of multiple layers of Transformer blocks, to obtain document-level representations. We choose to use DistilBERT as the main component for the clustering and generation models for two reasons. First, DistilBERT takes advantage of the Transformer's self-attention which captures long-range dependencies better than the recurrent architectures. Second, DistilBERT has a reduced size and comparable performance to BERT,

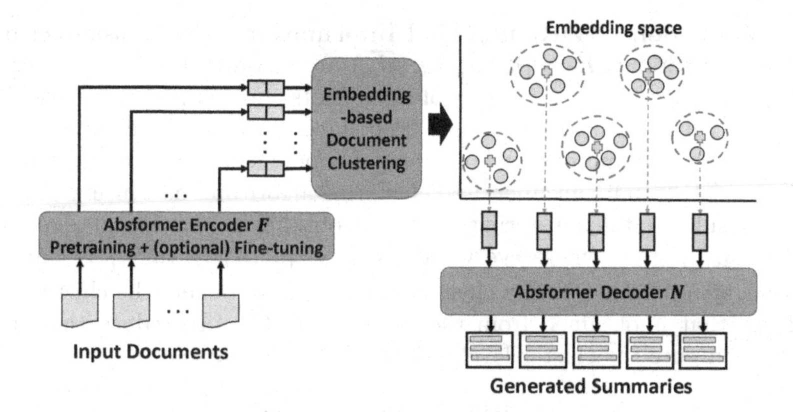

Fig. 1. Documents are clustered into semantically similar groups by pretraining the encoder F on a standard Masked Language Modeling (MLM) with an optional fine-tuning when labeling information is available. Then, a trained decoder N is used to generate summaries from the embeddings of the centers of clusters.

and our objective is to obtain both accurate document clusters and coherent summaries while keeping the time and memory complexity reasonable.

4.1 Pretraining of Encoder

We use a MLM objective with a masking rate of 15% sub-tokens in a document. After the pretraining phase, the pretrained DistilBERT is used as an encoder F to compute representations for documents, and the parameters $\boldsymbol{\theta}$ of DistilBERT are either jointly tuned with a multi-layer perceptron (MLP) if the labeling information is used during the document clustering, or kept frozen if unsupervised clustering is used on top of the computed document representations.

4.2 Embedding-Based Document Clustering

Our objective is to generate a summary for each group of documents that are semantically similar. The pretrained model F captures semantic matching signals by using the MLM objective as the pretraining task. Therefore, the document embeddings that are computed by F can be used as input to an unsupervised clustering algorithm to obtain clusters of semantically similar documents. The representation of a document d_i, denoted by $Rep(d_i)$, is given by:

$$Rep(d_i) = [\text{CLS}]x_1x_2 \ldots x_{n_i}[\text{SEP}] \tag{1}$$

where $d_i = x_1x_2 \ldots x_{n_i}$ is a sequence of n_i tokens, and [CLS] and [SEP] are DistilBERT special tokens that are added into the sequence similar to the single sentence classification. The encoder F takes as input the representation of a document d_i, denoted by $Rep(d_i)$, to compute the embedding $\overline{\boldsymbol{d_i}} \in \mathbb{R}^h$ that is extracted using the hidden state of the [CLS] token from the last Transformer block in the DistilBERT model F, where h is the dimension of the embedding.

Cluster Documents Without Label Information. The Transformer-based document embeddings $\overline{\boldsymbol{D}} = \{\overline{\boldsymbol{d_1}}, \overline{\boldsymbol{d_2}}, \ldots, \overline{\boldsymbol{d_n}}\}$ are forwarded to an unsupervised clustering step to obtain K clusters of documents. In this paper, we use the k-means clustering algorithm which has been shown to be both effective in terms of clustering results and efficient in terms of time and memory complexity. We leave investigating more unsupervised clustering algorithms as a future direction.

As a result of the clustering step, we obtain K clusters of documents $D = \{g_1, g_2, \ldots, g_K\}$, where every cluster $g_i = \{d_1^i, d_2^i, \ldots, d_{|g_i|}^i\}$ is composed of $|g_i|$ documents from D and a cluster center C_{g_i}. We denote the cluster center embedding that is obtained from the k-means by $\overline{\boldsymbol{C_{g_i}}}$. To reduce the outliers effect, we compute the cluster-level weight of every document:

$$w_{d_j^i} = \frac{\min_{d_k^i \in g_i} L_{norm}^2(\overline{\boldsymbol{C_{g_i}}}, \overline{\boldsymbol{d_k^i}})}{L_{norm}^2(\overline{\boldsymbol{C_{g_i}}}, \overline{\boldsymbol{d_j^i}})} \tag{2}$$

$L_{norm}^2(v_1, v_2)$ denotes the Euclidean norm between two vectors v_1 and v_2. Then, we update the center embedding $\overline{\boldsymbol{C_{g_i}}}$ of every cluster g_i using the cluster-level weights of documents:

$$\overline{\boldsymbol{C_{g_i}}} = \frac{\sum_{j=1}^{|g_i|} w_{d_j^i} \overline{\boldsymbol{d_j^i}}}{\sum_{j=1}^{|g_i|} w_{d_j^i}} \tag{3}$$

The embeddings of documents and centers that are obtained from the encoder F are forwarded to the summary generation step, and kept frozen during both summary generation training and inference.

Cluster Documents with Label Information. The unsupervised clustering step leads to clusters with semantically similar documents as a result of the MLM objective. There are cases where summaries should be generated for documents that share more common characteristics in addition to the semantic similarity. For example, if D is composed of product reviews, an interesting task is to generate a summary for similar products or similar ratings. In other words, the labeling information or the clustering criteria in this case is explicitly provided either by the product type or the rating score. Therefore, the pretrained encoder should be fine-tuned using the available labeling information to obtain clusters with semantically similar documents sharing common characteristics.

In the fine-tuning phase, each document $d_i \in D$ is associated with a label $l_i \in L$, with L is the set of labels (e.g. rating scores, or product types). The pretrained DistilBERT model is used as an encoder F, and its parameters $\boldsymbol{\theta}$ are jointly trained with a task-specific softmax layer W that is added on top of DistilBERT to predict the probability of a given label $l \in L$. The parameters of DistilBERT, denoted by $\boldsymbol{\theta}$, and the softmax layer parameters W are fine-tuned by maximizing the log-probability of the true label.

The fine-tuning phase results in $|L|$ clusters, with $|L|$ is the number of labels. Similar to clustering documents without labels, each cluster should be associated with a cluster center, and each document should belong to one cluster and have a membership weight to reduce the effect of outliers. For a given document d_i, the first step is to compute the probability distribution $p_i = \text{softmax}\left(W\overline{d_i}^{\theta}\right) \in \mathbb{R}^{|L|}$ of d_i over all the labels from L. The second step is to predict both the cluster of d_i, denoted by $c_i = \underset{l \in L}{\text{argmax}}\, p_i[l]$, and the membership weight, denoted by $w_i = \max_{l \in L} p_i[l]$. The cluster centers are computed similar to the case of clustering documents without label information using Eq. (3). Finally, we obtain $|L|$ clusters of documents $D = \{g_1, g_2, \ldots, g_{|L|}\}$, where every cluster $g_i = \{d_1^i, d_2^i, \ldots, d_{|g_i|}^i\}$ is composed of $|g_i|$ documents from D and a cluster center C_{g_i}.

For the rest of the paper, we denote the number of clusters by K regardless of the presence or absence of the label information during the clustering phase. For example, Fig. 1 depicts the case where $K = 5$. In conclusion, the document clustering phase results in (1) clusters $\{g_1, g_2, \ldots, g_K\}$ with semantically similar documents characterized by a cluster center embedding $\overline{\boldsymbol{C_{g_1}}}, \overline{\boldsymbol{C_{g_2}}}, \ldots, \overline{\boldsymbol{C_{g_K}}}$, respectively, and documents with membership weights $w_{d_i^{c(i)}}$, where $c(i)$ is the cluster associated with d_i, and (2) a trained DistilBERT-based encoder model F that is used to map documents into the embedding space. The outputs of the document clustering phase are forwarded to the summary generation phase.

5 Absformer: Summary Generation

Pretrained Transformer-based models have led to state-of-the art results in text summarization. For instance, PEGASUS [20] is an encoder-decoder Transformer-based model that is used for the supervised abstractive summarization. The ground truth summaries are given in the supervised abstractive summarization, so that both the encoder and decoder are jointly trained. In this paper, we address the unsupervised abstractive summarization so that end-to-end Transformer-based encoder-decoder models cannot be directly applied to generate summaries. We propose to train a Transformer-based decoder model N that generates the documents from the embeddings that are obtained from the frozen encoder F. Then, the trained decoder N is used to generate summaries from the embeddings of cluster's centers as shown in Fig. 1.

5.1 Encoder-Decoder Model

An important design choice in Absformer is that the encoder F should be kept frozen during the summary generation training in order to preserve the document clustering information in terms of document embeddings and cluster center embeddings as shown in the embedding space of Fig. 1. Therefore, we directly used the embeddings $\{\overline{\boldsymbol{d_1}}, \overline{\boldsymbol{d_2}}, \ldots, \overline{\boldsymbol{d_n}}\}$ as input to the decoder model. Compared to the traditional text-to-text encoder-decoder models, the decoder part of Absformer takes an embedding-level information as input instead of the token-level information, and it generates as output a token-level sequence.

Our proposed decoder has the same number of blocks as the DistilBERT-based encoder model F. Each block in F is composed of a self-attention layer and a feed forward layer. To incorporate the document embedding into the decoding process, we also include a cross-attention layer in the decoder of Absformer similar to T5 model [17]. In addition to the decoder blocks, the decoder N is composed of an embedding layer and a language model head.

Embedding Layer. In each timestamp, the decoder generates a token in order to either generate the document tokens from the input embedding, or the summary tokens from the embedding of the cluster's center. The token-level embedding is obtained from an embedding layer in the decoder N. This embedding layer is composed of word embeddings and position embeddings, and is initialized using the encoder's embedding layer to incorporate the token-level knowledge captured by the encoder F. The output of the embedding layer is forwarded to the first decoder block of N.

Decoder Block. Each block is composed of three components which are the self-attention head, cross-attention head, and feed forward layer.

Self-attention Head: This is a Transformer-based layer with 12 attention heads. There are 6 self-attention layers in total, and each layer is initialized using the corresponding self-attention from the frozen encoder F. The initialization from the encoder F leads to a faster convergence and a lower loss than initializing the self-attention heads from the vanilla DistilBERT. The Transformer-based self-attention heads capture long-range dependencies better than the recurrent architectures, and this leads to more accurate generated tokens even for long sequences. The deep contextualized embeddings, that are obtained at the timestamp t, are forwarded to the cross-attention head.

Cross-attention Head: The embeddings that are obtained from the self-attention head depend only on the generated tokens up to the timestamp t. To incorporate the encoder embeddings into the decoding process, we add a cross-attention head to compute both context- and encoder-aware embeddings. The cross-attention is also a Transformer-based layer with 12 attention heads. There are 6 cross-attention layers in total. Formally, in a given attention head from a cross-attention layer, three parametric matrices are introduced: a query matrix $Q \in \mathbb{R}^{h \times h}$, a key matrix $K \in \mathbb{R}^{h \times h}$, and a value matrix $V \in \mathbb{R}^{h \times h}$. For a given document d_i, suppose that the embeddings of the t generated tokens is denoted by $E_t \in \mathbb{R}^{t \times h}$, the cross-attention between the token embeddings E_t and the encoder embedding $\overline{\boldsymbol{d_i}}$ is given by:

$$\begin{aligned}
&\mathcal{Q} = E_t Q \in \mathbb{R}^{t \times h} \\
&\mathcal{K} = \overline{\boldsymbol{d_i}} K \in \mathbb{R}^{1 \times h} \\
&\mathcal{V} = \overline{\boldsymbol{d_i}} V \in \mathbb{R}^{1 \times h} \\
&Att(E_t, \overline{\boldsymbol{d_i}}, Q, K, V) = \text{softmax}\left(\frac{\mathcal{Q}\mathcal{K}^T}{\sqrt{h}}\right) \mathcal{V} \in \mathbb{R}^{t \times h}
\end{aligned} \tag{4}$$

So, the key output $\mathcal{K}$ and value output $\mathcal{V}$ are computed using the encoder embedding $\overline{\boldsymbol{d_i}}$, and the query output is computed using the generated tokens embeddings E_t. The context- and encoder-aware embeddings that are obtained from the cross-attention head are forwarded to the feed forward layer. We note that we also initialize the cross-attention heads using the self-attention parameters from the frozen encoder F.

Feed Forward Layer: This layer is composed of a feed forward network with 2 linear layers, and a normalization layer. These neural components are initialized using the corresponding feed forward layer of the encoder F. The output of the feed forward layer of a block l is forwarded to the self-attention head of the block $l+1$.

Language Model Head. The next token prediction is used to train the decoder N. In other words, in each timestamp t, the decoder N should output the token that corresponds to the position $t+1$. The language model head takes as input the embedding of the sequence of length t that is obtained from the last decoder block, and outputs a probability distribution over all the vocabulary of the decoder N (the same vocabulary as the encoder F). For faster convergence of the decoding training, we initialize the language model head of the decoder using the language model head of pretraining the encoder F on the MLM in the document clustering phase. The language model head of the decoder is composed of a first linear layer, GELU activation, normalization layer, and a second linear layer with a dimension of output layer that is equal to the size of the decoder's vocabulary.

5.2 Decoder Training

The decoder N is trained using the next token prediction task with teacher forcing. Instead of using a standard cross-entropy loss to reconstruct the original documents, we train our decoder N using a weighted cross-entropy loss in order to focus the efforts of training on reconstructing documents that are close to the cluster's center. Therefore, we reduce the effect of outliers on the generated summaries. Formally, the decoder loss function L_N of a batch of documents of size B is given by:

$$L_N = -\sum_{i \in B} \sum_{j=1}^{T_i} w_{d_i^{c(i)}} \times \log P\left(t_j \mid t_0^{j-1}\right) \tag{5}$$

where T_i is the length of document d_i, t_j is the j-th token of the document d_i, and t_0^{j-1} is the sequence of tokens from the document d_i up to token $j-1$.

5.3 Summary Decoding

After training the decoder N on reconstructing the documents from the embeddings that are obtained from the encoder F, the last step is to generate a summary for each cluster. Each cluster's center $\overline{C_{g_i}}$ is used as input to the decoder N as shown in Fig. 1. More specifically, the cross-attention head of each decoder block uses the cluster's center embedding $\overline{C_{g_i}}$ to compute the key and value matrices, so that we obtain a cluster-aware embedding for each timestamp as output from the last decoder block. The output embedding in each timestamp is forwarded to the language model head of the decoder N in order to compute the probability distribution over all the vocabulary.

Multiple methods [8,9] are proposed in the literature to select the next token from the computed probability distribution over the vocabulary. We experimentally found that a combination of top-p [9] and top-K [8] sampling methods lead to the best quality of generated summaries. We perform the summary decoding S times so that we obtain S different summaries for each cluster. Similar to Coavoux et al. [6], the S generated summaries for a given cluster g_i are ranked based on the cosine similarity between the embedding of each summary and the cluster's center embedding $\overline{C_{g_i}}$ in order to keep only the semantically similar summaries to each of the clusters.

6 Evaluation

6.1 Data Collections

Amazon Reviews. This dataset is composed of the Oposum corpus [2] which contains 3,461,603 Amazon reviews that are collected from 6 types of products (Bags and Cases, Bluetooth, Boots, Keyboards, TV, Vacuums). Each review is accompanied by a 5-star rating. For each product, there are 30 gold summaries that are used only to evaluate the generated summaries by our proposed model and baselines during the testing phase.

Yelp Reviews. This dataset includes a large number of reviews without gold-standard summaries. Each review is accompanied by a 5-star rating. Similar to Chu et al. [5], businesses with less than 50 reviews are removed, so that there are enough reviews to be summarized for each product. This dataset contains 1,297,880 reviews. We consider 7 types of products (Shopping, Home Services, Beauty & Spas, Health & Medical, Bars, Hotels & Travel, Restaurants). For each product, there are 20 gold summaries that are used only to evaluate the generated summaries by our model and baselines during the testing phase.

Ticket Data from Network Equipment. Our work is motivated by the business need to summarize ticket data from network equipment. When an issue occurs in mobile networks, a ticket is opened to describe the problem at various levels of detail. The ticket is usually resolved by a unit that is determined by the tester. To accelerate the process of resolving tickets, the tester can take advantage of the previous tickets to resolve common problems. However, it is a very time-consuming task for the tester to read the previous tickets in order to find similar issues. So, summarizing groups of similar tickets into a concise summary can help the tester to focus only on the important parts that are shared across a given group of tickets. Absformer will be used to generate summaries for groups of tickets, and these summaries will be used as an additional source of information for the tester to resolve tickets. The total number of tickets is 400k. For the confidentiality of the ticket data, we cannot show samples of tickets.

6.2 Baselines

We compare our proposed method Absformer against the following unsupervised abstractive MDS baselines: **Aspect + MTL** [6], **MeanSum** [5], **Copycat** [4], **DenoiseSum** [1], and **RecurSum** [11].

6.3 Experimental Setup

Our model is implemented using PyTorch, with NVIDIA RTX A6000. The dimension h is equal to 768. For each dataset, the encoder is pretrained using the MLM for 50 epochs with a batch size of 32. The learning rate is 1e-4 with a weight decay 0.01, and a warmup phase of 80,000 steps. For fine-tuning of the encoder using the labeling information, the model is trained for 10 epochs with a batch size 32. The learning rate is 3e-5 with a linear decrease that starts with a warmup period in which the learning rate increases. During the training step of phase II, we train the decoder for 40 epochs with a batch size of 8. The learning rate is 1e-4 with a weight decay 0.01, and a warmup phase of 20,000 steps. The maximum length of a sequence is 512. For summary decoding, the number of summaries S is equal to 10. The top-K is 50 and top-p is 0.95.

6.4 Experimental Results

We evaluate the performance of our proposed method and baselines using three ROUGE scores: ROUGE-1 (R-1), ROUGE-2 (R-2), and ROUGE-L (R-L).

Results on Amazon Dataset. Table 1(a) shows the performance of different approaches on the Amazon reviews. We show that our proposed method Absformer outperforms the baselines for R-1 and R-2. It is challenging to obtain a high R-L score because in our setting we summarize thousands of documents into a short summary, and it is unlikely to obtain a long sequence from the summary that exactly matches a sentence from ground truth summaries. In order to justify the importance of each component in our proposed method, we present the results of an ablation study for Absformer. In Absformer (w/o labeling information), we generate summaries for clusters of documents that are obtained from clustering the MLM-based embeddings using k-means. In Absformer (w/o decoder's initialization), the decoder is initialized from the vanilla DistilBERT instead of using the corresponding parameters from the encoder to initialize the embedding layer, the decoder block, and the language model head of the decoder. In Absformer (w/o pretraining), the encoder is fine-tuned directly on the target dataset. In Absformer (unweighted cross-entropy), all the weights $w_{d_i^{c(i)}}$ are equal to 1. Our full model Absformer outperforms all four system variations which supports the importance of (1) initializing the embedding

Table 1. ROUGE scores on publicly available benchmarks.

Method Name	ROUGE-1	ROUGE-2	ROUGE-L
Aspect + MTL [6]	30.00	5.00	17.00
MeanSum [5]	30.16	4.51	17.76
Copycat [4]	31.84	5.79	**20.00**
DenoiseSum [1]	34.82	6.12	18.58
RecurSum [11]	34.91	6.33	18.91
Absformer (w/o labeling information)	34.96	7.05	18.85
Absformer (w/o decoder's initialization)	35.85	7.12	19.14
Absformer (w/o pretraining)	35.52	7.08	19.11
Absformer (unweighted cross-entropy)	36.34	7.20	19.22
Absformer	**37.76**	**8.73**	**20.00**

(a) Amazon

Method Name	ROUGE-1	ROUGE-2	ROUGE-L
Aspect + MTL [6]	28.12	3.68	15.23
MeanSum [5]	28.66	3.73	15.77
Copycat [4]	28.95	4.80	17.76
DenoiseSum [1]	29.77	5.02	17.63
RecurSum [11]	33.24	5.15	**18.01**
Absformer (w/o labeling information)	33.75	6.12	16.23
Absformer (w/o decoder's initialization)	34.86	6.82	16.88
Absformer (w/o pretraining)	34.14	6.25	16.56
Absformer (unweighted cross-entropy)	35.92	7.12	17.36
Absformer	**36.87**	**8.05**	17.87

(b) Yelp

layer, the decoder block, and the language model head of the decoder with the corresponding parameters from the encoder, (2) the labeling information, (3) the pretraining of the encoder, and (4) the unweighted cross-entropy loss for training the decoder N. Compared to using BERT architecture in the encoder and decoder, the DistilBERT-based architecture of Absformer is **68%** faster in the generation phase, with no significant drop in the ROUGE scores of the testing phase (retaining **98%** of BERT performance in the summary generation).

Results on Yelp Dataset. Table 1(b) shows the performance of different approaches on the Yelp reviews collection. Consistent with Amazon dataset, our results on Yelp dataset shows that Absformer outperforms all the baselines and system variations for both R-1 and R-2.

Table 2. Surrogate metrics on the ticket data.

Model	$\mathbf{cosine}_{center}$	$\mathbf{cosine}_{top-50}$	$\mathbf{cosine}_{top-500}$	$\mathbf{cosine}_{top-5000}$
Aspect + MTL [6]	0.8126	0.7578	0.7252	0.7084
MeanSum [5]	0.8378	0.7723	0.7445	0.7273
Copycat [4]	0.8541	0.7996	0.7688	0.7356
DenoiseSum [1]	0.8596	0.8011	0.7852	0.7624
RecurSum [11]	0.8687	0.8135	0.7989	0.7783
Absformer (w/o labeling information)	**0.9034**	**0.8553**	**0.8453**	**0.8133**

Results on Ticket Data from Network Equipment. Unlike Amazon and Yelp datasets where there is a subset of ground truth summaries to report ROUGE scores in the testing phase, there are no ground truth summaries in the testing phase for the ticket data. Therefore, we cannot report ROUGE scores on this dataset to assess the quality of the generated summaries. Instead, surrogate metrics are used to evaluate the quality of the generated summaries for the ticket data. Similar to MeanSum [5], as a first surrogate metric, we report the average cosine similarity between the encoded generated summary and the embedding of a cluster for all K clusters ($cosine_{center}$). The second surrogate metric is the average cosine similarity between the encoded generated summary and the top-k tickets in each cluster for all K clusters ($cosine_{top-k}$). The labeling information is not available in the ticket data, so Absformer (w/o labeling information) is used to generate summaries. Table 2 shows that Absformer achieves better surrogate metrics than the baselines. The embedding of the generated summary by Absformer is close to the center's embedding with an average cosine similarity around 0.9. For top-k tickets in each cluster, when k increases, the average cosine similarity between the encoded summary and the embeddings of the top-k tickets decreases for both our method and baselines. This indicates the difficulty of summarizing a large number of documents in general, which simulates real case scenarios, and therefore is an important research direction.

Examples of Generated Summaries. In Fig. 2, we show examples of generated summaries from Absformer for the TV product from the Amazon dataset. We generate summaries for all 5 clusters, where each cluster represents a different predicted rating. Figure 2 highlights multiple aspects that are captured by our summarizer. For example, the summaries cover the *sound quality*, *picture quality*, *colors*, *adjustments*, *troubleshooting*, *price*, *menu button*, *setup*, etc. The summary provides a description about the functionality of each aspect, where low ratings focus on problems and malfunctions, and high ratings focus on the fully functioning aspects. The summaries also highlight the sentiments of the users where negative sentiment is expressed in low-rating clusters using words such as *disappointed*, and *terrible*, and positive sentiment is expressed in high-rating clusters using words such as *very excited*, and *good enough*. Our summaries are fluent, and capture the main points of the reviews.

Cluster of Predicted Rating = 1
i bought this tv as a christmas gift and it had problems. very disappointed. as others have noted the sound quality is terrible. there is a constant cut in the lower left side and bottom and the picture has to shut off repeatedly. no adjustments and no troubleshooting. the colors are dull and blurry and we still have to change channels manually. i have called vizio and they have been giving me pause to call back. after 2 calls they told us it was not available and could not find a tv.

Cluster of Predicted Rating = 2
i had a tv but it just quit working. i wish i'd gone with another brand. when i first got the unit home it started popping noise. when the tv started displaying a dark screen about 20 seconds before switching the input source. if you try to use a dvd player and play games in hd mode you'll notice a slight green hue in all four corners. it's annoying that there are lots of connections for tvs to attach to the connectors.

Cluster of Predicted Rating = 3
i bought this tv based on its great price. picture quality and color was great. but the menu button layout isn't very intuitive. when you get the hd channel you want to go to the av jack which gets annoying and requires a lot of tweak to get it to work. but if i just want a tv without a problem i'll probably look elsewhere.

Cluster of Predicted Rating = 4
i was very excited about this tv especially because the price was right. it arrived early and the unit is easy to setup and use. i took it out of the box and it worked fine. picture is clear and crisp. setup was easy and i had to do some tweak. the sound is not good but expected from an inexpensive tv. just plugged in my cable box via hdmi and the hd channels come in great. a dvd player from my laptop not fully hdtv works fine.

Cluster of Predicted Rating = 5
i have owned this tv for some time now. the picture quality is amazing. it is easy to set up and use and customize the software and control. colors are true black as well as other apps such as amazon video amazon prime and youtube. also netflix is also available for android's and ios. the price was right out of the box and the picture is perfect. set up was fairly easy just plug in and play. picture quality and sound are good enough for my purposes.

Fig. 2. Example of generated summaries for the TV product from Amazon. The light blue represents multiple aspects that are used to evaluate TVs. The red color represents either the description of aspects or the sentiment of the users. (Color figure online)

7 Conclusions

In this paper, we proposed a new unsupervised MDS method denoted by Absformer. We have shown that our Transformer-based two-phased model generates coherent and fluent summaries for clusters with semantically similar documents. Our Transformer-based decoder is trained by generating the documents from the embeddings that are obtained from the frozen encoder. Then, the trained decoder generates summaries from the centers of clusters. Improving the time and memory complexity, and achieving fast convergence are important points that are considered in our proposed model. An ablation study shows the importance of four components: (1) initializing the embedding layer, the decoder block, and the language model head of the decoder with the corresponding parameters from the encoder, (2) the labeling information, (3) the pretraining of the encoder, and (4) the unweighted cross-entropy loss.

References

1. Amplayo, R.K., Lapata, M.: Unsupervised opinion summarization with noising and denoising. In: Proceedings of the 58th Annual Meeting of the Association for Computational Linguistics, ACL 2020, pp. 1934–1945 (2020)

2. Angelidis, S., Lapata, M.: Summarizing opinions: aspect extraction meets sentiment prediction and they are both weakly supervised. In: Proceedings of the 2018 Conference on Empirical Methods in Natural Language Processing, pp. 3675–3686 (2018)
3. Brazinskas, A., Lapata, M., Titov, I.: Few-shot learning for opinion summarization. In: Proceedings of the 2020 Conference on Empirical Methods in Natural Language Processing, EMNLP 2020, pp. 4119–4135 (2020)
4. Brazinskas, A., Lapata, M., Titov, I.: Unsupervised opinion summarization as copycat-review generation. In: Proceedings of the 58th Annual Meeting of the Association for Computational Linguistics, ACL 2020, pp. 5151–5169 (2020)
5. Chu, E., Liu, P.J.: Meansum: a neural model for unsupervised multi-document abstractive summarization. In: Proceedings of the 36th International Conference on Machine Learning, ICML 2019. Proceedings of Machine Learning Research, vol. 97, pp. 1223–1232. PMLR (2019)
6. Coavoux, M., Elsahar, H., Gallé, M.: Unsupervised aspect-based multi-document abstractive summarization. In: Proceedings of the 2nd Workshop on New Frontiers in Summarization, pp. 42–47 (2019)
7. Devlin, J., Chang, M.W., Lee, K., Toutanova, K.: Bert: Pre-training of deep bidirectional transformers for language understanding. In: NAACL-HLT (2019)
8. Fan, A., Lewis, M., Dauphin, Y.N.: Hierarchical neural story generation. In: Proceedings of the 56th Annual Meeting of the Association for Computational Linguistics, ACL, pp. 889–898. Association for Computational Linguistics (2018)
9. Holtzman, A., Buys, J., Du, L., Forbes, M., Choi, Y.: The curious case of neural text degeneration. In: International Conference on Learning Representations (2020)
10. Iso, H., Wang, X., Suhara, Y., Angelidis, S., Tan, W.: Convex aggregation for opinion summarization. In: Findings of the Association for Computational Linguistics: EMNLP 2021, pp. 3885–3903. Association for Computational Linguistics (2021)
11. Isonuma, M., Mori, J., Bollegala, D., Sakata, I.: Unsupervised abstractive opinion summarization by generating sentences with tree-structured topic guidance. Trans. Assoc. Comput. Linguist. **9**, 945–961 (2021)
12. Jin, H., Wang, T., Wan, X.: Multi-granularity interaction network for extractive and abstractive multi-document summarization. In: Proceedings of the 58th Annual Meeting of the Association for Computational Linguistics, ACL 2020, pp. 6244–6254. Association for Computational Linguistics (2020)
13. Li, P., Lam, W., Bing, L., Guo, W., Li, H.: Cascaded attention based unsupervised information distillation for compressive summarization. In: Proceedings of the 2017 Conference on Empirical Methods in Natural Language Processing, EMNLP 2017, pp. 2081–2090. Association for Computational Linguistics (2017)
14. Liu, Y., Lapata, M.: Hierarchical transformers for multi-document summarization. In: Proceedings of the 57th Conference of the Association for Computational Linguistics, ACL 2019, pp. 5070–5081 (2019)
15. Ma, C., Zhang, W.E., Guo, M., Wang, H., Sheng, Q.Z.: Multi-document summarization via deep learning techniques: a survey. ACM Comput. Surv. **55**, 1–37 (2022)
16. Nguyen, T.N.A., Shen, M., Hovsepian, K.: Unsupervised class-specific abstractive summarization of customer reviews. In: Proceedings of The 4th Workshop on e-Commerce and NLP, pp. 88–100 (2021)
17. Raffel, C., et al.: Exploring the limits of transfer learning with a unified text-to-text transformer. J. Mach. Learn. Res. **21**, 140:1-140:67 (2020)
18. Sanh, V., Debut, L., Chaumond, J., Wolf, T.: DistilBERT, a distilled version of BERT: smaller, faster, cheaper and lighter. CoRR abs/1910.01108 (2019)

19. Vaswani, A., et al.: Attention is all you need. In: Advances in Neural Information Processing Systems, vol. 30, pp. 5998–6008 (2017)
20. Zhang, J., Zhao, Y., Saleh, M., Liu, P.J.: PEGASUS: pre-training with extracted gap-sentences for abstractive summarization. In: Proceedings of the 37th International Conference on Machine Learning, ICML 2020, vol. 119, pp. 11328–11339. PMLR (2020)

A Comparison of Demographic Attributes Detection from Handwriting Based on Traditional and Deep Learning Methods

Fahimeh Alaei(✉) and Alireza Alaei

Faculty of Science and Engineering, Southern Cross University, Gold Coast, Australia
{fahimeh.alaei,ali.alaei}@scu.edu.au

Abstract. Analyzing handwritten documents and detecting demographic attributes of writers from handwritten samples has received enormous attention from various fields of research, including psychology, computer science and artificial intelligence. Automatic detection of age, gender, handedness, nationality, and qualification of writers based on handwritten documents has several real-world applications, such as forensics and psychology. This paper proposes two simple but effective methods to detect the demographic information of writers from offline handwritten document images. The proposed methods are based on traditional and deep learning approaches. In the traditional machine learning method, the Rank Transform feature extraction method is used for measuring the intensity in handwriting images. The extracted handcrafted features are then fed into Support Vector Machine based classifiers to predict the demographical attributes of writers. In the deep learning method, a Convolutional Neural Network model based on the ResNet architecture with a fully connected layer, followed by a softmax layer is used to provide probability scores to facilitate demographic information detection. To evaluate the proposed methods and compare the results with the results in the literature, a comprehensive set of experiments was conducted on a frequently used benchmark database, KHATT. Both methods performed relatively well in predicting different demographic attributes. However, considering the settings in our experiments, the results obtained from the traditional model indicated better demographic detection compared to the deep learning models in all the tasks.

Keywords: Demographic Attributes · Detection Models · Handwritten Document Images · Rank Transform · ResNet Architecture

1 Introduction

Handwriting, by its nature, accommodates several vital demographic characteristics, such as age, gender, handedness, qualification, and nationality. Handwriting analysis and recognition play significant roles in revealing such characteristics in various applications in the fields of science, including psychology, neurology, forensics, medicine, and computer. The problem of demographic attribute detection from handwritten document images has received a lot of attention in real-world applications, such as information retrieval from historical documents, bank checks verification, suspect crimes, immigrant classification, and disease prediction in recent years.

M. Coustaty and A. Fornés (Eds.): ICDAR 2023 Workshops, LNCS 14194, pp. 167–179, 2023.
https://doi.org/10.1007/978-3-031-41501-2_12

Similar to any automatic detection system, a demographic attributes detection system has at least two main components: feature extraction and classification. To develop an accurate demographic detection system, extracting a set of discriminative features that can differentiate the variation in handwriting is crucial. Features can be extracted using traditional or deep learning based methods from different components of handwritten document images, such as characters, words, lines, paragraphs, patches, or even whole handwriting samples. Considering the traditional approach, the gradient of image pixels and its effectiveness for detecting demographic attributes of writers from handwritten samples were discussed in the literature [1–3]. In [2], multi-gradient directional (MGD) features using Canny edge detection, histograms of angles, and frequencies were extracted to perform gender identification from handwritten document images. The multi-gradient directional features can provide the direction of dominant pixels. A gender identification rate of 86.5% was reported when they used their own database for experiments. The authors further introduced adaptive multi-gradient features (AMGF) [3] to slightly improve gender identification results (87.0%) using the same database for experiments. Gradient features, such as histogram of oriented gradients (HOG) and gradient local binary patterns (GLBP), were further used in [1] for the detection of age, gender, and handedness from handwriting. The extracted features were fed into a Support Vector Machine (SVM) classifier, and accuracies of 55.55%, 71.11%, and 76.78% were reported for age, gender, and handedness detection tasks, respectively.

In [4], GoogleNet and ResNet architectures, as deep learning based feature extraction approaches, were used to characterize handwritten documents in the proposed gender detection. The extracted features were then fed into an SVM classifier to detect gender. These methods were evaluated using a database owned by the authors, and gender detection accuracies of 80.05% and 83.32% were reported for GoogleNet and ResNet, respectively [4]. These two methods were developed and examined for only a two-class problem. AlexNet architecture was also investigated to detect the gender of writers in [5]. The bilinear convolutional neural network (B-CNN) with ResNet blocks was proposed for age and gender detection in [22]. In this method, two generated parallel CNN blocks are concatenated into one bilinear vector and fed to the classification layers. The proposed method was evaluated using three databases, and the accuracies of 76.17% for gender detection and 78.17% for two age classes on KHAT databases were reported. The AlexNet was used in the feature extraction phase, and then Linear Discriminant Analysis (LDA) was employed for detection/classification purposes. The proposed method was evaluated using the QUWI database, and the best result of 70.08% correct gender detection was reported. In another study, Illouz introduced a neural network architecture consisting of four convolutional layers, a single fully-connected layer, and a softmax output layer for the gender detection task [6]. For classification, the majority vote and average softmax were considered. The Hebrew database was used to evaluate this method. The highest gender detection accuracy of 82.89% was reported with the majority vote classification.

From the literature, we noted that although there has been a noticeable amount of research on detecting two demographic attributes- age and gender- from handwriting, it is still considered a challenging research problem with several open research questions. For example, neither human nor automatic computerized systems could acquire high detection accuracy in age and gender detection [1, 2, 7, 8]. In addition, little research on detecting handedness, nationality, and qualification, as other demographic attributes, from handwriting can be found in the literature [7, 9]. It is also worth noting that a two-class problem, but not a multi-class problem, was considered in most of the literature for age detection. For instance, male or female in the gender detection task, right-handed and left-handed in the handedness detection task, and age categories below 35 or above 35 years old in the age detection task were considered in the literature [10].

Therefore, in this paper, we investigate detecting five demographic attributes using traditional and deep learning methods as two different feature extraction approaches in this domain. The usefulness of the rank transforms feature extraction method, as a traditional feature extraction method, has not been investigated for any demographic attributes detection methods. Thus, the rank transform method is considered for feature extraction. The extracted features using the rank transform method are then fed into SVM classifiers for multi-class problems of age, nationality, and qualification detection from handwritten documents, in addition to two-class problems of gender and handedness detection.

Moreover, in the deep learning approach, the usefulness of ResNet architecture for extracting features from handwriting samples and a fully connected layer for classification is also investigated in this research work. The ResNet architecture is used for deep feature extraction, as it preserves the gradient information present in handwritten documents. The performance of the proposed methods is examined for three-class in age and qualification and nine-class in nationality detection problems. Experimental results obtained from the proposed method show a better performance than the methods in the literature. Lastly, the obtained results from both methods are compared.

The rest of the paper is organized as follows. Section 2 demonstrates the proposed approach. Experimental results and comparative analysis are discussed in Sect. 3. Section 4 concludes the paper and provides some future research directions.

2 Proposed Approach

The block diagram of the proposed traditional and deep learning based approaches is provided in Fig. 1. The proposed methods include pre-processing, feature extraction using either deep learning or traditional method, and classification components.

There are 2 phases in both methods, training and testing. In the training phase of the traditional method, document images are first resized into 100×400 and then converted into grayscale in the pre-processing step. The rank transform method is then applied on the pre-processed images to extract features. The extracted rank transform features are used to train an SVM with a polynomial kernel in the learning process of our traditional approach. In the testing phase, the SVM classifier is used to predict the demographic attributes of writers from handwriting document images.

In the deep learning approach, the input images are first converted into RGB format and resized into 100×400 dimensions in the pre-processing steps. A data augmentation technique is further used to increase the number of training samples and to improve the learning process. By training the ResNet using document images, appropriate weights are computed for all neurons at different layers. The extracted features from testing samples

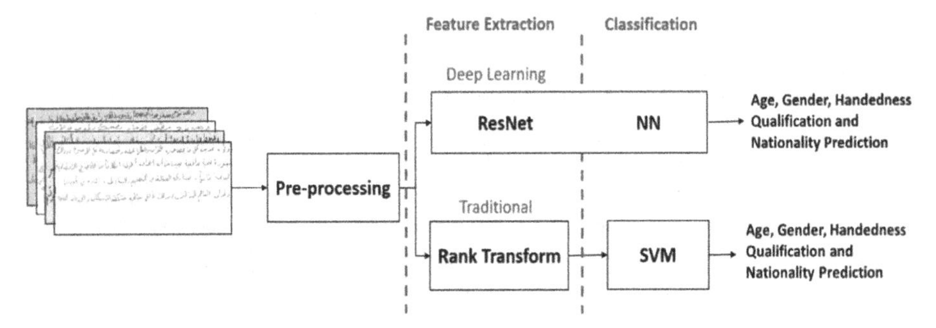

Fig. 1. The block diagram of the proposed method.

are then fed into the trained ResNet model with a fully connected neural network layer, followed by a softmax layer as an activation layer to finally predict the age category, gender, handedness, nationality, and qualification of the writers of the given handwritten document images.

2.1 Traditional Method

Rank Transform Feature Extraction. It has been proved that there is intensity variation, disparity, discrete points, and discontinuity in the handwriting of people of different genders, ages, and backgrounds [11, 12]. Since each person has distinct handwriting attributes with coherent parameters, parametric measures may not perform well in identifying demographic attributes from handwriting. This refers to factionalism. Thus, using a local transform with significant local variation and tolerating factionalism in the images is crucial for extracting discriminative features from handwriting document images. To address this, we have used the rank transform, a non-parametric local transform method that can better characterize the demographic attributes of writers from their handwriting. The rank transform method measures the local intensity of images and ranks the relative ordering of intensities, not the intensity values [13].

Rank transform (RT) [13] is a non-parametric texture-based feature extraction method used to measure the local intensity of the pixels. Consider P as a pixel of image I, $I(P)$ is intensity, and $N(P)$ is a set of pixels surrounding the pixel P with a diameter d in the square neighborhood. The non-parametric transforms are obtained by comparison of intensities of P and $N(P)$ pixels, for example, $I(P\prime)$ based on the following function,

$$\varepsilon(P, P') = \begin{cases} I(P') < I(P), & 1 \\ Otherwise & 0. \end{cases} \quad (1)$$

The number of pixels in the region that have less intensity compared to the intensity of the center pixel is called the rank transform $R(P)$ and is computed as follows:

$$R(P) = \|\{P\prime \epsilon N(P) | I(P\prime) < (P)\}\|. \quad (2)$$

It should be noted that $R(P)$ is an integer in the range of $\{0, \ldots, d^2 - 1\}$ and depending on the sign of the comparison, the integers are transformed and stored in the histogram. As a result, 9 different patterns acquire using the RT method.

SVM Classifier. Support Vector Machine (SVM) [14] has originally been developed based on a combination of methods used for supervised learning and binary classification. It has later been adapted for multi-class problems and has shown its superiority in various applications [15]. It is, therefore, considered in our proposed traditional method for classification purposes. Considering the definition of SVM, we have a set of training samples as $(u_i, v_i) \in R^M \times \{\pm 1\}$, where i is the number of samples per class, M is the data dimension, and $f : R^M \rightarrow \{\pm 1\}$ are set of functions. The function f is selected during the training of the SVM in order to increase the margin between the classes so that the testing data are classified more accurately using the following function:

$$\mathrm{f(z)} = \mathrm{sign}\left(\sum_{\mathrm{j}=1}^{\mathrm{SV}} \alpha_{\mathrm{j}} \mathrm{y_j} \mathrm{K}\left(\mathrm{u_j}, \mathrm{u_k}\right) + \mathrm{b}\right) \tag{3}$$

where SV is the number of support vectors, and b is a bias parameter. In Eq. (3), any mathematical function can be used as kernel $K\left(u_j, u_k\right)$. The polynomial kernel was selected in our proposed method, as it provided the best results in our experiments. The equation of the polynomial kernel is defined below, and l is the degree of the polynomial.

$$K\left(u_j, u_k\right) = \left(u_j.u_k + 1\right)^l \tag{4}$$

The SVM was fitted to a multi-class classification for tasks with more than two classes.

2.2 Deep Learning Method

ResNet Neural Network based Feature Extraction. Residual Networks or ResNet [16], as one of the robust deep neural networks in the literature, has been applied in several domains and applications, including object detection, image segmentation, and image classification [17, 18]. It comprises convolutional, pooling, and fully connected layers followed by a softmax output layer. Although the importance of gradient features in the traditional feature extraction approach for detecting demographic attributes from handwriting has been explored in the literature [1–3], the gradient characteristic may be omitted in deep learning methods when applied to document images. In deep learning models, more hidden layers are generally stacked for better learning. Stacking the layers will, however, vanish the edges and gradients of the handwriting images. Thus, the absence of the gradient information of the image pixels may result in poor prediction/detection of demographic information, as curvature and gradients are vital characteristics in people's handwriting [1, 16]. This issue has been addressed in the ResNet architecture by considering identity connection [16, 19]. The identity connection can protect the network from exploding or vanishing gradients. Therefore, a modified architecture of the ResNet is considered for detecting demographic attributes in this research work.

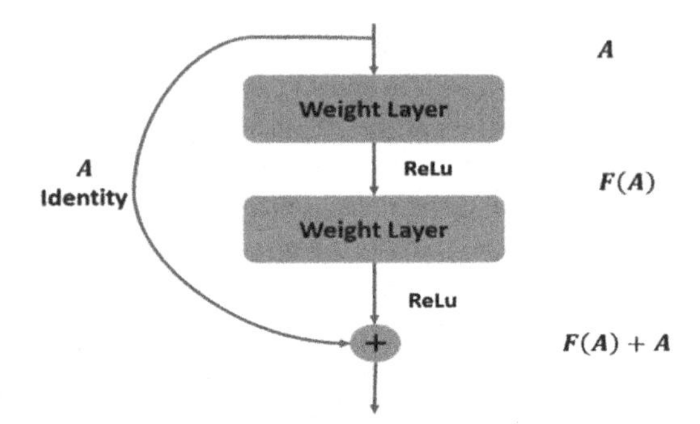

Fig. 2. Design of a residual block (adapted from [16]).

Adding the layers and creating deeper architecture may increase the training error, resulting in lower classification accuracy. To address this issue, the layers explicitly consider a residual mapping. The assumption is that optimizing residual mapping is easier than original mapping [16]. In Fig. 2, the identity connection or shortcut connection works similarly to a bridge between the input and the end of the residual block. It should be noted that there are weight layers and activation functions in the network, but they are not illustrated in Fig. 2. The layers are stacked to learn a residual mapping in residual networks.

In Fig. 2, A is considered as the input to the first layer, and $F(A) + A$ is the output from the identity connection. The mapping function, however, is defined as follows:

$$H(A) = F(A) + A \tag{5}$$

and in the same way, the residual function is denoted by

$$F(A) = H(A) - A. \tag{6}$$

As is evident in Fig. 2, the stacked layers learn the residual function $F(A)$, instead of learning the original function $H(A)$. Thus, in ResNet, residual learning is considered rather than feature learning. The model can continue to train and go deeper by passing the residual to the next layer. This has helped us extract residual features, such as curvature and gradient, from handwritten document images that are important characteristics of different handwritings. In our experiments, we used the convolutional window of size 3×3, and following the depth of networks, the number of filters increased from 64 to 2048. The networks include 50-layer including 48 convolutional layers, one max-pool layer, and one average pool layer. We considered the learning rate of 1e–4 for the experiments. In addition, the Adam optimizer was used as it can improve the problem with gradient features and maintain the parameter learning rates.

Fully Connected Neural Networks. For the classification in the deep learning method, a fully connected layer [20] is used in our proposed deep learning based demographic attributes detection method. This layer is based on a traditional neural network. In the proposed fully connected or densely connected layer, every input of the input vector impacts every output of the output vector. The output size in this layer is the number of available classes in each task. The output of this layer is sent to a softmax layer. The softmax function converts the raw output of the fully connected layer into a probability score using the following equation:

$$\mathrm{softmax}(t)_y = \frac{e^{t_y}}{\sum_{s=1}^{N} e^{t_s}} \tag{7}$$

where t is the vector output from the previous layer, and the output of the softmax equation is always a positive value between 0 and 1. A cross-entropy based classification layer is finally included after the softmax layer to obtain the class labels of the testing samples.

3 Experimental Results and Discussions

3.1 Databases and Evaluation Metrics

The KHATT database [21] was chosen for our experiments since it has the ground truth for most demographic attributes we considered in this research work. This database is composed of 4000 documents written by 1000 Arabic-speaking writers with 4 different age categories, 3 qualification backgrounds, and 9 nationalities. The handwritten document images were saved at different resolutions, such as 200, 300, and 600 DPI. Of the 4000 document images, 2000 samples contain the same text content, and the rest, 2000 samples, have different text. We considered all samples in our experiments.

Table 1 shows more details about the KHAT database and the number of samples in each class based on different demographic attributes. From Table 1, it is noticeable that the database is imbalanced concerning all demographic attributes. This makes the detection of demographic attributes more challenging. To get an idea of document images in the KHAT database, a few samples are shown in Fig. 3. Figures (a1) and (a2) are the same paragraph copied by all the writers. Figure (a1) was written by a right-handed male in the age category of 16–25 years old with high school qualification from Saudi Arabia, and (a2) was written by a left-handed male in the age category of 26–50 years old and with a university qualification from Morocco. Figures (b1) and (b2) are different paragraphs written by choice of writers. Figure (b1) was written by a right-handed female in the age category of 16–25 years old with a university qualification from Morocco, and (b2) was written by a left-handed female with < 15 years old and high school qualification from the USA.

Most demographic detection tasks in the literature used accuracy as a metric for the evaluation of methods. Accordingly, we considered the accuracy for evaluating our proposed methods in this research work. Accuracy is defined as the correct detection of age, gender, handedness, nationality, and qualification over the actual number of test handwriting document images.

As mentioned, due to the insufficient samples in the database to train the deep learning method, random rescaling as a data augmentation technique is considered to increase the number of samples for training purposes. For the deep learning experiments, document images are resized to 100×400 dimensions. This size was decided experimentally in line with what has been considered in the literature. The document images are then converted

Table 1. Number of samples in each class in the KHATT database.

Attribute	Class	Number of samples
Age	< 15	504
	16–25	2572
	26–50	860
	> 50	64
Gender	Male	2708
	Female	1292
Handedness	Right-handed	3712
	Left-handed	288
Qualification	Elementary school	352
	High School	2148
	University	1500
Nationality	Saudi Arabia	2704
	Morocco	360
	Jordan	316
	Yemen	180
	Palestine, USA, etc	116, 64,...

ذهب نوح بظفر ضرغام بصحبة رؤوف بن لؤي رائق ظافر مطهوط وهلال خازن [illegible] الحج
بدأت قوافل الحجيج حاج اثر اخر يلي. عند وصولنا وسماع شيخ كان جاري في الخيمة
يتكلم وهو نائم بكلمات لا افهمها مثل انقض [illegible] له الضابط لزمتك. سألت
راجح هل بلغ أصحابنا ظ ع ك، ث ح ض، ص غ ه أننا في الحج هل تعلم
فائدة الكلمات التالية لهذا النص: [illegible]

(a1)

ذهب نوح مظفر ضرغام بصحبة رؤوف بن لؤي
رائق ظافر مطهوط وهلال خازن للحج. بدأت قوافل
الحجيج حاج اثر آخر يلي. عند وصولنا طفنا وسعينا
مع شيخ. كان جاري في الخيمة يتكلم وهو نائم
بكلمات لا افهمها مثل انقض يغلبها لم [illegible] لزمتك.
سألت راجح هل بلغ اصحابنا ظ ع ك خ ض ب س ش
ص غ ه اننا في الحج. هل تعلم فائدة الكلمات التالية
لهذا.

(a2)

أولاً - تحديث كل ما يخص عمارة التراث وتطويرا عملية الاستعادة، قبل كل شيء، وذلك
بصورة علمية واضحة بعيدة عن أية أهداف أخرى، انطلاقا من الأوضاع الاقتصادية
العالمية. ثانياً - المشاركة الفعالة في التصميم السليم لهذه المادة في أوروبا
وبلدان العالم الثالث، وذلك لأجل معالجة مشكلة الإسكان والحرمان العامة

(b1)

وكلها عالية من أرضها من جهة البر وتتصل الى سلسلة جبال السراة وتكثر فيها الأودية التي تنحدر
من الهضبة وتأخذ بعضها الاتجاهات من الشمال الشرقي أو من الشمال الغربي الى سواحل البحر
حيث تشكل اتجاهات المياه والسيول وتكون أعلى ارتفاع لسلسلة الجبال الجنوبية في أقصى الجنوب
الشرقي من الجزيرة.

(b2)

Fig. 3. Four different samples of handwritten document images in the KHATT database.

into color RGB images suitable to ResNet architecture. However, for the traditional experiments as the pre-processing step, the images only are resized and converted to gray for all the conducted tasks. Since a smaller number of samples are available in the age category of > 50, we combined those samples with the age category of 26–50 for the age detection task. Therefore, experiments in the age detection task are conducted with 3 subclasses.

For each round of experiments, 70% of the samples were randomly selected for training purposes, 20% for validation, and 10% for testing. The cross-validation strategy was used in these experiments, and after repeating the experiments 5 times, the statistical median of the obtained accuracies was calculated and reported in this paper.

3.2 Results and Discussion

The results obtained from the traditional method using the SVM classifier with different kernels, including Gaussian, linear, and polynomial, are presented in Table 2. It is worth noting that the ANOVA test was conducted, and the results were statistically significant. In the age, gender, and qualification detection task, the obtained accuracies using polynomial kernels were considerably higher than the Gaussian and linear kernels. In handedness and nationality detection tasks, the accuracies obtained with different kernels were similar. Since the polynomial kernel performed better than the other two kernels in most demographic attribute detection tasks, the results obtained from this kernel have been considered for further comparisons.

Table 2. The obtained results using different kernel functions in SVMs.

Tasks	Gaussian	Linear	Polynomial
Age	81.45%	82.37%	**83.65%**
Gender	83.84%	83.09%	**84.35%**
Handedness	93.51%	93.51%	**93.51%**
Nationality	**84.34%**	**84.34%**	84.30%
Qualification	63.34%	62.86%	**64.19%**

The results obtained from both the traditional and deep learning methods in different tasks of demographic attribute detection are demonstrated in Table 3. From Table 3, it is evident that, in the age detection task, the accuracies obtained from both methods were comparably high. Though, the traditional method, with an accuracy of 83.65%, performed better than the deep learning method, with an accuracy of 79.08%.

In the gender detection task, however, the difference between the traditional and deep learning methods was highly significant, as the accuracy obtained using the traditional method was almost 12% higher than the deep learning approach.

In addition, although the accuracy obtained using the traditional method for handedness detection was slightly (0.71%) higher than the deep learning method, the difference is not noticeable.

It is also noticed that, among all the demographic attribute detection problems, traditional and deep learning models performed significantly different in nationality detection. The results obtained in the nationality detection task indicated more than a 14% difference between the accuracy achieved by the traditional approach using the RT method and the deep learning approach using the ResNet method for feature extraction. The highest accuracy of 84.30% was obtained by the traditional method.

Table 3 further indicates a difference of 2.69% between the results obtained in the qualification detection task, where the traditional method accomplished higher accuracies compared to the deep learning approach.

Table 3. The obtained results using deep learning and traditional models.

Tasks	ResNet	RT
Age	79.08%	83.65%
Gender	72.02%	84.35%
Handedness	92.80%	93.51%
Nationality	70.20%	84.30%
Qualification	61.50%	64.19%

In summary, from the results obtained from traditional and deep learning methods (Table 3), we can conclude that traditional models performed better in all tasks, with higher accuracies between 0.71%, and 14.10%. The lowest and highest differences, with 0.71% and 14.10%, were in handedness and nationality detection, respectively. We further noted that both methods performed comparably well when they were used for age, qualification, and handedness detection, of which the handedness detection models based on both approaches achieved the highest accuracy among other demographic attributes. However, the traditional approach using RT features and SVM provided relatively better results in gender and nationality detection problems compared to the deep learning method. The first reason for this can be because of a low number of samples for learning in some classes of the KHAT database. The second reason can be the difference in the number of samples between females and males and between people from Saudi and other nationalities (imbalanced data), as demonstrated in Table 1. The SVM and traditional classifiers can generally handle this problem better than deep neural networks.

3.3 Comparative Analysis

For a fair comparison analysis of the results obtained from the proposed methods in this research work and the results reported in the literature, the methods in the literature that used the KHATT database were considered for comparison. Other studies in the literature used their own databases or databases that were not publicly available. Thus, we did not consider them for comparison. Moreover, we mainly noticed age and gender detection using handwriting have been studied in the literature compared to other demographic attributes, which have got attraction in recent years. Therefore, we compared the results obtained from our experiments using age and gender detection models. For this evaluation, we applied similar conditions discussed in [10, 11]. In our experiments, two-thirds of the data was considered for training and the remaining for testing. It should also be noted that similar to [10], only two age categories, 16–25 and 26–50, were used in the experiments. The results obtained from our models and the results from the literature are listed in Table 4. From the results shown in Table 4, it is clear that our proposed traditional approach based on rank transform features and SVM classifier could acquire the highest age detection results in the literature. Our proposed method performed significantly better, with 73.48% accuracy, compared to the performance of methods in the literature. The accuracy obtained from our proposed method is around 9% higher than the reported accuracy in [11]. The results obtained from our proposed traditional gender detection method are comparable to those obtained by Navya et al. [3]. However, the result (74.95%) obtained from our method is higher than the reported results in [10] using HOG and GLBP feature extraction methods and the SVM classifier. We believe the RT feature extraction method needs a reasonable number of samples in each class for training to capture the required variations to perform better.

Table 4. The comparison of age and gender detection with the literature methods.

Attribute	Study	Feature Extraction	Classification	Accuracy
Age	[10]	HOG GLBP	SVM	48.15% 55.55%
	[11]	Disconnectedness features	K-means	64.44%
	Our method	RT	SVM	**73.48%**
Gender	[10]	HOG GLBP	SVM	68.89% 74.44%
	[3]	AMGF	Correlation between text lines	75.6%
	Our method	RT	SVM	74.95%

4 Conclusion and Future Work

This paper presented two methods, traditional and deep learning based, for detecting demographic attributes from offline handwritten document images. Deep learning models were developed based on the ResNet architecture. The proposed traditional methods were designed using the rank transform feature extraction technique and SVM classifier. We have developed five different models for age, gender, handedness, nationality, and qualification detection. The experiments were conducted on the KHATT dataset for the evaluation of all proposed models. The proposed traditional methods obtained better results compared to the deep learning methods in all the problems. Our classification results were comparable to previous methods in the literature. In particular, our results for age detection were significantly better on the same dataset compared to the state-of-the-art methods. In the future, we plan to investigate the combination of two or more demographic attribute detection based on a single model using handwriting samples.

References

1. Bouadjenek, N., Nemmour, H., Chibani, Y.: Histogram of Oriented Gradients for writer's gender, handedness and age prediction. In: Symposium on Innovations in Intelligent Systems and Applications, pp.1–5 (2015). https://doi.org/10.1109/INISTA.2015.7276752
2. Navya, B.J., et al.: Multi-gradient directional features for gender identification. In: Proceedings of the International Conference on Pattern Recognition, pp.3657–3662 (2018). https://doi.org/10.1109/ICPR.2018.8546033
3. Navya, B.J., et al.: Adaptive multi-gradient kernels for handwritting based gender identification. In: Proceedings of the International Conference on Frontiers in Handwriting Recognition, pp.392–397 (2018). https://doi.org/10.1109/ICFHR-2018.2018.00075
4. AL-Qawasmeh, N., Suen, C.Y.: Gender detection from handwritten documents using concept of transfer-learning. In: Lu, Y., Vincent, N., Yuen, P.C., Zheng, W.-S., Cheriet, F., Suen, C.Y. (eds.) ICPRAI 2020. LNCS, vol. 12068, pp. 3–13. Springer, Cham (2020). https://doi.org/10.1007/978-3-030-59830-3_1
5. Moetesum, M., Siddiqi, I., Djeddi, C., Hannad, Y., Al-Maadeed, S.: Data driven feature extraction for gender classification using multi-script handwritten texts. In: Proceedings of the International Conference on Frontiers in Handwriting Recognition, pp. 564–569 (2018). https://doi.org/10.1109/ICFHR-2018.2018.00104
6. Illouz, E., (Omid) David, E., Netanyahu, N.S.: Handwriting-based gender classification using end-to-end deep neural networks. In: Kůrková, V., Manolopoulos, Y., Hammer, B., Iliadis, L., Maglogiannis, I. (eds.) ICANN 2018. LNCS, vol. 11141, pp. 613–621. Springer, Cham (2018). https://doi.org/10.1007/978-3-030-01424-7_60
7. Al Maadeed, S., Hassaine, A.: Automatic prediction of age, gender, and nationality in offline handwriting. EURASIP J. Image Video Process. **2014**(1), 1–10 (2014). https://doi.org/10.1186/1687-5281-2014-10
8. Alaei, F., Alaei, A.: Gender detection based on spatial pyramid matching. In: Lladós, J., Lopresti, D., Uchida, S. (eds.) ICDAR 2021. LNCS, vol. 12824, pp. 305–317. Springer, Cham (2021). https://doi.org/10.1007/978-3-030-86337-1_21
9. Morera, Á., Sánchez, Á., Vélez, J.F., Moreno, A.B.: Gender and handedness prediction from offline handwriting using convolutional neural networks. Complexity 2018 (2018)
10. Bouadjenek, N., Nemmour, H., Chibani, Y.: Age, gender and handedness prediction from handwriting using gradient features. In: Proceedings of the International Conference on Document Analysis and Recognition, pp. 1116–1120 (2015). https://doi.org/10.1109/ICDAR.2015.7333934
11. Basavaraja, V., Shivakumara, P., Guru, D.S., Pal, U., Lu, T., Blumenstein, M.: Age estimation using disconnectedness features in handwriting. In: Proceedings of the International Conference on Document Analysis and Recognition, pp. 1131–1136 (2019). https://doi.org/10.1109/ICDAR.2019.00183
12. Alaei, F., Alaei, A.: Handwriting analysis: Applications in person identification and forensic. In: Daimi, K., Francia, G., III., Encinas, L.H. (eds.) Breakthroughs in Digital Biometrics and Forensics, pp. 147–165. Springer, Cham (2022). https://doi.org/10.1007/978-3-031-10706-1_7

13. Zabih, R., Woodfill, J.: Non-parametric local transforms for computing visual correspondence. In: Eklundh, J.-O. (ed.) ECCV 1994. LNCS, vol. 801, pp. 151–158. Springer, Heidelberg (1994). https://doi.org/10.1007/BFb0028345
14. Cortes, C., Vapnik, V.: Support-vector networks. Mach. Learn. **20**(3), 273–297 (1995). https://doi.org/10.1007/BF00994018
15. Weston, J., Watkins, C.: Support vector machines for multi-class pattern recognition. Esann **99**, 219–224 (1999)
16. He, K., Zhang, X. Ren, S., Sun, J.: Deep residual learning for image recognition. In: Proceedings of the Conference on Computer Vision and Pattern Recognition, pp. 770–778 (2016). https://doi.org/10.1109/CVPR.2016.90
17. Pan, T.-S., Huang, H.-C., Lee, J.-C., Chen, C.-H.: Multi-scale ResNet for real-time underwater object detection. SIViP **15**(5), 941–949 (2020). https://doi.org/10.1007/s11760-020-01818-w
18. Fan, Z., Liu, Y., Xia, V, Hou, J., Yan, F., Zang, Q.: ResAt-UNet: a U-shaped network using ResNet and attention module for image segmentation of urban buildings. Select. Topics Appl. Earth Observ. Remote Sens. **16,** 1–20 (2023). https://doi.org/10.1109/JSTARS.2023.3238720
19. He, K., Zhang, X., Ren, S., Sun, J.: Identity mappings in deep residual networks. In: Leibe, B., Matas, J., Sebe, N., Welling, M. (eds.) ECCV 2016. LNCS, vol. 9908, pp. 630–645. Springer, Cham (2016). https://doi.org/10.1007/978-3-319-46493-0_38
20. Saxe, A.M., McClelland, J.L., Ganguli, S.: Exact solutions to the nonlinear dynamics of learning in deep linear neural networks. ArXiv Prepr. ArXiv13126120 (2013)
21. Mahmoud, S.A., et al.: KHATT: an open Arabic offline handwritten text database. Pattern Recognit. **47**(3), 1096–1112 (2014)
22. Rabaev, I., Alkoran, I., Wattad, O., Litvak, M.: Automatic gender and age classification from offline handwriting with bilinear ResNet. Sensors **22**(24), 9650 (2022). https://doi.org/10.3390/s22249650

A New Optimization Approach to Improve an Ensemble Learning Model: Application to Persian/Arabic Handwritten Character Recognition

Omid Motamedisedeh[1], Faranak Zagia[2], and Alireza Alaei[3](✉)

[1] Faculty of Industrial and Systems Engineering, Tarbiat Modares University, Tehran, Iran
[2] Center Tehran Branch, Islamic Azad University, Tehran, Iran
[3] Faculty of Science and Engineering, Southern Cross University, Gold Coast, Australia
alireza20alaei@gmail.com

Abstract. Due to the advancement of technology, handwriting recognition has become more important than ever. As a result, several methods for document image recognition have been developed in the literature. This paper presents a new ensemble model based on the Feedforward Neural Networks (FFNN) to accurately recognize Persian and Arabic handwritten characters. As training and optimizing FFNN models have a significant role in obtaining optimal results, two optimization algorithms are integrated into the proposed handwritten recognition method. The Particle Swarm Optimization algorithm is integrated into the proposed model to improve the Neural Networks learning process. The FFNN architectures are further optimized using the League Championship Algorithm. The ensemble model is fed by a set of handcrafted features, including directional and intersection features, extracted from handwritten text. The proposed model is evaluated using three different datasets. Results obtained from the proposed models demonstrate higher accuracies compared to the state-of-the-art models.

Keywords: Handwriting Recognition · Ensemble Model · Neural Network · League Championship Algorithm · Particle Swarm Optimization

1 Introduction

Despite technological advances, handwritten documents still play an important role in our day-to-day life. Handwriting is still a natural, effective and convenient way of representing, keeping, exchanging, and communicating information. Many people use it while capitalizing on computers and other available smart devices. In the past, many practical applications, such as the automation of processes in post or bank offices, have been developed in the literature [1].

Recently, the number of smart handheld device users has increased significantly. This resulted in increasing demand for automatic handwritten document processing, enabling people to use these devices effectively. Thus, new research and applications in this area are still viable [2].

M. Coustaty and A. Fornés (Eds.): ICDAR 2023 Workshops, LNCS 14194, pp. 180–194, 2023.
https://doi.org/10.1007/978-3-031-41501-2_13

The research on recognizing handwritten document images, including characters and numerals, for most scripts, such as Latin-based scripts, Chinese, Japanese, and Korean, is mature, and several high-performance methods are available in the literature [3, 10, 11, 13]. There are also several studies about Persian (Urdu and Arabic as similar scripts) handwritten document image recognition in the literature [1, 2, 4–9, 14]. These methods can be grouped into two main categories, conventional and deep learning based approaches. The literature of the conventional Persian/Arabic handwritten character recognition approach is enriched with the application of several feature extraction methods, such as projection profiles, fractal code, moment, run-length, fuzzy elastic patterns, Gradient operator, wavelet, and directional, followed by classification techniques, including Neural Networks (NN) and Vector Quantization, Hidden Markov Model, SVM and K-Nearest Neighbor [2, 6]. Deep learning approaches have recently been applied to recognize Persian/Arabic handwritten documents [7, 8]. Table 1 provides an overview of some recent works in this context.

Despite rich literature in Persian and Arabic handwritten analysis and recognition, the reported accuracies are lagging compared to other scripts [2, 3, 14]. Apart from cursiveness, the existence of several characters with similar shapes and outlines, multiple forms for each character concerning its position in words and more classes compared to Western languages [2], lack of model optimization is also crucial in obtaining lower recognition accuracies. Therefore, in this research work, we introduced the use of some novel optimization algorithms to improve the recognition accuracies of Persian/Arabic handwritten document images. An ensemble model based on three independent neural networks with different architectures is proposed to perform the recognition process. The number of hidden layers and weight of each connection in the proposed NNs are obtained automatically using the Particle Swarm Optimization (PSO) method [15] and the League Championship Algorithm (LCA) [12]. The LCA automatically finds the optimal number of the hidden layers and neurons in each layer of different NNs, and the PSO is applied to find the optimum weights in the proposed NNs.

Moreover, as in the Persian and Arabic alphabet sets, there are several characters with similar shapes and more than two parts; we introduced five new features, including the number of connected graphs, horizontal, vertical, diagonal, and off-diagonal junction points, in addition to undersampled bitmaps technique and modified chain-code direction frequencies used in [2]. Results obtained from the proposed model, when applied to publicly available handwritten datasets, revealed the superiority of the proposed method compared to the state-of-the-art methods. The main contribution of this research work is not only the improvement of the recognition results of handwritten document images but also proposing a new optimization solution for designing neural networks in ensemble models by using the LCA and PSO methods.

The rest of the paper is organized as follows: The proposed model is described in Sect. 2. Section 3 provides experimental results and comparative analysis. Finally, Sect. 4 concludes the paper.

Table 1. An overview of a few handwritten document image recognition methods in the literature.

Method	Year	Model	Language	Application
Parvaz and Sabri [4]	2013	Dynamic programming	Arabic	In the proposed model, text lines are divided into words and words are recognized by a fuzzy polygon matching algorithm
Halavati and Shouraki [5]	2007	Elastic fuzzy pattern recognition	Persian	A new fuzzy model is presented to recognize elastic patterns of Persian words. The accuracy of the proposed model on the test data (including 1250 words) is about 78%
Shahmoradi and Shouraki [6]	2018	Fuzzy elastic matching machine	Persian	A fuzzy Elastic Matching Machine, as a flexible mathematical structure, is presented to solve the problem for the Persian language. The proposed model is applied to Kabir's Persian datasets, and its accuracy is evaluated at about 75%
Mersa et al. [7]	2019	CNN	Persian	A combination of CNN and SVM is used to solve the problem. The GPDS-Synthetic of the proposed model is about 6%
Safarzadeh and Jafarzadeh [8]	2020	CNN and RNN-CTC	Persian	In the proposed model, convolutional neural networks and recurrent neural networks are used to solve the problem and connectionist temporal classification is applied to eliminate the segmentation step required in the conventional methods

(continued)

Table 1. *(continued)*

Method	Year	Model	Language	Application
Mowlaie et al. [9]	2003	Support vector machines (SVMs)	Persian	The SVM is used to recognize the handwriting pattern of written addresses on postal cards
Farlynda et al. [3]	2021	Backpropagation algorithm	English	A neural network with a hidden layer and 64 neurons is by backpropagation learning algorithm is used to solve the problem
Manocha et al. [10]	2021	Deep-Learning	Devanagari	The paper has been focused on Devanagari Handwritten Recognition Systems background, challenges, opportunities and future research trends
Retsinas et al. [11]	2021	Weight Oracle labels	English	They proposed a model with a 2% error rate for recognizing the characters and a 5% error rate for recognizing the Words

2 Proposed Method

The proposed method comprises two main components, pre-processing and feature extraction, and constructing the proposed ensemble model based on the NNs followed by fine-tuning its parameters. Details of each component are discussed in the following subsections.

2.1 Preprocessing and Feature Extraction

Pre-processing is an essential part of a recognition/classification method. In this research work, simple pre-processing methods, such as image binarization and normalization, are performed to simplify the process and obtain normalized features. Each input image is first binarized using Otsu's algorithm, and the extracted bounding of each character is then normalized into 49 × 49 pixels. We used Otsu's algorithm as one of the literature's most common binarization techniques. The size is decided based on experimental analysis [2].

One of the difficulties of Persian handwritten document image recognition is that several characters are composed of a main shape and one or more components, such as dots and diacritical marks. To handle this challenge, in addition to the directional features extracted in [2], five more simple but effective features, including the number of intersections between horizontal, vertical, diagonal and off-diagonal lines and contours of the image, and the number of connected graph features, are extracted and included in the feature set. In order to not repeat the entire feature extraction process presented in [2], the pre-processing and feature extraction step is depicted in Table 2 and briefly summarized as follows.

Table 2. Feature extraction process.

	Result	Step	Result	Parameters
Original image	ض	49 non-overlappig blocks		
Contour extraction		Intersection and number of connected components features		b_1=4 b_2=4 b_3=4 b_4=2 b_5=8
Normalized image of 49×49 pixels		Directional features of the block (i=23)		$a_{23\,1}$=15 $a_{23\,2}$=0 $a_{23\,3}$=18 $a_{23\,4}$=10

i. Binarize the image and find the boundary box for the input image
ii. Normalize the boundary box into an image of 49 × 49 pixels and then extract the contours of the images.
iii. Divide the normalized contour image into 49 independent/non-overlapped blocks of size 7 × 7 pixels, numbered from 1 (top left) to 49 (bottom right).
iv. Extract four directional features (a_{i1} to a_{i4}) for block i, where i = 1 to 49. This process results in 196 (49 × 4) directional features [2].
v. Compute features b_1, b_2, b_3, b_4, and b_5 as the number of connected graphs (connected components), and find the number of intersections between horizontal, vertical, diagonal and off-diagonal lines and the contours of the character.
vi. As a result, a feature set of size 201 (196 + 5) features [b_1, b_2,... b_5, a_{11}, a_{12}, a_{13}, a_{14}, a_{21}, a_{22}, a_{23}, a_{24}, ..., $a_{49\,1}$, $a_{49\,2}$, $a_{49\,3}$, $a_{49\,4}$] is obtained.

2.2 The Ensemble Model

Ensemble models are a machine learning technique to combines multiple individual models to produce a more accurate and robust prediction or classification model than any single model alone. The individual models, also known as base models, can be of different or the same types and trained on different subsets of the data [18]. In ensemble models, including bagging, stacking, and boosting, different models are used and trained separately based on the same or different datasets, and then an aggregation function (such as averaging, weighted voting or even machine learning models) is used to combine the results [18].

We have used neural networks to build our ensemble model. The structure of the proposed ensemble model includes three independent neural networks, whose results are aggregated by a weighted average based on their efficiency, as shown in Fig. 1. The number of NNs will also be decided during the optimization process.

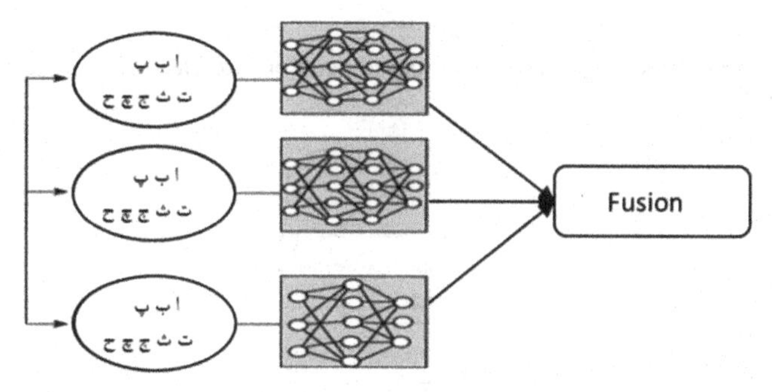

Fig. 1. The block diagram of the proposed ensemble model.

2.3 Neural Networks

This research work proposes three NN with different numbers of hidden layers and neurons to perform Persian handwritten character recognition. Figure 2 demonstrates a general design of an NN containing an input layer, two hidden layers, and an output layer. In the proposed model, the input layer is defined by $X_n = [b_1, b_2, \ldots b_5, a_{11}, a_{12}, a_{13}, a_{14}, \ldots, a_{49\,4}]$, and the output layer is represented by y (the labels of the Persian characters). One of the challenges in designing a NN is defining the number of hidden layers and fine-tuning the weights between neurons at different layers. In order to solve this problem, LCA and PSO are used in this research work. In the proposed model, the LCA [12] is used to optimize the NNs' architecture and the optimal number of hidden layers and neurons in each layer, and the PSO [15] is further used to optimize the weighted matrix in the NNs. Figure 3 shows a pictorial description of the proposed ensemble model and its optimization process. Details of the LCA and PSO methods are presented in the following subsections.

2.4 League Championship Algorithm

The League Championship Algorithm (LCA) is a stochastic population-based algorithm proposed for the continuous global optimization problem [12]. The LCA considers a few scenarios and simulates a championship environment in an imaginary league. The scenario (team) with a higher fitness function value in each competition will win, and others will lose. The loser team changes its parameters for the next round of competition.

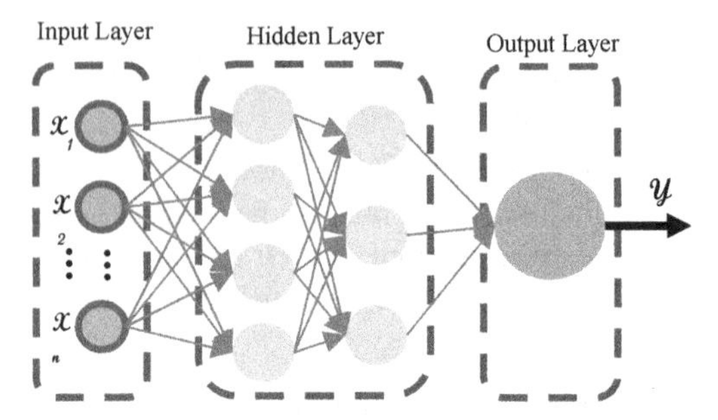

Fig. 2. The general block diagram of the proposed Feedforward Deep Neural Network model.

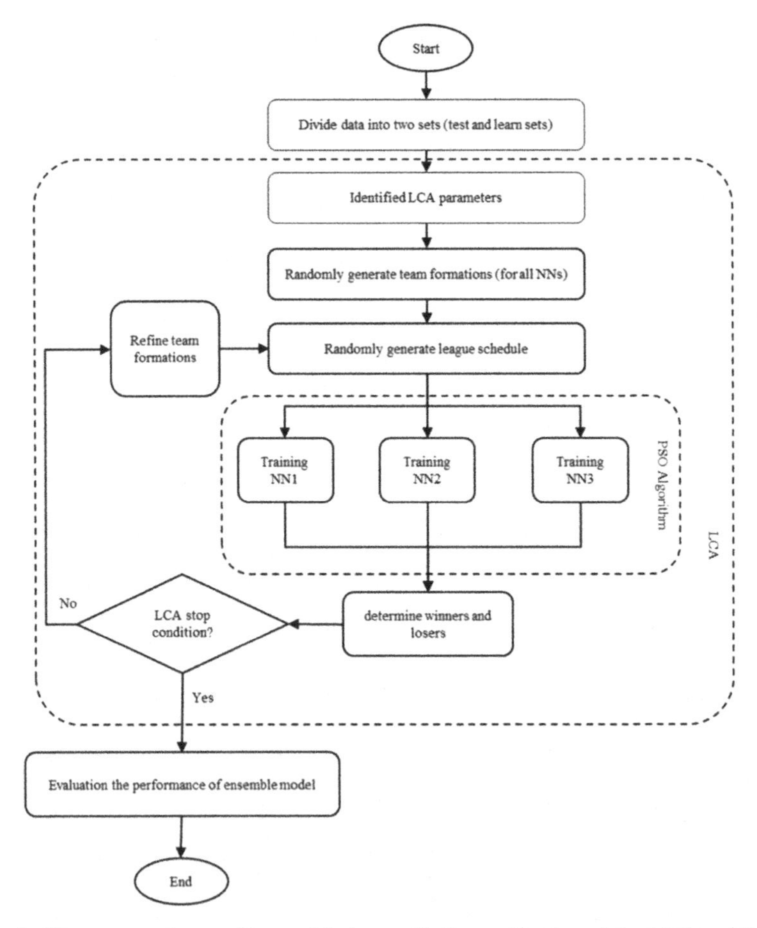

Fig. 3. The proposed ensemble model along with the application of the LCA and PSO.

This process is iterated until all scenarios are compared with each other. At the end of the process, the scenario with the highest scores indicates the winner or the optimized solution (global) [12]. To avoid getting stuck in the local optimal solution, after each iteration (week of competition), each team devises the required changes in its formation (generation of a new solution) for the contest in the next week [12].

As mentioned, one of the challenges in NN is finding the number of hidden layers to obtain optimal results. In this research, we applied the LCA as a heuristic model to optimize the structure of the proposed NN models. The application of LCA to the proposed model is demonstrated in Fig. 3. As shown in Fig. 3, in each iteration, the structures of NNs in the ensemble model are considered as the teams' structures, and in each competition, the accuracy rate of two different structures are compared, and the structure of NNs with a lower accuracy rate will be changed based on the structure of the next team for the next competition.

As shown in Fig. 3, to optimize the NN architecture, the LCA is applied considering the following steps:

Step I: set $t=1$ and determine league size (L) and the number of seasons (S).
Step II: initial team formations (determining the structures of three NNs) are generated randomly, as initially l_i (an integer number) are generated randomly for the team number i to determine the number of hidden layers, and then $n_{i,1}$, $n_{i,2}$,..... $n_{i,li}$ are randomly generated to determine the number of neurons in each layer.
Step III: the fitness value of each formation is assessed by running the ensemble model.
Step IV: league schedule is generated.
Step V: teams compete with each other based on the schedule, and winners and losers will be determined.
Step VI: if Mod(t,L-1) <> 0, go to *Step IX;* else *continue.*
Step VII: for each pair of teams, a member of the loser team is replaced by the same member of the winner team, and the solutions will be refined.
Step VIII: set t=t+1 and set up a new team formation for each team based on the artificial match analysis process.
Step IX: If t>S(L-1), the whole process is stopped; otherwise, go to *Step III*.

2.5 Particle Swarm Optimization Algorithm

PSO is a heuristic algorithm that finds the optimum solution in Np-hard (nondeterministic polynomial time) problems based on birds' behavior in the process of finding food [15]. Each solution is considered a bird trying to find food (a fitness function evaluates food quality). The best experience of each bird is considered *pbest,* and the best solution among all birds is considered *gbest*. In each iteration, the birds move toward *pbest* and *gbest* by specific velocity.

This research work uses the PSO method to optimize the weight vectors between different layers of each NN, as shown in Fig. 4.

The NNs with different architectures are trained by employing the following steps:

Step I. The initial particles that represent the weight matrix and biases are generated randomly. The ith particles for an NN by one hidden layer will be donated by $P_i^0 = \{W_{1i}^0, W_{2i}^0, a_{1i}^0\}$, where $W^0{}_{1i}$ is the weights matrix of the ith particle from the input layer to a hidden layer, $W^0{}_{1i}$ is the weight matrix of the ith particle from a hidden layer to the output layer, and a_{1i} is a bias for the ith particle.

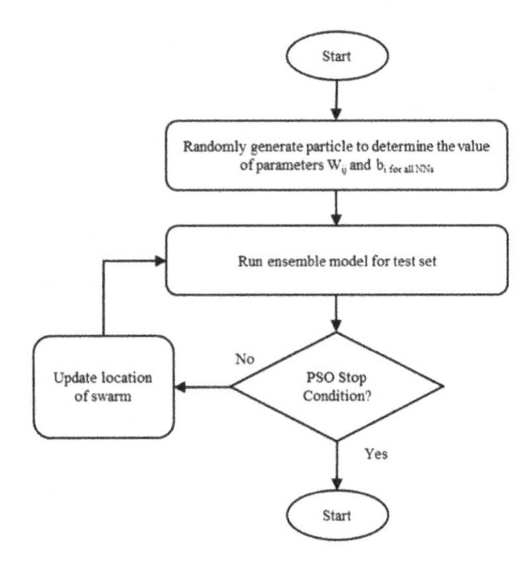

Fig. 4. Particle Swarm Optimization method used in our proposed ensemble model.

Step II. Run the NN for all particles and calculate the fitness function for all particles.
Step III. Update *pbest* and *gbest* based on the fitness function values.
Step IV. Check the stop condition and stop the algorithm if it is true, else go to the next step.
Step V. Evaluate velocity values for the ith particles.
Step VI: Calculate new locations for all particles by Eq. (1).

$$P_i^{j+1} = P_i^j + V_i^{j+1} \tag{1}$$

Which V_i^{j+1} is evaluated based on the below formula.

$$V_i^{j+1} = V_i^j + r\alpha[P_{best,i}^j - P_i^j] + s\beta[g_{best}^j - P_i^j] \tag{2}$$

where $P_{best,i}^j$ is the best position for the ith particle in the jth iteration, P_{best}^j is the best position for all particles in the jth iteration, α and β are random parameters between 0 and 1 and r and s are constants in Eq. (8).

$$P_i^j = \{W_{1i}^j, W_{2i}^j, a_{1i}^j\} \tag{3}$$

$$g_{best}^j = \{W_{gbest_1}^j, W_{gbest_2}^j, a_{gbest_1}^j\} \tag{4}$$

$$P_{best,i}^j = \{W_{pbest,i_1}^j, W_{pbest,i_2}^j, a_{pbest,i_1}^j\} \tag{5}$$

$$V_i^j = \left\{V_{1i}^j, V_{2i}^j, V_{1i}^{'j}\right\} \tag{6}$$

$$P_i^{j+1} = \{W_{1i}^j + V_{1i}^{j+1}, W_{2i}^j + V_{2i}^{j+1}, a_{1i} + V_{1i}^{'j+1}\} \tag{7}$$

So particle location will be calculated as Eq. (8).

$$P_i^{j+1} = \{W_{1i}^j + V_{1i}^j + r\alpha\left[W_{pbest,i_1}^j - W_{1i}^j\right] + s\beta\left[W_{gbest_1}^j W_{1i}^j\right], W_{2i}^j + V_{2i}^j + r\alpha\left[W_{pbest,i_2}^j - W_{2i}^j\right] + s\beta\left[W_{gbest_2}^j - W_{2i}^j\right], a_{1i}^j + V_{1i}^{\prime j+1} + r\alpha\left[a_{pbest,i_1}^j - a_{1i}^j\right] + s\beta\left[a_{gbest_1}^j - a_{1i}^j\right]\} \quad (8)$$

Step VII: Go to Step *II*

3 Experimental Results and Comparative Analysis

We considered two Persian handwritten character datasets to evaluate the performance of the proposed model. The first one is the IFHCDB [16], an offline handwritten dataset of 52,020 samples populated in 32 classes [16]. The second one is an online Persian handwritten character dataset including 5,843 samples in 32 classes [17]. The online data (as a noisy sample) has been converted into images to evaluate the performance of the proposed model on the images obtained from an online dataset [17]. In our experiments, we used the 10-fold cross-validation protocol and accuracy as an evaluation metric. In all our experiments, both the architecture and the weight matrix of NNs were optimized using the training set, and then the performance of the model was computed using the test set. The results obtained from our experiments on the IFHCDB using different numbers of NNs, hidden layers and neurons are presented in Table 3. As shown in Table 3, the best results were obtained based on an ensemble model composed of three NNs, of which the first NN contains two hidden layers of 9 and 3 neurons, the second NN has two hidden layers of 6, and 5 neurons and the third on contains three hidden layers of 6, 4, and 2 neurons.

Considering the IFHCDB [16] for experiments, in addition to the proposed ensemble model, we employed several other classifiers, including the original NN without the application of the proposed optimization process, Nearest Neighbore (1-NN) and K-NN (K = 3 and 5), and "simple ensemble" model without considering new features for comparison purposes. The results obtained from each fold of 10-fold cross-validation based on different models are shown in Fig. 5. As can be seen from Fig. 5, in all cases, the proposed model has outperformed other methods. The average recognition results obtained from different models are further presented in Table 4. From Table 4, it is evident that a recognition accuracy of 98.1% has been obtained from the proposed model on the IFHCDB. Moreover, the recognition rate obtained from the proposed model on the IFHCDB has been improved by 5.6% compared to the original NN. This improvement in accuracy is higher compared to other models.

We further performed several experiments on the online dataset [17]. The results obtained based on 10-fold cross-validation from different models are shown in Fig. 6. From Fig. 6, it is clear that similar to the results obtained from the IFHCDB dataset, the performance of the proposed method is higher than all other models. Moreover, the average recognition results obtained from all models are provided in Table 4. From Table 4, it is noted that the best recognition accuracy of 95.2% has been obtained from the proposed model considering the online dataset [17]. The improvement of accuracy obtained from our proposed model on the online dataset compared to the simple NN is

Table 3. Comparison of results obtained from NNs with different architectures considering the IFHCDB for experimentation.

Number of NNs	Optimum Architecture (Number of hidden layers and neurons)	Accuracy
2	[(5,7), (6,3,4,6,3)]	93.3%
3	**[(9,3), (6,5), (6,4,2)]**	**98.1%**
4	[(5,5,4,5), (5,6), (5,4,8,4,7), (5,6,3,4,7)]	97.9%
5	[(6,3,2,4), (3,3,5,2,2), (5,6), (3,3,5), (3,3)]	97.6%
6	[(2,7,2), (6,5,5,7,7), (5,2), (6,5), (3,4,8,7,2), (3,5)]	95.3%
7	[(7,4,2), (7,6,4,4), (7,6,3), (7,6,4), (6,4,5,5), (4,5), (6,5)]	97.2%
8	[(6,5), (8,5), (5,4,4), (8,5,3), (5,5,4,3,6), (4,3,2), (4,3,5,3), (2,2)]	97.4%
9	[(8,4,8,2), (7,3), (4,8), (7,3,3), (5,4), (4,8,3,4,4), (7,8,4), (3,8), (5,2,4,4)]	96.1%
10	[(8,4,2,8), (4,6), (8,5,2,4,4), (4,6,4,3,3), (3,3,5,3,4), (5,5,4), (3,6,5,4,8), (4,5), (5,6,2,8), (2,2,6,2,5)]	97.6%

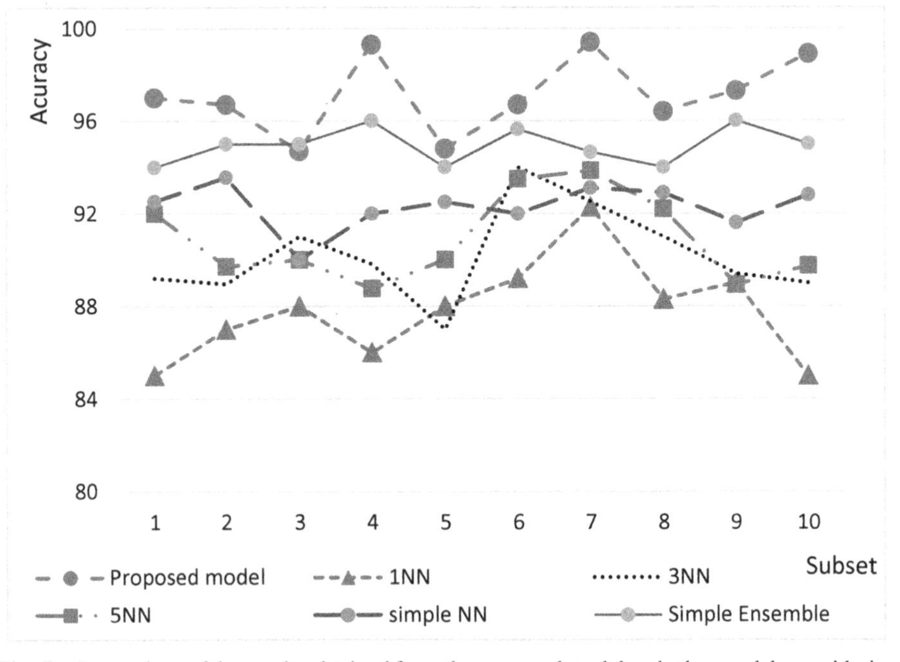

Fig. 5. Comparison of the results obtained from the proposed model and other models considering the IFHCDB for experimentation.

Table 4. Comparison of results obtained from different models using both Persian datasets.

Method	Dataset	
	Recognition Accuracy (%)	
	IFHCDB dataset [16]	Online dataset [17]
1-NN	88.4	83.6
3-NN	90.2	88.4
5-NN	90.9	87.2
Simple NN	92.5	90.8
Simple Ensemble	94.92	91.8
Proposed Model	**98.1**	**95.2**

consistent with the IFHCDB. However, the results obtained from the proposed model on the online dataset were lower than those on the IFHCDB model. The main reason for these lower accuracies is the smaller number of samples in the online dataset compared to the IFHCDB, making the training and convergence of the proposed model challenging.

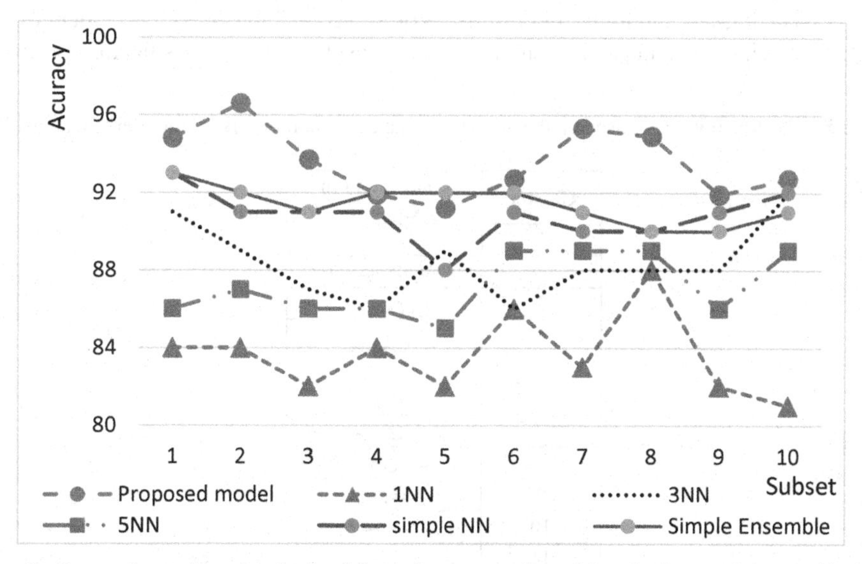

Fig. 6. Comparison of results obtained from the proposed model and other models considering the online dataset for experimentation.

To compare the performance of the proposed model with other models in the literature, the Kaggle database, including 3,410 offline handwritten English characters, was further considered for experiments. The same protocol used as before for the evaluation and results are provided in Fig. 7. Figure 7 shows that the proposed model outperformed the simple NN and provided higher accuracies than the simple NN model.

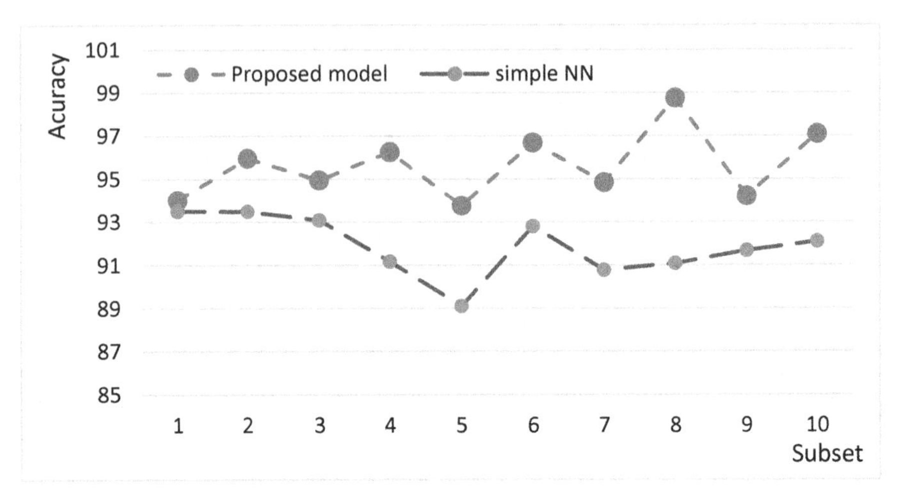

Fig. 7. Comparison of the results obtained from the proposed model and other models considering the English dataset for experimentation.

To get an idea of the erroneous results where the proposed model could not correctly recognize the characters, we performed a simple clustering of the erroneous cases in both Persian datasets and found 11 clusters, as presented in Table 5. As it is evident from Table 5, the characters in each cluster are similar, resulting in misclassification results.

Table 5. Clusters obtained based on the analysis of the confusion matrix of both Persian datasets.

Cluster	Characters in Cluster
1	آ
2	ب, پ, ت, ث
3	ج, چ, خ, غ
4	ح, ع, ق
5	ر, د, و
6	ذ, ز, ژ
7	س, ص, ی, ل, ن
8	ش, ض, ف
9	ط, ظ
10	ک, گ
11	م, ه

4 Conclusion and Future Work

We proposed a novel ensemble of NNs, in conjunction with the LCA and PSO, to automatically design and construct a new model for recognizing Persian Handwritten documents. In the proposed model, the NNs were considered independently, and results were aggregated by a weighted average based on their efficiency. The LCA was applied to optimize the NN architectures, and the PSO was used to optimize the training process of NNs. As we had a small number of data samples and limited computing resources, we extracted some conventional features instead of feeding the entire images to the model for training and testing. The results obtained from the proposed model on three different datasets revealed that the model provided higher recognition accuracies compared to other models.

In future, we plan to extend our model to other recognition tasks and apply deep feature learning methods instead of using handcrafted features to improve the accuracies of Persian handwritten document recognition.

References

1. Elleuch, M., Jraba, S., Kherallah, M.: The effectiveness of transfer learning for arabic handwriting recognition using deep CNN. J. Inf. Assur. Secur. **16**(2) (2021)
2. Alaei, A., Nagabhushan, P., Pal, U.: A new two-stage scheme for the recognition of persian handwritten characters. In: Proceedings of the 12th International Conference on Frontiers in Handwriting Recognition (2010)
3. Farlinda, S., et al.: Application of backpropagation algorithm for handwriting recognition. In: Journal of Physics: Conference Series (2021)
4. Tanvir Parvez, M., Mahmoud, S.A.: Arabic handwriting recognition using structural and syntactic pattern attributes. Pattern Recognit. **46**(1), 141–154 (2013)
5. Halavati, R., Shouraki, S.B.: Recognition of Persian online handwriting using elastic fuzzy pattern recognition. Int. J. Pattern Recognit. Artif. Intell. **21**(03), 491–513 (2007)
6. Shahmoradi, S., Bagheri Shouraki, S.: Evaluation of a novel fuzzy sequential pattern recognition tool (fuzzy elastic matching machine) and its applications in speech and handwriting recognition. Appl. Soft Comput. **62**, 315–327 (2018)
7. Mersa, O., Etaati, F., Masoudnia, S., Araabi, B.N.: Learning representations from Persian handwriting for offline signature verification, a deep transfer learning approach. In: Proceedings of the 4th International Conference on Pattern Recognition and Image Analysis (IPRIA) (2019)
8. Safarzadeh, V.M., Jafarzadeh, P.: Offline Persian handwriting recognition with CNN and RNN-CTC. In: Proceedings of the 25th International Computer Conference (2020)
9. Mowlaei, A., Faez, K.: Recognition of isolated handwritten Persian/Arabic characters and numerals using support vector machines. In: Proceedings of the IEEE XIII Workshop on Neural Networks for Signal Processing (2003)
10. Manocha, S.K., Tewari, P.: Deep learning approaches for Devanagari handwriting recognition. In: Proceedings of the 9th International Conference on Reliability, Infocom Technologies and Optimization (Trends and Future Directions) (2021)
11. Retsinas, G., Sfikas, G., Nikou, C.: Iterative weighted transductive learning for handwriting recognition. In: Lladós, J., Lopresti, D., Uchida, S. (eds.) ICDAR 2021. LNCS, vol. 12824, pp. 587–601. Springer, Cham (2021). https://doi.org/10.1007/978-3-030-86337-1_39
12. Husseinzadeh Kashan, A.: League championship algorithm (LCA): an algorithm for global optimization inspired by sport championships. Appl. Soft Comput. **16**, 171–200 (2014)
13. Ruiz-Parrado, V., Heradio, R., Aranda-Escolastico, E., Sánchez, Á., Vélez, J.F.: A bibliometric analysis of off-line handwritten document analysis literature (1990–2020). Pattern Recognit. **125** (2022)
14. Mozaffari, S., Soltanizadeh, H.: ICDAR handwritten Farsi/Arabic character recognition competition. In: Proceedings of the10th ICDAR, pp.1413–1417 (2009)

15. Kennedy, J., Eberhart, R.: Particle swarm optimization. In: Proceedings of the IEEE International Conference on Neural Networks, IV. pp. 1942–1948 (1995)
16. Mozaffari, S., Faez, K., Faradji, F., Ziaratban, M., Golzan. S.M.: A comprehensive isolated Farsi/Arabic character database for handwritten OCR research. In: Proceedings of the10th IWFHR, pp. 385–389 (2006)
17. Razavi, S.M., Kabir, E.: A data base for online Persian handwritten recognition. In: Proceedings of the 6th Conference on Intelligent Systems, Kerman (2004)
18. Guo, H., Liu, Y., Yang, D., Zhao, J.: Offline handwritten Tai Le character recognition using ensemble deep learning. Vis. Comput. **38**, 1–14 (2021). https://doi.org/10.1007/s00371-021-02230-2

BN-DRISHTI: Bangla Document Recognition Through Instance-Level Segmentation of Handwritten Text Images

Sheikh Mohammad Jubaer[1], Nazifa Tabassum[1], Md Ataur Rahman[1](✉), and Mohammad Khairul Islam[2]

[1] Department of CSE, Premier University, Chittagong, Bangladesh
jubaer.puc@gmail.com, nazifa.puc@gmail.com, ataur.cse@puc.ac.bd
[2] Department of CSE, University of Chittagong, Chittagong, Bangladesh
mkislam@cu.ac.bd

Abstract. Handwriting recognition remains challenging for some of the most spoken languages, like Bangla, due to the complexity of line and word segmentation brought by the curvilinear nature of writing and lack of quality datasets. This paper solves the segmentation problem by introducing a state-of-the-art method (BN-DRISHTI (**Code and Demo:** https://github.com/crusnic-corp/BN-DRISHTI)) that combines a deep learning-based object detection framework (YOLO) with Hough and Affine transformation for skew correction. However, training deep learning models requires a massive amount of data. Thus, we also present an extended version of the BN-HTRd dataset comprising 786 full-page handwritten Bangla document images, line and word-level annotation for segmentation, and corresponding ground truths for word recognition. Evaluation on the test portion of our dataset resulted in an F-score of 99.97% for line and 98% for word segmentation. For comparative analysis, we used three external Bangla handwritten datasets, namely BanglaWriting, WBSUBNdb_text, and ICDAR 2013, where our system outperformed by a significant margin, further justifying the performance of our approach on completely unseen samples

Keywords: Handwritten Text Recognition (HTR) · Data Annotation · Image Segmentation · Computer Vision · Deep Learning

1 Introduction

Line and word segmentation are one of the most fundamental parts of handwritten document image recognition. As the field of deep learning is maturing at an unprecedented speed, the choice for solving this sort of task employing off-the-shelf deep learning frameworks is getting popular nowadays for its efficiency. However, few attempts have been made to utilize this approach for Bangla handwritten recognition task due to the scarcity of datasets in this domain. Our previous endeavors involved an initial dataset-making process named BN-HTRd (v1.0), comprising of Bangla handwritten document images and only line-level

M. Coustaty and A. Fornés (Eds.): ICDAR 2023 Workshops, LNCS 14194, pp. 195–212, 2023.
https://doi.org/10.1007/978-3-031-41501-2_14

annotations and ground truths for word recognition. However, that dataset was incomplete due to the missing word-level annotation. Therefore, to have a more comprehensive and useable handwritten recognition dataset, we have extended the BN-HTRd (v4.0) dataset[1] by integrating word-level annotations and necessary improvements in the ground truths for the word recognition task.

As segmentation plays a vital role in recognizing handwritten documents, another pivotal *contribution* of this paper is the conglomeration of a state-of-the-art method for segmenting lines and words from transcribed images. Our approach treats the segmentation task as an object detection problem by identifying the distinct instances of similar objects (i.e., lines, words) and demarcating their boundaries. Thus in a way, we are performing `instance-level segmentation` as it is particularly useful when homogeneous objects are required to be considered separately. To do so, we partially rely on the YOLO (You Only Look Once) framework. However, the success of our method is more than just the training of the YOLO algorithm. In order to get the perfect words segmented from possibly complex curvilinear text lines, we had to improvise our approach to retrieve the main handwritten text lines correctly by removing other unnecessary elements. For that, we used a combination of the Hough and Affine transform methods. The Hough transform predicts the skew angles of the main handwritten text lines, and the Affine transform rotates them according to the expected gradients, making them straight horizontally. Therefore, the word segmentation approach provides much better results compared to the segmentation on skewed lines. Thus, the main contributions of this paper are threefold:

1. Introducing a straightforward **`novel hybrid approach`**, for instance-level handwritten document segmentation into corresponding lines and words.
2. Achieved *state-of-the-art* (`SOTA`) scores on three different prominent Bangla handwriting datasets for line/word segmentation tasks.
3. Set a new `benchmark` for the BN-HTRd dataset. Also, `extended`[2] it to be one of the largest and the most comprehensive Bangla handwritten document image segmentation and recognition dataset by adding `200k+` annotations.

2 Related Work

CMATERdb [21] is one of the oldest character-level datasets consisting of 150 Bangla handwritten document images distributed among two versions. Another prominent character-level dataset having 2000 handwritten samples named **BanglaLekha-Isolated** [5] contains 166105 handwritten characters written by an age group of 6 to 28. **Ekush** [15], which is a multipurpose dataset, contains 367,018 isolated handwritten characters written by 3086 individual writers. The authors also benchmarked the dataset using a multilayer CNN model (**EkushNet**) for character classification, achieving an accuracy of 97.73% on their dataset while scoring 95.01% in the external **CMATERdb** dataset.

[1] **Extended Dataset:** https://data.mendeley.com/datasets/743k6dm543.
[2] **Changes:** https://data.mendeley.com/v1/datasets/compare/743k6dm543/4/1.

A paragraph-level dataset that resembles our dataset in terms of word-level annotation is the **BanglaWriting** [12] dataset, which includes single-page handwriting comprising 32,787 characters, 21,234 words, and 5,470 unique words produced by 260 writers of different ages and personalities. Another paragraph-level unannotated dataset **WBSUBNdb_text** [10], consisting of 1383 handwritten Bangla scripts having around 100k words, was collected from 190 transcribers for the writer identification task. While in terms of a document-level dataset, mostly resembling our own, **ICDAR 2013** [22] handwriting segmentation contests dataset comes with 2649 lines and 23525 word-level annotations for 50 handwritten document images on Bangla.

Segmenting handwritten document images in terms of lines and words is the most crucial part when it comes to end-to-end handwritten document image recognition. In ***Projection-based*** methods [8,9,13,14], the handwritten lines are obtained by computing the average distance between the peaks of the projected histogram. A method based on the skew normalization process is proposed in [3]. ***Hough-based*** methods [9] represent geometric shapes such as straight lines, circles, and ellipses in terms of parameters to determine geometric locations that suggest the existence of the desired shape. The author of [8] presented a skew correction technique for handwritten Arabic document images using their optimized randomized Hough transform, followed by resolving the primary line for segmentation. For layout analysis, ***Morphology-based*** approaches [7,9] have been used along with piece-wise painting (PPA) algorithms [2], to segment script independent handwritten text lines. In contrast, ***Graph-based*** approaches [9,11,23] compactly represent the image structure by keeping the relevant information on the arrangement of text lines. ***Learning-based*** techniques recently became popular for segmenting handwritten text instances. The authors of [4,19,20,24] used a Fully Convolutional Network (FCN) for this purpose. A model based on the modified multidimensional long short-term memory recurrent neural networks (**MDLSTM RNNs**) was proposed in [6]. An unsupervised ***clustering*** approach [16] was utilized for line segmentation which achieved an F-score of 81.57% on the BN-HTRd dataset.

A series of consistent recent works on **Bangla handwriting segmentation** [1,17,18] is carried out by a common research team that also developed the WBSUBNdb_text dataset. Their technique predominantly relies on the projection profile method and connected component analysis. They initially worked on a tri-level (line/word/character) segmentation [17] while their latest works are focused solely on word [1] and line segmentation [18]. Moreover, in [18], the method serves the line segmentation on multi-script handwritten documents while the other two research only work for the Bangla scripts.

Our work can be categorized as a **Hybrid Approach** for segmenting lines and words. Our supervised models employ YOLO deep learning framework to predict lines and words from handwritten document images. We used the Hough Line Transform to measure the segmented line's skew angle, then corrected it with Affine Transform. These combinations were never used in the literature for Bangla handwritten recognition tasks.

3 Dataset

Data annotation is one of the most crucial parts of the dataset curation process where supervised learning is concerned. As a primary text source, we considered the BBC Bangla News platform since it does not require any restrictions and has an open access policy. Hence, we downloaded various categories of news content as files in TEXT and PDF format, renamed files according to the sequence of 1 to 237, and put them in separate folders. We distributed those 237 folders among 237 writers of different ages, disciplines, and genders. They were instructed to write down the text file's contents in their natural writing style and to take pictures of the pages afterward. This resulted in 1,591 handwritten images in total. Due to the complexity of the task, we were only able to recruit a total of 75 individuals to annotate lines of assigned handwritten images using an annotation tool called `LabelImg`. As a result, we were only able to annotate a maximum of 150 folders. The resultant annotation produced YOLO and PASCAL VOC formatted ground truth for line segmentation. These 150 folders of handwritten images and their line annotations were included in the first version of the BN-HTRd dataset [16]. For the purpose of word segmentation, we have extended the dataset (v4.0) by adding bounding-box annotations of individual words for all the annotated lines. We also organized each word of the text file into separate rows in Excel in order to create the ground truth Unicode representation of the corresponding word's images for recognition purposes in the future.

We used this extended BN-HTRd dataset containing annotations in 150 folders to develop and test our system. It contains a total of 786 handwritten images comprising 14,383 lines and 1,08,181 words. The rest of the unannotated 87 folders were automatically annotated using our system, resulting in an additional 14,836 lines and 1,06,135 words, which we denoted as Automatic Annotations. For the purpose of experimental evaluation, we split the 150 folders into two subsets and took one image from each of the folders for either validation or testing (resulting in 75 images for each subset). The rest of the 636 images were used for training purposes. Table 1 below shows this subdivision.

Table 1. Distribution of extended BN-HTRd (v4.0) dataset for experimentation. (**Splitted Dataset:** https://doi.org/10.57967/hf/0546)

Type	Purpose	Train	Valid	Test	Total
Doc. Images	Line Segmentation	636	75	75	786
Line Images	Word Segmentation	11,471	1,515	1,397	14,383
Word Images	Word Recognition	86,055	11,712	10,414	1,08,181

4 Proposed Methodology

We have broken down our overall system architecture in Fig. 1, which consists of six parts. Those six parts cover the overall process of how our system functions.

Before dissecting those parts in detail in the later sections (4.1–4.5), we will provide a brief overview in the following:

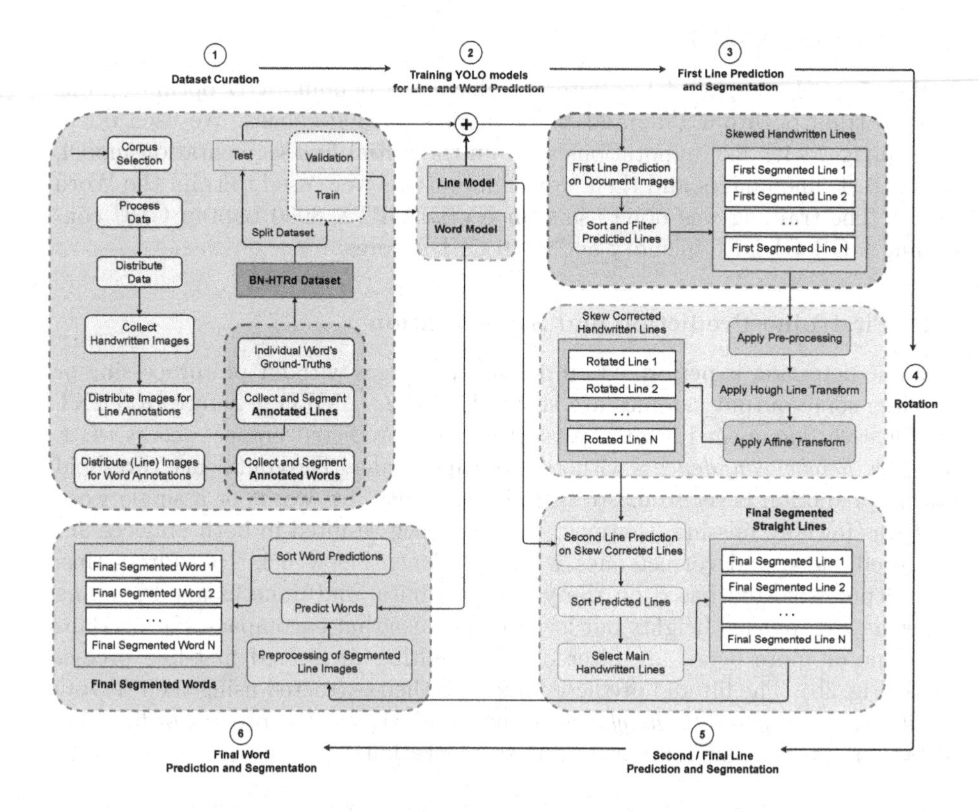

Fig. 1. Overall System Architecture for BN-DRISHTI.

- Our efforts in making and extending the BN-HTRd dataset involved various development processes such as distributing the data to the writers, manual annotations, and making it compatible with supervised learning methods such as ours (details in Sect. 3).
- Although training the models is a crucial part of any supervised system, it was not enough in our case despite YOLO being one of the best frameworks. It was predicting redundant lines, which we had to eliminate in order to get better segmentation scores (details in Sects. 4.1 and 4.2).
- As we were also getting some unnecessary lines along with the target line, a better line segmentation method is essential to segment the words correctly. To remove them, we rotated the curvilinear lines using the Hough-Affine transformation and corrected their skewness (details in Sect. 4.3).
- We applied the final/second YOLO line prediction on the skew-corrected lines, followed by some post-processing in order to extract the main handwritten line (details in Sect. 4.4).

– Finally, word prediction and segmentation are performed on skew-corrected final segmented line images using the word model (details in Sect. 4.5).

4.1 Training Models

YOLOv5x (XLarge) model architecture having a default SGD optimizer was used to train both our Line and Word models for 300 epochs. We used document images with line annotations to train the initial line segmentation model. In contrast, line images and their word annotations were used to train the Word model. The training was done using an NVIDIA RTX 3060 Laptop GPU containing 6 GB GDDR6 memory and 3840 CUDA cores.

4.2 First-Line Prediction and Segmentation

The line detection is performed on document images without pre-processing or resizing; some output samples are shown in Fig. 2a. YOLO generates a TEXT file for each document image, representing each predicted line as <*class_id, x, y, width, height, confidence*> without particular order. The confidence threshold during prediction is set to 0.3 to include lines with few words or a single word that was initially missing. However, this approach resulted in both unnecessary line predictions and correct ones with confidence below 0.5. To address this, the output is sorted based on the y-axis attribute, and unnecessary bounding boxes having unusual heights but lower confidence that encompasses or overlaps with one or more boxes are filtered out, resulting in filtered first-line predictions (Fig. 2b). The filtered predicted lines are then extracted using their YOLO attributes: $< x, y, width, height >$. Figure 2c illustrates the process of first-line detection, filtering, and corresponding segmentation.

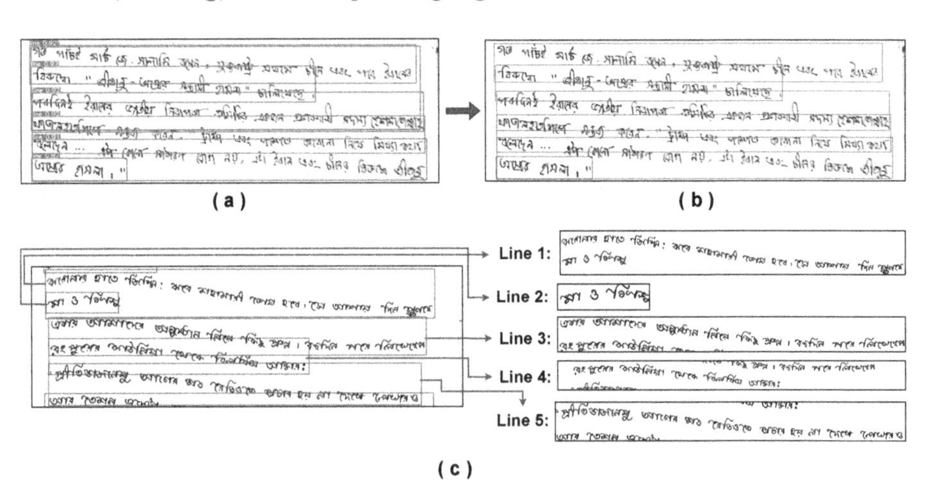

Fig. 2. Representation of First-line prediction and segmentation, where a) sample image with first-line prediction containing multiple unnecessary predictions, b) filtered first-line prediction, and c) another sample image with filtered first-line prediction and segmentation for curvilinear handwriting.

4.3 Rotation (skew Estimation and Correction)

After analyzing our first segmented line images, we found out that, with the main handwritten line, we are also getting some unwanted lines at the top or bottom due to the skewness of the lines and the rectangular shape of the predicted bounding box. Therefore, the skew correction over the first line prediction is important in order to retrieve the main handwritten line. We denoted this process as *Rotation*, which is performed by applying the Hough line and Affine transform. We have represented the overall rotation process in Fig. 3.

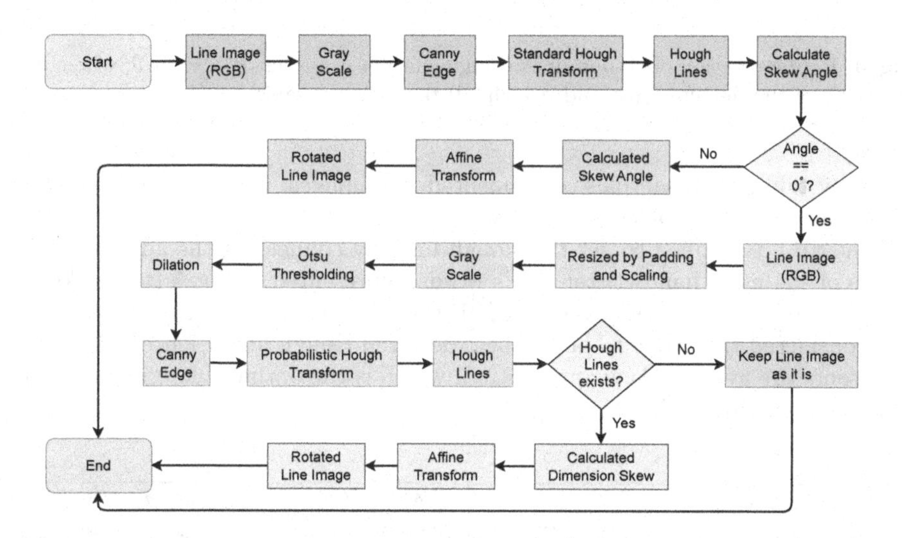

Fig. 3. Flowchart of skew estimation and correction over the first predicted lines.

4.3.1 Skew Estimation: We categorized handwritten lines' skew into two types: Positive and Negative (shown in Fig. 4). The skew angle estimation is performed in two phases:

1. Line Skew (LSkew) Estimation: where we applied the Standard Hough Transform (SHT).
2. Dimension Skew (DSkew) Estimation: where we applied the Probabilistic Hough Transform (PHT).

LSkew: In the Bangla writings, each word consists of letters and the letters are often connected by a horizontal line called 'mātrā'. By connecting those horizontal lines above the words using SHT, we construct straight lines, which we denote as Hough lines. Using those Hough lines, we estimate the skew angle of the main handwritten line. In terms of the representation of LSkew (Fig. 4), if the detected Hough lines have positive skew, the estimated skew angle will be negative; otherwise positive. We illustrate this LSkew estimation process in

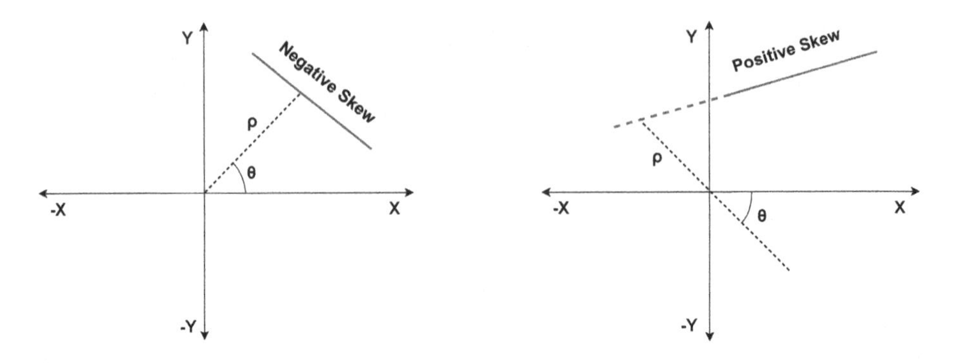

Fig. 4. Representation of Hough lines using equation ρ = x*cos θ + y*sin θ; where θ is the angle of the detected line and ρ is the distance from x-axis.

Fig. 5 by taking two samples of segmented line images, where one got positive skew, and the other got negative skew.

The SHT is applied to get the Hough lines by connecting the adjacent edge points of the main handwritten line's words, represented in Fig. 5 (top). Consequently, we calculated the average of all the detected Hough lines' parameters and considered this value to be the best detected Hough line. Figure 5 (bottom) represents the average skew angle (θ_{avg}), which is the optimal skew angle of our best detected Hough line.

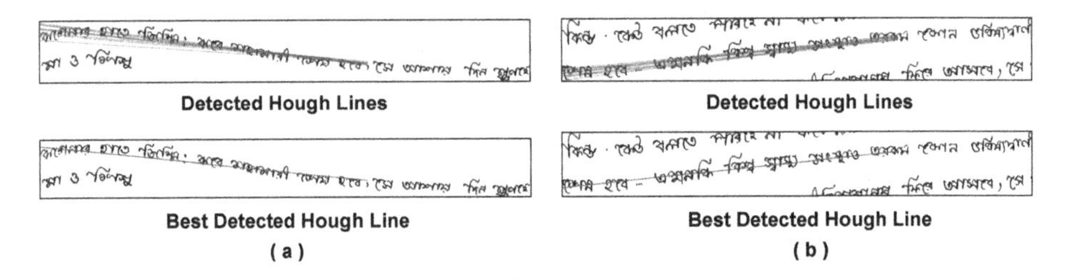

Fig. 5. Detected (top) and the Best Detected (bottom) Hough Lines, where the main handwritten line contains, a) Positive Skew, and b) Negative Skew.

DSkew: In some cases, SHT fails to detect the Hough lines, despite the main handwritten line on those segmented images being well skewed. We identified that the dimension of those failed images is too small compared to the standard dimension of the line images where SHT works. Moreover, in most cases, those line images contain only a few words, in such cases, not requiring any skew correction. Therefore, we opt for the DSkew process by applying PHT. We perform up-scaling on those failed images by preserving the aspect ratio before applying PHT (shown in Fig. 6).

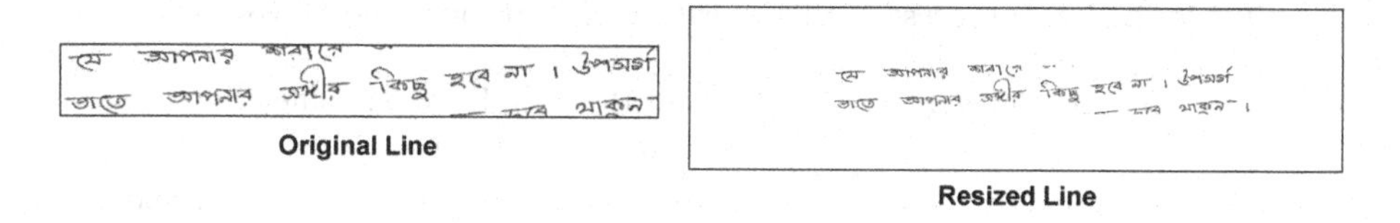

Fig. 6. Changing the dimension of nonstandard line image before applying PHT.

We apply some preprocessing steps such as image binarization and morphological operation with a 3×3 kernel to make the objects' lines and overall shape thicker and sharper. Finally, the canny edge detection method is applied before we can use the PHT. The output of preprocessing steps can be seen in Fig. 7.

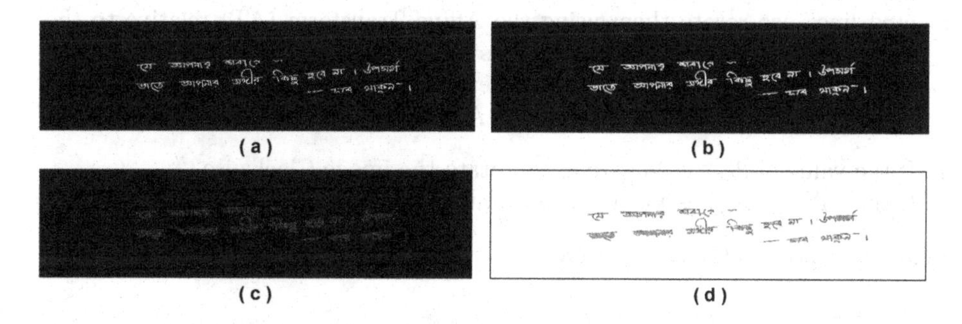

Fig. 7. Preprocessing and Hough line detection of sample resized line image represented in Fig. 6; where a) Binarization, b) Morphological Dilation, c) Canny edge detection, and d) Detected Hough lines using PHT.

The PHT not only joins the 'matra' of words but also connects any subsets of the points of each word edges individually if there is any potential Hough line. We named it dimension skew or DSkew, as each word component in the image takes part in the skew estimation process. Like SHT, we also get the typical Hough line parameters such as (x_1, y_1), (x_2, y_2), and (ρ, θ) in PHT. Hence, we applied the PHT in the edge detected image (of Fig. 7c) and got the Hough lines detected, shown in Fig. 7d. As the process detects multiple Hough lines for almost every word, therefore, each line has many θ, which we denote as *Degree*. To obtain the optimal skew angle of that image, we perform a voting process by dividing the xy space into six cases to determine where the maximum detected Hough lines had fallen. We then take an average of those lines' parameters to find the average of Degree ($Degree_{avg}$) and consider this as the skew angle of the detected Hough lines by PHT. The six cases of the voting process and their outcomes are given in Table 2:

Table 2. Voting process of DSkew with their categories and outcomes.

Based On	Voting Categories	Detected Hough Line Types	Final outcome as an average of degrees
Coordinates	x_1 equals x_2	Vertical	Return $Degree_{avg}$ as 90°
	y_1 equals y_2	Straight	Return $Degree_{avg}$ as 0°
Quadrants	$-45° \leq$ Degree $\leq 0°$	Positive Skew	Return $Degree_{avg}$
	$-90° \leq$ Degree $<-45°$	Negative Skew	Return $Degree_{avg}$
	$0° <$Degree $\leq 45°$	Negative Skew	Return $Degree_{avg}$
	$45° <$Degree $\leq 90°$	Positive Skew	Return $Degree_{avg}$

4.3.2 Skew Correction: In order to correct the estimated skew of our segmented lines, we rotate them using the Affine Transform (AT) relative to the center of the image. The rotation process for LSkew and DSkew is as follows:
LSkew: After estimating the optimal skew angle (θ_{avg}) using LSkew, we rotate the image with that skew angle through AT using the following two conditions:

1. If the value of θ_{avg} is Negative, we rotate the image Clockwise.
2. If the value of θ_{avg} is Positive, we rotate the image Anti-Clockwise.

Figure 8 illustrates the skew-correction for the segmented lines of Fig. 5.

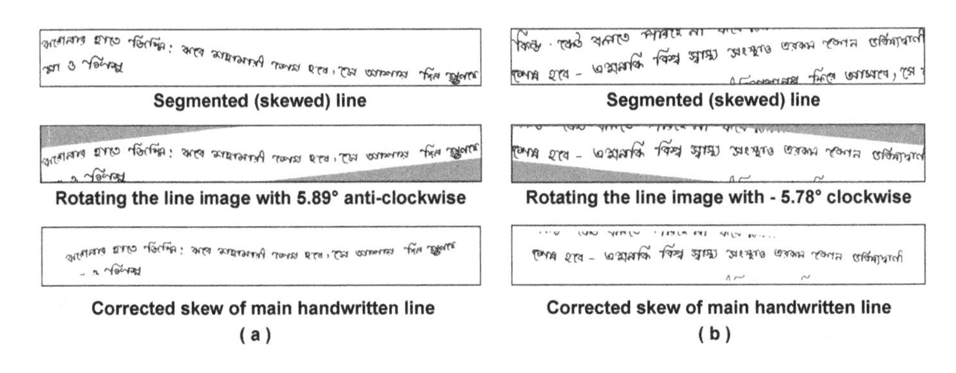

Fig. 8. Skew correction of segmented lines using AT where the original line was, a) Negative Skewed (rotated anti-clockwise); and b) Positive Skewed (rotated clockwise).

DSkew: The rotation for DSkew correction is similar to the rotation for LSkew correction, but the process of finding the optimal degree for rotation is different. Here, we calculate the optimal skew angle (θ_{avg}) based on the estimated $Degree_{avg}$ from DSkew. Then according to θ_{avg}, we rotate the image using AT by following the four conditions listed in Table 3:

Table 3. Conditions for skew correction for the process of DSkew.

No.	Conditions	Optimal Skew (θ_{avg})	Rotation
1	$-45° \leq Degree_{avg} \leq 0°$	$\theta_{avg} = Degree_{avg}$	Clockwise
2	$-90° \leq Degree_{avg} < -45°$	$\theta_{avg} = Degree_{avg} + 90°$	Anti-clockwise
3	$0° < Degree_{avg} \leq 45°$	$\theta_{avg} = Degree_{avg}$	Anti-clockwise
4	$45° < Degree_{avg} \leq 90°$	$\theta_{avg} = Degree_{avg} - 90°$	Clockwise

4.4 Final/Second Line Prediction and Segmentation

Final or second line prediction is applied on the skew-corrected lines to retrieve the main handwritten lines by eliminating the unwanted lines. Before that, we trim down each side of the DSkewed line image by a little portion to avoid unnecessary word prediction. Here, we consider a confidence threshold of 0.5. We also follow a selection process when we have multiple lines even after the second line prediction, as described below:

1. **The number of line predictions is one:** In this case, we segment the line with the given bounding box attributes, like in Fig. 9. If the width of the predicted line is less than 40% of the image width, we keep it as it is.

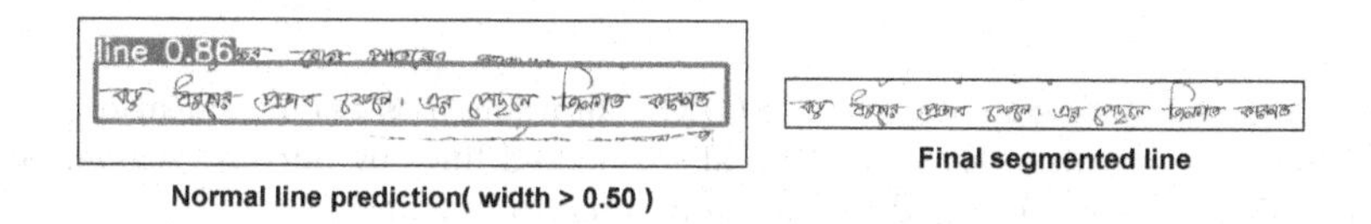

Fig. 9. Line image with single line prediction and segmentation.

2. **The number of line predictions is two:** In this case, normally, we segment the line prediction with maximum widths, like in Fig. 10. But, if both the predicted line's width is less than 50% of the image width, then we check their confidence and segment the line with maximum confidence. Otherwise, we keep the image as it is.

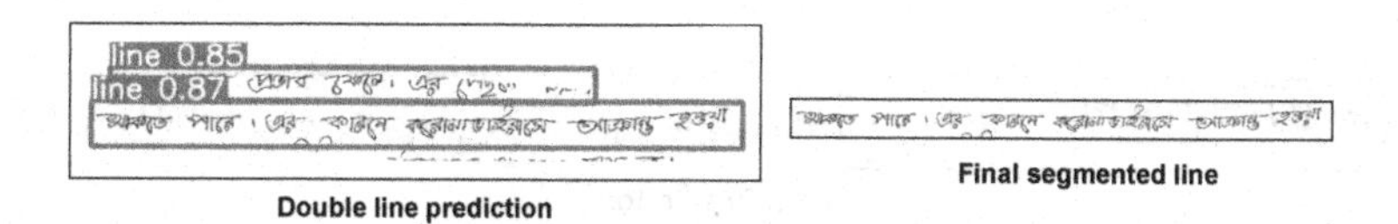

Fig. 10. Line image with two line prediction and segmentation (usual case).

3. **The number of line predictions is three:** In this case, we segment the line which stays in the middle like in Fig. 11.

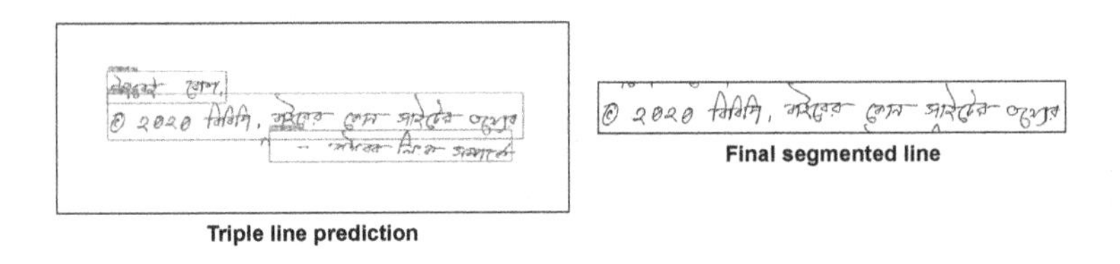

Fig. 11. Line image with two line prediction and segmentation.

4. **The number of line predictions is more than three:** Unseen cases where we select and segment the line having maximum width.

As the segmented lines have passed through the pre-processing, rotation, and final line segmentation process, we now have our final lines segmented from the handwritten document images. Note that, we also keep track of the predicted *line numbers* within the document for future recognition purposes. Figure 12 illustrates the resultant final line segmentation of the lines represented in Fig. 5.

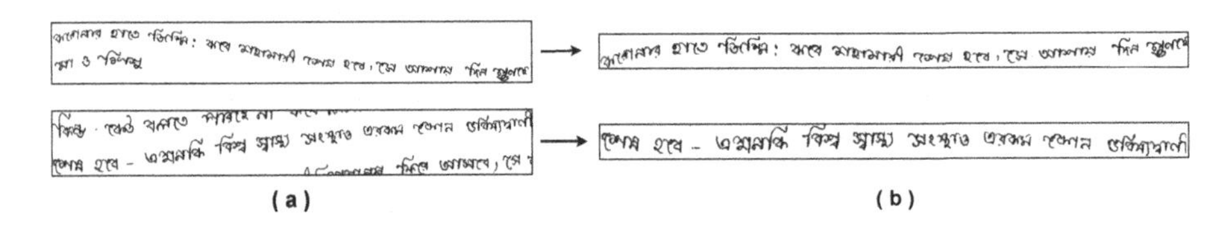

Fig. 12. a) Initial line segmentation by YOLO containing mostly curvilinear or skewed handwritten lines with noises, and b) Final segmented lines by our line segmentation approach, which are straight and without any unnecessary lines.

4.5 Word Prediction and Segmentation

We perform word prediction on the Final segmented lines by directly employing our custom YOLO word model, where we set the confidence threshold to be 0.4. We also sort the predictions based on the horizontal axis of the lines in order to get the position of a particular word in that line for future recognition purposes. Figure 13 illustrates word prediction and segmentation from the running example.

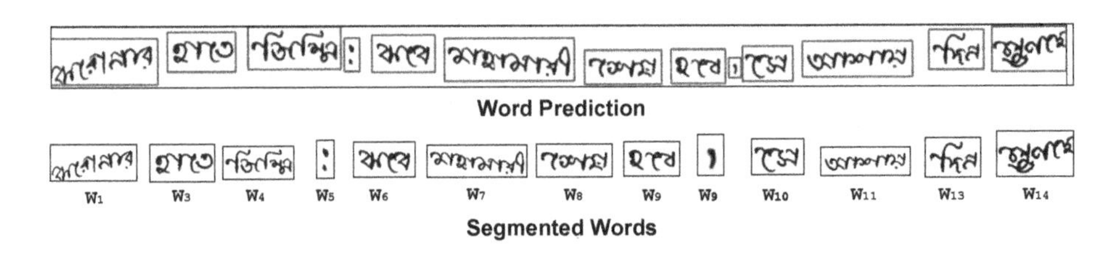

Fig. 13. Word prediction and segmentation on skew corrected final segmented lines; where W_i is the ith word within the line.

5 Experimental Results

In this section, we evaluate the efficiency of our line and word segmentation approach on the `BN-HTRd` dataset. We will also compare our results with an unsupervised line segmentation approach of `BN-HTR_LS` system [16].

5.1 Evaluation Matrices

Two bounding boxes (lines) are considered a one-to-one match if the total matching pixels exceed or equal the evaluator's approved threshold (T_a). Let N be the number of ground-truth elements, M be the count of detected components, and $o2o$ be the number of one-to-one matches between N and M; the Detection Rate (DR) and Recognition Accuracy (RA) are equivalent of Recall and Precision. Combining these, we can get the final performance metric FM (similar to F-score) using the equation below:

$$DR = \frac{o2o}{N}, \quad RA = \frac{o2o}{M}, \quad FM = \frac{2DR * RA}{DR + RA} \tag{1}$$

5.2 Line Segmentation

For the evaluation of our `BN-DRISHTI` line segmentation approach, we first did the **Quantitative analysis** on the test set of 75 handwritten document images from the BN-HTRd dataset containing 1397 (N) manually annotated ground truth lines. Our segmentation approach's final line prediction was 1396 (M). Among those, the number of $o2o$ matches was 1314. However, by using only YOLO trained model, we got 1433 (M) which implies that YOLO predicted 37 more redundant lines as compared to our approach, making our approach much superior. These results are listed in rows 2–3 of Table 4.

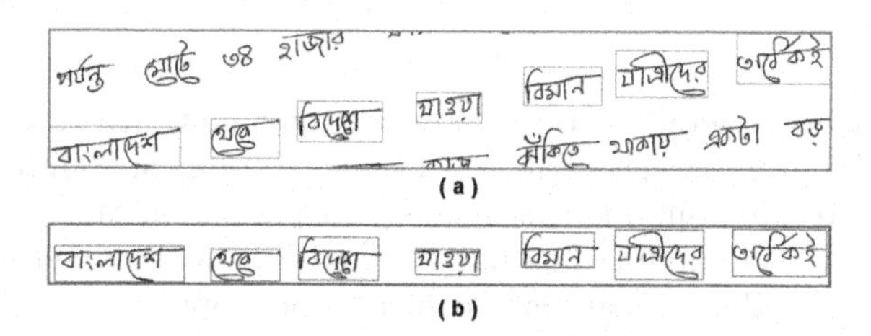

Fig. 14. a) Ground truth annotation on skewed line; Vs. b) Prediction on straight line.

After analyzing the line's ground truth and prediction bounding boxes visually (see Fig. 14), we came to the conclusion that the overlap between them for each line is not quite accurate since we performed skew correction before segmenting the line images. Thus, in automatic or quantitative evaluation the results we are getting are not as significant as we were expecting, since almost every line of the document images was segmented perfectly. Hence, we decided to do a **Qualitative evaluation** by going through all the ground truth and predictions manually to find the *o2o* for each handwritten document. And the overall *o2o* match was 1396, which is equal to the final line predictions we were getting. In Table 4 we put together the relative performance of our line segmentation approach's (`BN-DRISHTI`) quantitative and qualitative analysis as compared to the unsupervised approach of `BN-HTR_LS` system[3] [16] where they only performed line segmentation on the same dataset.

Table 4. Comparison of line segmentation results on BN-HTRd test sets.

Approaches	N	M	o2o	DR(%)	RA(%)	FM(%)
BN-HTR_LS [16]	2915	3437	2591	88.88	75.38	81.57
YOLO line model	1397	1433	1314	94.06	91.7	92.86
BN-DRISHTI (Quantitative)	1397	1396	1314	94.06	94.13	94.09
BN-DRISHTI (Qualitative)	1397	1396	1396	**99.93**	**1.00**	**99.97**

5.3 Word Segmentation

For this experiment, we used 10,414 manually annotated ground truth words within the line images of the test set's 75 handwritten documents. Our word model predicted 10,348 words. Table 5 shows the score of **Quantitative** analysis.

Table 5. Quantitative evaluation of our word segmentation on BN-HTRd test sets.

Ground Truths	Prediction	DR (%)	RA (%)	FM (%)
10,414	10,348	15.2	17.7	16.0

Again for the same aforementioned reason, the quantitative evaluation does not do justice to our approach's true word segmentation capabilities. Hence, we visually compared the ground truths against our predictions and found that the position of the words bounding box has changed drastically due to the changes in image dimension during our line segmentation approach, as illustrated in Fig. 14. This occurred because the original ground truth annotation was on the skewed lines, and our word prediction was done on the skew-corrected straight lines. Thus, after analyzing the ground truth and prediction bounding boxes, we came to the conclusion that the evaluation will not be fair if done automatically. Therefore, we again opt for a manual **Qualitative** analysis. We show both the quantitative and qualitative results in Table 6.

[3] **BN-HTR_LS Codebase:** https://github.com/shaoncsecu/BN-HTR_LS.

Table 6. Results of our word segmentation approach on the original ground truth (skewed) vs. skew-corrected (straight) lines from the BN-HTRd test sets.

Analysis	Word prediction on	N	M	DR	RA	FM
Quantitative	First segmented (Skewed) line	10,414	10,383	0.39	0.45	0.42
Quantitative	Final segmented (Straight) line	10,414	10,348	0.15	0.17	0.16
Qualitative	Final segmented (Straight) line	10,414	10,348	0.98	0.98	**0.98**

In Table 6, the qualitative analysis results perfectly justify our systems word segmentation capabilities. We also emphasize that word segmentation is far more precise when combined with our skew correction strategy.

5.4 Comparative Analysis

ICDAR 2013 Dataset[22]**:** This handwriting segmentation contests dataset contains 50 images for Bangla. As ground truth (N), we got 879 lines and 6,711 words; against which our system segmented 874 lines and 6,667 words (M). We choose team Golestan-a, Golestan-b, and INMC for performance comparison, as the Golestan method outperforms all other contestants with an overall score (SM) of 94.17%. And for Line segmentation, the INMC method was on the top with a 98.66% FM score. The comparison in Table 7 indicates that our system outperforms Golestan and INMC team's SM scores by a good margin. While our word segmentation results absolutely smashed the competitors, the line segmentation score was only second to INMC by a narrow margin.

Table 7. Comparison among top teams of ICDAR 2013 and our BN-DRISHTI system.

Systems	Class	N	M	o2o	DR (%)	RA (%)	FM (%)	SM (%)
Golestan-a	Lines	2649	2646	2602	98.23	98.34	98.28	94.17
	Words	23525	23322	21093	89.66	90.44	90.05	
Golestan-b	Lines	2649	2646	2602	98.23	98.34	98.23	90.06
	Words	23525	23400	21077	89.59	90.07	89.83	
INMC	Lines	2649	2650	2614	98.68	98.64	**98.66**	93.96
	Words	23525	22957	20745	88.18	90.36	89.26	
BN-DRISHTI	Lines	879	874	863	98.18	98.74	98.46	**96.65**
	Words	6711	6677	6348	98.74	95.07	**94.83**	

BanglaWriting Dataset [12]**:** It comprises 260 full-page Bangla handwritten documents and only the words ground truth. We manually evaluated the word segmentation results using randomly selected 50 document images from this dataset, as the word annotation was done directly over the document without any intermediate line annotation. Those selected 50 images contain 4409 words, and our system correctly segmented 4186 words against them. Table 8 indicates how our system performed on the BanglaWriting dataset.

Table 8. Word segmentation results on fifty images of BanglaWriting dataset.

Task	N	M	o2o	DR (%)	RA (%)	FM (%)
Word Segmentation	4409	4219	4186	94.9	99.2	97.0

WBSUBNdb_text Dataset [10]**:** This publicly available dataset has been used by two of the most prominent line [18] and word [1] segmentation methods for evaluation. As it contains 1352 Bangla handwriting without any ground truth, we only performed a qualitative analysis similar to the settings mentioned in those papers. We positioned our approach against these systems in Table 9.

Table 9. Comparison of segmentation results based on WBSUBNdb_text dataset.

Systems	Class	DR (%)	RA (%)	FM (%)
WBSUBNdb	Lines [18]	96.99	97.07	97.02
	Words [1]	86.96	93.25	90.0
BN-DRISHTI	Lines	99.27	99.44	**99.35**
	Words	96.85	97.18	**97.01**

6 Conclusions

The main contribution of this research is the significant improvement in line and word segmentation for Bangla handwritten scripts, which lays the foundation of our envisioned Bangla Handwritten Text Recognition (HTR). To alleviate the shortage of Bangle document-level handwritten datasets for future researchers, we have extended our BN-HTRd dataset. Currently, it is the largest dataset of its type with line and word-level annotation. Moreover, keeping the recognition task in mind, we have stored the words' Unicode representation against their position in the ground truth text. The main recipe behind our approach's overwhelming success is a two-layer line segmentation technique combined with an intricate skew correction in the middle. Our proposed line segmentation approach has achieved a near-perfect benchmark evaluation score in terms of F measure (99.97%) compared to the unsupervised approach (81.57%) of BN-HTR_LS [16]. The word segmentation technique also achieved an impressive score (98%) on the skew-corrected lines by our system compared to the skewed lines. Furthermore, we have compared our method against the previous SOTA systems on three of the most prominent Bangla handwriting datasets. Our approach outperformed all those methods by a significant margin, making our `BN-DRISHTI` system a new state-of-the-art for Bangla handwritten segmentation task. We aim to expand our work by integrating supervised word recognition to build an "End-To-End Bangla Handwritten Image Recognition system".

References

1. Agarwal, K., Mantry, A., Halder, C.: Word segmentation of offline handwritten Bangla text lines. In: Mandal, J.K., Buyya, R., De, D. (eds.) Proceedings of International Conference on Advanced Computing Applications. AISC, vol. 1406, pp. 551–560. Springer, Singapore (2022). https://doi.org/10.1007/978-981-16-5207-3_46
2. Alaei, A., Pal, U., Nagabhushan, P.: A new scheme for unconstrained handwritten text-line segmentation. Pattern Recogn. **44**(4), 917–928 (2011)
3. Bal, A., Saha, R.: An improved method for text segmentation and skew normalization of handwriting image. In: Sa, P.K., Sahoo, M.N., Murugappan, M., Wu, Y., Majhi, B. (eds.) Progress in Intelligent Computing Techniques: Theory, Practice, and Applications. AISC, vol. 518, pp. 181–196. Springer, Singapore (2018). https://doi.org/10.1007/978-981-10-3373-5_18
4. Barakat, B., Droby, A., Kassis, M., El-Sana, J.: Text line segmentation for challenging handwritten document images using fully convolutional network. In: 2018 16th International Conference on Frontiers in Handwriting Recognition (ICFHR), pp. 374–379. IEEE (2018)
5. Biswas, M., et al.: Banglalekha-isolated: a multi-purpose comprehensive dataset of handwritten Bangla isolated characters. Data Brief **12**, 103–107 (2017)
6. Bluche, T.: Joint line segmentation and transcription for end-to-end handwritten paragraph recognition. In: Advances in Neural Information Processing Systems, vol. 29 (2016)
7. Boudraa, O., Hidouci, W.K., Michelucci, D.: An improved skew angle detection and correction technique for historical scanned documents using morphological skeleton and progressive probabilistic hough transform. In: 2017 5th International Conference on Electrical Engineering-Boumerdes (ICEE-B), pp. 1–6. IEEE (2017)
8. Boukharouba, A.: A new algorithm for skew correction and baseline detection based on the randomized Hough transform. J. King Saud Univ. Comput. Inf. Sci. **29**(1), 29–38 (2017)
9. Fernández-Mota, D., Lladós, J., Fornés, A.: A graph-based approach for segmenting touching lines in historical handwritten documents. Int. J. Doc. Anal. Recog. (IJDAR) **17**(3), 293–312 (2014). https://doi.org/10.1007/s10032-014-0220-0
10. Halder, C., Obaidullah, S.M., Santosh, K., Roy, K.: Content independent writer identification on Bangla script: a document level approach. Int. J. Pattern Recogn. Artif. Intell. **32**(09), 1856011 (2018)
11. Kumar, J., Kang, L., Doermann, D., Abd-Almageed, W.: Segmentation of handwritten textlines in presence of touching components. In: 2011 International Conference on Document Analysis and Recognition, pp. 109–113. IEEE (2011)
12. Mridha, M.F., Ohi, A.Q., Ali, M.A., Emon, M.I., Kabir, M.M.: Banglawriting: a multi-purpose offline Bangla handwriting dataset. Data Brief **34**, 106633 (2021)
13. Mullick, K., Banerjee, S., Bhattacharya, U.: An efficient line segmentation approach for handwritten Bangla document image. In: 2015 Eighth International Conference on Advances in Pattern Recognition (ICAPR), pp. 1–6. IEEE (2015)

14. Nicolaou, A., Gatos, B.: Handwritten text line segmentation by shredding text into its lines. In: 2009 10th International Conference on Document Analysis and Recognition, pp. 626–630. IEEE (2009)
15. Rabby, A.K.M.S.A., Haque, S., Islam, M.S., Abujar, S., Hossain, S.A.: Ekush: a multipurpose and multitype comprehensive database for online off-line Bangla handwritten characters. In: Santosh, K.C., Hegadi, R.S. (eds.) RTIP2R 2018. CCIS, vol. 1037, pp. 149–158. Springer, Singapore (2019). https://doi.org/10.1007/978-981-13-9187-3_14
16. Rahman, M.A., Tabassum, N., Paul, M., Pal, R., Islam, M.K.: BN-HTRd: A Benchmark Dataset for Document Level Offline Bangla Handwritten Text Recognition (HTR) and Line Segmentation. In: Computer Vision and Image Analysis for Industry 4.0, pp. 1–16. CRC Press 6000 Broken Sound Parkway NW, Suite 300, Boca Raton, FL 33487–2742 (2023)
17. Rakshit, P., Halder, C., Ghosh, S., Roy, K.: Line, word, and character segmentation from Bangla handwritten text-a precursor toward Bangla HOCR. Adv. Comput. Syst. Secur. **5**, 109–120 (2018)
18. Rakshit, P., Halder, C., Sk, M.O., Roy, K.: A generalized line segmentation method for multi-script handwritten text documents. Expert Syst. Appl. **212**, 118498 (2023)
19. Renton, G., Chatelain, C., Adam, S., Kermorvant, C., Paquet, T.: Handwritten text line segmentation using fully convolutional network. In: 2017 14th IAPR International Conference on Document Analysis and Recognition (ICDAR), vol. 5, pp. 5–9. IEEE (2017)
20. Renton, G., Soullard, Y., Chatelain, C., Adam, S., Kermorvant, C., Paquet, T.: Fully convolutional network with dilated convolutions for handwritten text line segmentation. Int. J. Docu. Anal. Recogn. (IJDAR) **21**(3), 177–186 (2018). https://doi.org/10.1007/s10032-018-0304-3
21. Sarkar, R., Das, N., Basu, S., Kundu, M., Nasipuri, M., Basu, D.K.: Cmaterdb1: a database of unconstrained handwritten Bangla and Bangla-English mixed script document image. Int. J. Docu. Anal. Recogn. (IJDAR) **15**, 71–83 (2012)
22. Stamatopoulos, N., Gatos, B., Louloudis, G., Pal, U., Alaei, A.: ICDAR 2013 handwriting segmentation contest. In: 2013 12th International Conference on Document Analysis and Recognition, pp. 1402–1406. IEEE (2013)
23. Surinta, O., Holtkamp, M., Karabaa, F., Van Oosten, J.P., Schomaker, L., Wiering, M.: A path planning for line segmentation of handwritten documents. In: 2014 14th International Conference on Frontiers in Handwriting Recognition, pp. 175–180. IEEE (2014)
24. Vo, Q.N., Lee, G.: Dense prediction for text line segmentation in handwritten document images. In: 2016 IEEE International Conference on Image Processing (ICIP), pp. 3264–3268. IEEE (2016)

Text Line Detection and Recognition of Greek Polytonic Documents

Panagiotis Kaddas[1,2](✉), Basilis Gatos[1], Konstantinos Palaiologos[1,3], Katerina Christopoulou[1,4], and Konstantinos Kritsis[5]

[1] Computational Intelligence Laboratory, Institute of Informatics and Telecommunications, National Center for Scientific Research Demokritos, GR-153 10, Agia Paraskevi, Athens, Greece
{pkaddas,bgat,k.palaiologos,achristopoulou}@iit.demokritos.gr

[2] Department of Informatics and Telecommunications, University of Athens, 157 84 Athens, Greece

[3] Hellenic Institute, Royal Holloway, University of London, Egham Hill, Egham, Surrey TW20 0EX, UK

[4] Harokopio University, School of Environment, Geography and Applied Economics, Department of Economics and Sustainable Development, 17676 Athens, Greece

[5] Department of English Studies, University of Cyprus, 1678 Nicosia, Cyprus
kritsis.konstantinos@ucy.ac.cy

Abstract. In this work we highlight the significance of Text Line Detection in documents. By utilizing a well-known Deep Neural Network and by proposing some simple but efficient modifications applied during training of such models, we can achieve very accurate results in different datasets of high diversity. Moreover, such models can be robust even when trained with few data. Our focus is on Greek polytonic documents (typewritten and handwritten) and we provide a new dataset to the public (GTLD-small) for text line detection. We evaluate our method through scenarios applied to the detection and recognition tasks, while demonstrating promising results when compared to popular commercial or open-source systems.

Keywords: Deep Neural Networks · Object Detection · Text Line Detection · Text Line Recognition · YOLOv5 · Circular Smooth Label

1 Introduction

Over the last few decades, there has been a growing interest in studying and examining images of historical and modern documents, largely driven and motivated by the need to investigate material stored in libraries and archives [4]. There are millions of pages that have been digitized and are accessible as images, but in order to facilitate efficient work with such documents by humanists, historians and the general public, there is ongoing research and academic discourse aimed at understanding and converting the content of these documents [17] in a digital form. This is equal to recognize and comprehend the textual and non-textual content through an automated procedure.

M. Coustaty and A. Fornés (Eds.): ICDAR 2023 Workshops, LNCS 14194, pp. 213–225, 2023.
https://doi.org/10.1007/978-3-031-41501-2_15

Recently, the research of document image analysis has investigated utilizing deep learning for the segmentation and recognition of typewritten and handwritten text. At this point, it is worth noting that images of historical documents are distinct from images of natural scenes, where deep learning techniques are already applied widely [8]. As deep learning techniques advance, the automated processing and understanding of documents has become increasingly efficient, concerning text recognition, while highlighting two key steps as crucial towards success [4]: Text line detection and text line recognition. Text line detection is a technique used in document processing and computer vision to detect and extract text lines from a scanned document [21]. It is often being applied as a pre-processing step for Optical Character Recognition (OCR) and Layout Analysis, while having great impact on the accuracy of each task [6].

In this work, by utilizing a variation of the well-known YOLOv5 ([13]) Deep Neural Network model (YOLOv5-OBB[1]) and by proposing some simple but efficient modifications applied during training, we can achieve very accurate text line detection results on different datasets of high diversity. Moreover, we demonstrate the efficiency of such workflow on the text recognition task.

The contributions of this paper are as follows: a) We introduce a new dataset, named GTLD dataset, with annotated text line quadrilateral polygons and a smaller subset of 1.642 documents (including annotations on the Tobacco-3482 dataset) is publicly available[2]. b) We show in the experiments that the introduced oriented quadrilateral polygons can improve accuracy on the text line detection task, especially when text line boxes overlap. c) We provide promising results on text line recognition using an end-to-end workflow.

The rest of the paper is organized as follows. Section 2 presents related works, Sect. 3 introduces the proposed method, Sect. 4 demonstrates experimental scenarios and results and Sect. 5 presents the conclusion of this work.

2 Related Work

Text Line Detection. Text line detection methods have been proposed for many years [3,16,18,22], where most techniques focus on historical documents because of their diverse nature. In addition, segmentation of unconstrained handwritten manuscripts is usually the most challenging task to comprehend [7]. In such cases, text line detection is sometimes equivalent to the detection of baselines.

In [10], a two-stage workflow is proposed to detect baselines, where the first stage is a pixel labeling (or goal-oriented binarization) deep hierarchical neural network (ARU-Net), which detects foreground baseline pixels and separators. Finally, second stage clusters extracted superpixels in order to build baselines. In [2], dhSegment is proposed, where a generic neural network (U-Net [20] based on ResNet50 [12]) enables mutli-task detection for page segmentation, layout analysis, line detection and ornament extraction. A similar technique has been

[1] https://github.com/hukaixuan19970627/yolov5_obb.
[2] https://doi.org/10.5281/zenodo.8020403.

proposed in [5], where Doc-UFCN utilizes a light U-shaped Fully Convolutional Network (FCN) without any dense layers for page segmentation and text line detection. In [1], text line detection is achieved when document image binarization is applied as a pre-processing step and connected component analysis enables the localization of text blocks and lines. Finally, in [8] the authors propose a Mask-RCNN [11] based architecture for detecting text lines in historical Arabic handwritten documents, where they group labeled results after detecting text lines on image patches.

Text Line Recognition. Recognizing the text in documents has been frequently considered as a subsequent step to the text line detection task. This paper focuses on recognition of Greek polytonic documents, where variations in writing styles and accents is a very challenging task [9]. In [23], the authors propose a workflow for the recognition of Greek printed polytonic scripts, using a character segmentation step combined with intensity-based feature extractor and $k-$nearest neighbors (KNN) classification. Following the neural network logic and by using only Long Short-Term Memory (LSTM) Networks, in [24], the authors demonstrate some interesting results for Greek polytonic scripts.

Over the years, many methods have been proposed for the Handwriting Text Recognition (HTR) problem, where most approaches focus on Deep Learning techniques [25]. In [17], an attention-based sequence-to-sequence (Seq2Seq) model was proposed, where a Convolutional Neural Network (CNN) with Bidirectional LSTM (BLSTM) networks encode the image into a feature representation. Then, a decoder that utilizes a recurrent layer interprets this information and an attention mechanism enables focusing on the most relevant encoded features.

3 Proposed Method

3.1 Object Detection Model

The overall proposed architecture is provided in Fig. 1. Inspired by the success of the well-known YOLO [19] object detectors, we use a variation of the YOLOv5 model [13], utilized for the detection of oriented quadrilateral polygons [26], (named YOLOv5-OBB for convenience).

As shown in Fig. 1, the CSP-Darknet53 is the Darknet53 backbone architecture, combined the Cross Stage Partial (CSP) network strategy. CSP is applied in order to overcome the problem of redundant gradients by truncating the gradient flow, while preserving the advantage of residual and dense blocks used in the backbone in order to flow information to the deepest layers and to overcome the vanishing gradient problem. Using CSP, inference speed is increased (less FLOPS) while the number of parameters is being reduced. YOLOv5 has a major change in the Neck of the architecture, when compared to it's predecessors. A variant of Spatial Pyramid Pooling layer feeds a Path Aggregation Network (PANet) in which the CSP strategy has been incorporated via the

BottleNeckCSP layer. Finally, the Head of the network is composed from 3 convolution layers for predicting the location of bounding boxes, as long as the object classes with their scores and the angle estimation.

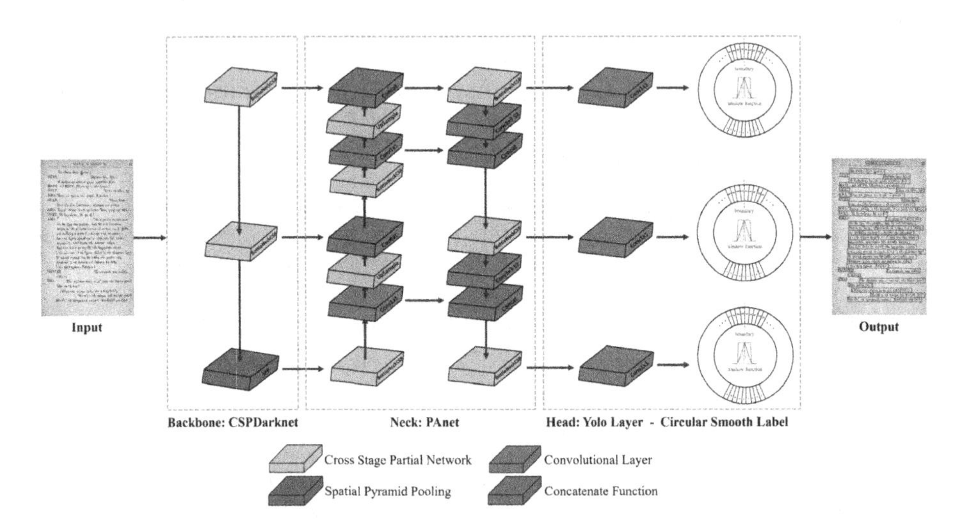

Fig. 1. Architecture of the proposed YOLOv5-OBB model: The **CSP-Darknet53** is the backbone feature extractor and a variant of **Spatial Pyramid Pooling (SPP)** layer feeds a **Path Aggregation Network (PANet)**. The Head of the network is composed from 3 convolution layers combined with the **Circular Smooth Label** technique in order to classify the quadrilateral's angle and to accurately detect the text lines in the document. In **BottleNeckCSP** convolutional layers, the **Cross Stage Partial (CSP)** technique has been incorporated in order to overcome the problem of redundant gradients.

3.2 Circular Smooth Label for Angle Estimation

In addition to the YOLOv5 model in order to localize text lines with orientation, this work follows the Circular Smooth Label (CSL) technique proposed in [26] in order to classify the rotation angle of the object's bounding polygon. The use of regression methods can cause issues as ideal predictions fall outside the defined range. The result of a prediction can be controlled more effectively if the object angle is approached as a classification problem, where the angle is treated as a category label, and the number of categories is determined by the angle range. However, changing the method from regression to one-hot encoded classification can result in a slight decrease of accuracy.

With the CSL method, the angles are encoded in a circular/repeating pattern (Fig. 1), while windowing functions (pulse, rectangular, triangle and Gaussian) can be used to exterminate boundary problems in order to obtain better prediction of the angle's class.

3.3 Extended Polygonal Labeling

In many cases, ground-truth text lines are strictly annotated and the polygonal edges almost overlap with the characters. Even when this is not the case, we observe that LSTM layers in a text line recognition system handle edge characters better, as demonstrated in the experiments Section. So, by loosening the enclosing polygon, both line detection and recognition systems improve. For this reason and during training of the YOLOv5-OBB model, we shift every point $p_i = (x_i, y_i), i \in 1, 2, 3, 4$ of the $n-th$ text line quadrilateral of width w_n over the horizontal axis, resulting to the new point $p_i^{'} = (x_i^{'}, y_i^{'})$, following Eqs. 1, 2:

$$x_i^{'} = x_i + a \times k \times w_n \tag{1}$$

$$y_i^{'} = \frac{(y_j - y_i)}{(x_j - x_i)} \times (x_i^{'} - x_i) + y_i \tag{2}$$

where $j = (i+1) \bmod 4$, $a = -1$ for the two leftmost points and $a = +1$ for the two rightmost points. We empirically use $k = 0.03$ as a shift factor relative to the width of the quadrilateral, since we observe that resulting polygons do not overlap, while starting and ending characters are recognized properly from the text line recognition system. A visualization is given in Fig. 2.

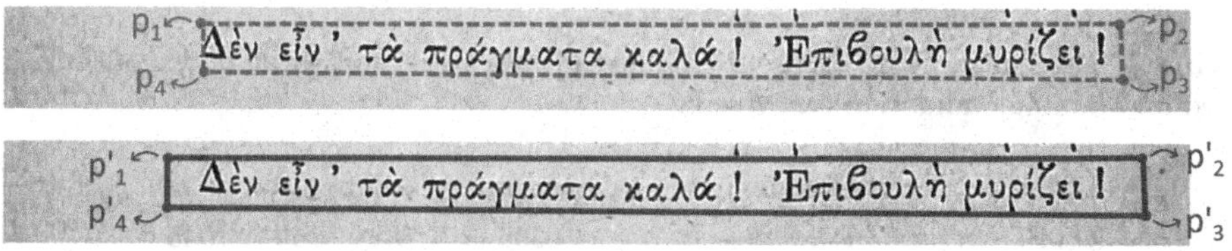

Fig. 2. Extended Polygonal Labeling: Ground-truth text lines are sometimes annotated strictly and the polygonal edges almost overlap with the characters (upper image). By loosening the enclosing polygon over the horizontal axis, both line detection and recognition systems improve (lower image).

4 Experiments

4.1 Datasets

Since the detection of text lines can be considered in general as a language-independent segmentation task, we combine two datasets with a totaling number of 17.133 images. We perform our experiments on the Tobacco-3482 dataset [14] along with a new Greek polytonic dataset named GTLD. The latter is a combination of two heterogeneous collections, with the first subset containing also handwritten documents. The overview of the datasets is given in Table 1. Smaller portions of each dataset (denoted with a *"small"* suffix in Table 1) with a total number of 1.642 images are available along with the corresponding annotations at a text line level (See footnote 2).

Tobacco-3482 Dataset. The Tobacco-3482 [14] dataset consists of document images belonging to 10 classes such as letter, form, email, resume, memo, etc. The dataset has 3482 images and we provide annotations at a text line level.

Greek Text Line Detection Dataset (GTLD). We introduce a new dataset, named GTLD (See footnote 2), which contains a total of 13.651 images acquired from two collections: The ShakeIT collection contains images of Greek polytonic documents that originate from almost 140 books of Greek translations of Shakespeare's plays published between 1842 and 1950. For this work, we use 28 books with a total of 3955 randomly selected pages. The PIOP collection consists of almost 100.000 digital documents from the Historical Archives of the Piraeus Bank Group Cultural Foundation (PIOP)[3]. This archive includes a significant number of collections starting from the early 20th century. In this paper, we use a total of 9696 randomly selected pages. Both collections contain mostly Greek polytonic characters, with rare cases of English/German. Exemplar images from both collections are shown in Fig. 3.

Table 1. Overview of the datasets included in this work and the number of images used for training, validation and testing.

Dataset	Collection	#Total	#train	#val	#test
GTLD	PIOP	9.696	6.782	969	1945
	ShakeIT	3.955	2.754	394	807
GTLD-small	PIOP-small	950	672	90	188
	ShakeIT-small	357	264	27	66
Tobacco-3482	Tobacco-3482	3.482	2.433	345	704
	Tobacco-3482-small	335	240	30	65

4.2 Experimental Results

Text Line Detection: For the text line detection task, we train all models shown in the following for 80–150 epochs (more epochs when using smaller subsets) using SDG algorithm with a starting learning rate of 0.001 and batch size equal to 4. We also apply common augmentation techniques during training (random affine transformations, color jitter, flipping, etc.) over the input images. For the Circular Smooth Label, we use Gaussian kernel with a radius of 4. As a

[3] https://www.piop.gr/en/istoriko-arxeio.aspx.

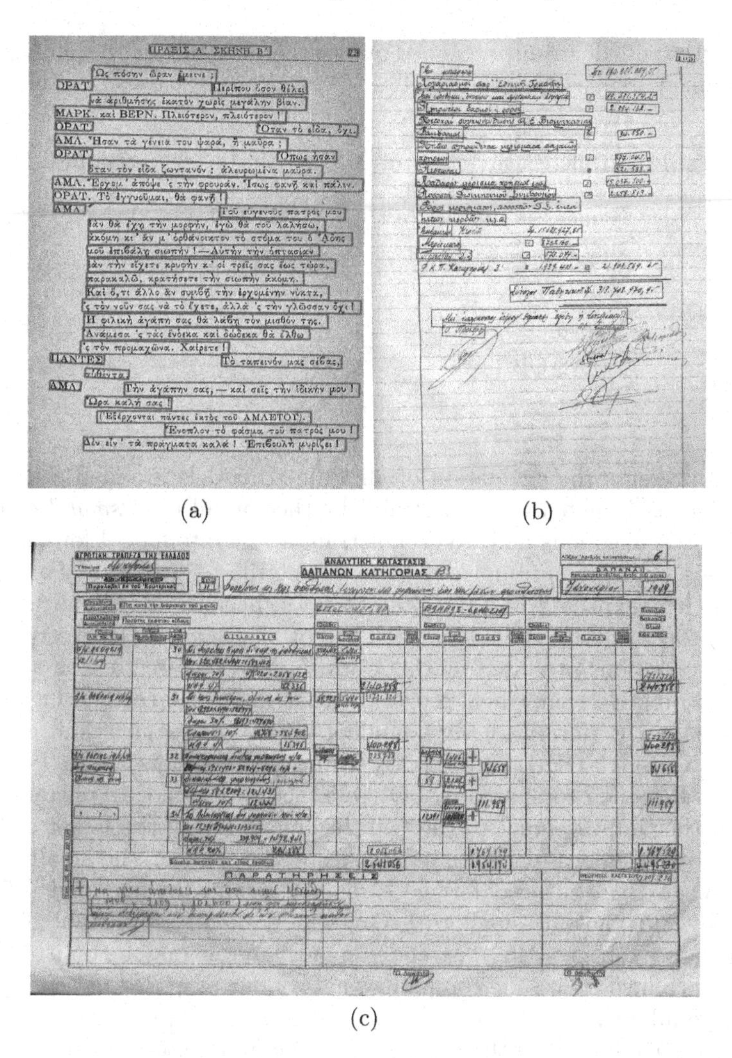

Fig. 3. Exemplar documents of the GTLD dataset: a): typewritten text (ShakeIT), b) handwritten text (PIOP), c) "Hybrid" text (PIOP).

YOLOv5 backbone, we use the YOLOv5-large[4] (YOLOv5l) model, pre-trained on MSCOCO dataset [15]. We keep the instance of the model that performed best on the validation set. As evaluation metrics, we use recall (R), precision (P), f1-score (F1) and mean Average Precision (mAP), evaluated at mAP@.5 and mAP@.5:.95, following the best evaluation practices in the domain of object detection. Table 2 shows the different architectures used in the following experiments, where "Rect" means rectangular polygons, "Quad" means quadrilateral

[4] https://github.com/ultralytics/yolov5.

polygons, "Extended" means extended quadrilateral or rectangular polygons and "CSL" means Circular Smooth Label

Table 2. Different architectures for the text line detection task.

Name	Rect	Quad	Extended	CSL
YOLOv5-rect	✓			
YOLOv5-rect-ext	✓		✓	
Yolov5-OBB		✓		✓
Yolov5-OBB-ext		✓	✓	✓

Table 3 presents the evaluation of all trained models for each collection. It is noted that each model has been trained either on all the *"small"* collections or the *"large"* collections and we do not train a separate model for each collection. At this point, it is important to highlight the following: a) For all small collections, it is expected to have lower metrics, because we use fewer training samples in contrast with the larger ones. b) For all setups, YOLOv5-rect is inferior to the variations proposed by this work. c) YOLOv5-rect-ext is superior to the YOLOv5-rect version, which proves the contribution of considering extended bounding polygons (for this case, bounding boxes). d) Application of the CSL technique (YOLOv5-OBB) significantly improves detection results (+7.5% **on mAP@.5:.95)** when compared to YOLOv5-rect. e) Extended Polygonal Labeling is a further improvement to the CSL technique for all the larger collections (+0.1%, +0.3%, +0.4% on mAP@.5:.95 respectively). For the smaller ones it is useful when there is need for more refined results (higher mAP@.5:.95 on PIOP-small and ShakeIT-small with Greek polytonic scripts), but if detail is not important (Tobacco-3482), extended quadrilaterals may be omitted.

In order to investigate the competitiveness of our proposed method, we evaluate two popular systems on our datasets. The first one, Google Cloud Vision API[5], is commercial, while the second one is the well-known open-source OCR engine Tesseract[6] (version 5.0.0). In both cases we evaluate the systems using each one's default setup. The evaluation results are given in Table 4. It is observed that the Tesseract engine fails to process all collections. One reason is that Tesseract provides only bounding rectangles instead of quadrilaterals and predictions suffer from the problems mentioned above. Google Cloud Vision API seems to perform better and gives more refined results on ShakeIT-small, were the layout of the documents is simpler. On the other hand, Our method is superior to all over the more diverse collections of PIOP-small and Tobacco-3482-small.

[5] https://cloud.google.com/vision.
[6] https://github.com/tesseract-ocr/tesseract.

Table 3. Evaluation results (%) for the GTLD (PIOP, ShakeIT) and Tobacco-3482 datasets

Collection	Model	P	R	mAP@.5	mAP@.5:.95
PIOP-small	YOLOv5-rect	**85.1**	59.4	68.9	34.5
	YOLOv5-rect-ext	82.7	58.0	69.8	37.2
	YOLOv5-OBB	84.7	**65.2**	**74.0**	39.9
	YOLOv5-OBB-ext	84.5	64.8	73.5	**41.4**
ShakeIT-small	YOLOv5-rect	95.5	95.4	95.5	53.8
	YOLOv5-rect-ext	95.7	95.3	95.6	52.5
	YOLOv5-OBB	**99.2**	**99.5**	**99.4**	65.4
	YOLOv5-OBB-ext	99.0	98.8	99.2	**68.1**
Tobacco-3482-small	YOLOv5-rect	87.6	79.7	83.9	42.9
	YOLOv5-rect-ext	85.8	81.0	84.5	42.3
	YOLOv5-OBB	88.5	**86.1**	**88.3**	**50.3**
	YOLOv5-OBB-ext	**89.3**	85.6	86.4	47.1
PIOP	YOLOv5-rect	87.0	74.3	81.1	47.6
	YOLOv5-rect-ext	**89.2**	78.3	85.7	50.4
	YOLOv5-OBB	88.4	79.6	86.5	54.4
	YOLOv5-OBB-ext	88.7	**80.9**	**87.1**	**54.5**
ShakeIT	YOLOv5-rect	95.2	94.9	96.6	65.8
	YOLOv5-rect-ext	99.3	99.2	99.4	68.0
	YOLOv5-OBB	99.3	**99.2**	**99.4**	71.9
	YOLOv5-OBB-ext	**99.4**	99.1	**99.4**	**72.2**
Tobacco-3482	YOLOv5-rect	85.5	82.7	85.4	50.9
	YOLOv5-rect-ext	90.5	86.0	89.5	57.6
	YOLOv5-OBB	**91.1**	87.3	90.6	58.8
	YOLOv5-OBB-ext	91.0	**88.8**	**91.3**	**59.2**

Text Line Recognition: Finally, order to signify the the robustness of the proposed method, we evaluate our systems indirectly over the subsequent task of text line recognition on Greek polytonic documents. For this reason, we perform the following steps:

Table 4. Comparison (mAP %) of the proposed method against other popular commercial/open-source systems.

Collection	Model	mAP@.5	mAP@.5:.95
PIOP-small	Google Cloud Vision API	52.0	31.9
	Tesseract 5.0.0	23.3	9.9
	YOLOv5-OBB-ext	**73.5**	**41.4**
ShakeIT-small	**Google Cloud Vision API**	95.2	**75.4**
	Tesseract 5.0.0	23.3	9.93
	YOLOv5-OBB-ext	**99.2**	68.1
Tobacco-3482-small	Google Cloud Vision API	67.6	31.3
	Tesseract 5.0.0	28.1	10.1
	YOLOv5-OBB-ext	**86.4**	**47.1**

- We train a text line recognition module, by utilizing the Calamari-OCR[7] engine. We use about 100.000 annotated text lines (OCR level) from the PIOP and ShakeIT collections. We choose the *"htr+"* architecture ([17]).
- We apply text line detection using four models (YOLOv5-rect, YOLOv5-rect-ext, YOLOv5-OBB and YOLOv5-OBB-ext) trained on the smaller collections.
- In order to construct the full page OCR from the predictions, we sort the detected text lines using Density-based spatial clustering (DBSCAN) and the coordinates of the extracted clusters.
- We consider the full page OCR of the 66 ShakeIT-small test images for evaluation. We choose this collection because it has simpler layout in order to minimize reading order errors from the previous step.
- As evaluation metrics, we consider the well-known Character Error Rate (CER) and Word Error Rate (WER).

From the results of Table 5 it is demonstrated again that the CSL method with extended quadrilaterals yields the best results and assists significantly to the text line recognition task. CER and WER results also validate the robustness of our text line recognition module. Another observation is that the text line recognition workflow that has been proposed above can be considered as an end-to-end workflow without any prior processing steps except for the line detection module combined with an grouping algorithm like DBSCAN for line ordering.

[7] https://github.com/Calamari-OCR/calamari.

Table 5. Evaluation of page OCR results using text line detection models on the ShakeIT-small test images.

Model	WER(%)	CER(%)
YOLOv5-rect	19.1	6.6
YOLOv5-rect-ext	17.2	5.3
YOLOv5-OBB	12.4	3.5
YOLOv5-OBB-ext	**9.8**	**3.2**

5 Conclusion

In this work, text line detection and recognition have been considered, with a focus on Greek polytonic documents. It was proposed that by utilizing YOLOv5 with some easy and simple techniques like the CSL method and extended polygonal labeling, there is a significant improvement on the text line detection task and accuracy results outperform popular commercial and open-source systems. Moreover, by applying this logic to the subsequent task of performing OCR even at a page level, similar improvements are noted and text line detection (with a line ordering technique) can be the only step before OCR in an end-to-end workflow. Finally, a new dataset was introduced (GTLD), with a smaller subset being publicly available for future works.

Acknowledgment. This research has been partially co-financed by the European Union and Greek national funds through the Operational Program Competitiveness, Entrepreneurship and Innovation, under the call RESEARCH-CREATE-INNOVATE, project Culdile (Cultural Dimensions of Deep Learning, project code: T1EDK-03785), the Operational Program Attica 2014-2020, under the call RESEARCH AND INNOVATION PARTNERSHIPS IN THE REGION OF ATTICA, project reBook (Digital platform for re-publishing Historical Greek Books, project code: ATTP4-0331172) and the project "Corpus-assisted drama translation research: Shakespeare In Translation - ShakeIT", under the call of internal research project funding of University of Cyprus (UCY https://www.ucy.ac.cy/directory/en/).

References

1. Ahn, B., Ryu, J., Koo, H.I., Cho, N.I.: Textline detection in degraded historical document images. EURASIP J. Image Video Process. **2017**(1), 82 (2017)
2. Ares Oliveira, S., Seguin, B., Kaplan, F.: dhSegment: a generic deep-learning approach for document segmentation. In: 16th International Conference on Frontiers in Handwriting Recognition (ICFHR), 2018, pp. 7–12. IEEE (2018)
3. Basu, S., Chaudhuri, C., Kundu, M., Nasipuri, M., Basu, D.: Text line extraction from multi-skewed handwritten documents. Pattern Recogn. **40**(6), 1825–1839 (2007)
4. Boillet, M., Kermorvant, C., Paquet, T.: Robust text line detection in historical documents: learning and evaluation methods. Int. J. Doc. Anal. Recog. (IJDAR) **25**(2), 95–114 (2022)

5. Boillet, M., Kermorvant, C., Paquet, T.: Multiple document datasets pre-training improves text line detection with deep neural networks. In: 2020 25th International Conference on Pattern Recognition (ICPR), pp. 2134–2141 (2021)
6. Diem, M., Kleber, F., Sablatnig, R.: Text line detection for heterogeneous documents. In: 2013 12th International Conference on Document Analysis and Recognition (ICDAR), pp. 743–747 (2013)
7. Diem, M., Kleber, F., Sablatnig, R., Gatos, B.: cBAD: ICDAR 2019 competition on baseline detection. In: 2019 International Conference on Document Analysis and Recognition (ICDAR), pp. 1494–1498 (2019)
8. Droby, A., Kurar Barakat, B., Alaasam, R., Madi, B., Rabaev, I., El-Sana, J.: Text line extraction in historical documents using mask r-CNN. Signals **3**(3), 535–549 (2022). https://doi.org/10.3390/signals3030032
9. Gatos, B., et al.: GRPOLY-DB: an old Greek polytonic document image database. In: 2015 13th International Conference on Document Analysis and Recognition (ICDAR), pp. 646–650 (2015)
10. Grüning, T., Leifert, G., Strauß, T., Michael, J., Labahn, R.: A two-stage method for text line detection in historical documents. Int. J. Docu. Anal. Recogn. (IJDAR) **22**(3), 285–302 (2019)
11. He, K., Gkioxari, G., Dollár, P., Girshick, R.: Mask R-CNN. In: 2017 IEEE International Conference on Computer Vision (ICCV), pp. 2980–2988 (2017)
12. He, K., Zhang, X., Ren, S., Sun, J.: Deep residual learning for image recognition. In: 2016 IEEE Conference on Computer Vision and Pattern Recognition (CVPR), pp. 770–778 (2016)
13. Jocher, G., et al.: ultralytics/yolov5: v7.0 - YOLOv5 SOTA Realtime Instance Segmentation (2022). https://doi.org/10.5281/zenodo.7347926
14. Kumar, J., Ye, P., Doermann, D.: Structural similarity for document image classification and retrieval. Pattern Recogn. Lett. **43**, 119–126 (2014). iCPR2012 Awarded Papers
15. LIn, T.-Y., et al.: Microsoft COCO: common objects in context. In: Fleet, D., Pajdla, T., Schiele, B., Tuytelaars, T. (eds.) ECCV 2014. LNCS, vol. 8693, pp. 740–755. Springer, Cham (2014). https://doi.org/10.1007/978-3-319-10602-1_48
16. Louloudis, G., Gatos, B., Pratikakis, I., Halatsis, C.: Text line detection in handwritten documents. Pattern Recogn. **41**(12), 3758–3772 (2008)
17. Michael, J., Labahn, R., Grüning, T., Zöllner, J.: Evaluating sequence-to-sequence models for handwritten text recognition. In: 2019 International Conference on Document Analysis and Recognition (ICDAR), pp. 1286–1293 (2019)
18. Nicolas, S., Paquet, T., Heutte, L.: Text line segmentation in handwritten document using a production system. In: 9th International Workshop on Frontiers in Handwriting Recognition, pp. 245–250 (2004)
19. Redmon, J., Divvala, S., Girshick, R., Farhadi, A.: You only look once: unified, real-time object detection. In: 2016 IEEE Conference on Computer Vision and Pattern Recognition (CVPR), pp. 779–788 (2016)
20. Ronneberger, O., Fischer, P., Brox, T.: U-Net: convolutional networks for biomedical image segmentation. In: Navab, N., Hornegger, J., Wells, W.M., Frangi, A.F. (eds.) MICCAI 2015. LNCS, vol. 9351, pp. 234–241. Springer, Cham (2015). https://doi.org/10.1007/978-3-319-24574-4_28
21. Sahare, P., Dhok, S.B.: Review of text extraction algorithms for scene-text and document images. IETE Tech. Rev. **34**(2), 144–164 (2017). https://doi.org/10.1080/02564602.2016.1160805

22. Shi, Z., Govindaraju, V.: Line separation for complex document images using fuzzy runlength. In: Proceedings of the First International Workshop on Document Image Analysis for Libraries (DIAL 2004), p. 306. DIAL 2004, IEEE Computer Society, USA (2004)
23. Sichani, A.M., Kaddas, P., Mikros, G.K., Gatos, B.: OCR for Greek polytonic (multi accent) historical printed documents: development, optimization and quality control. In: Proceedings of the 3rd International Conference on Digital Access to Textual Cultural Heritage, pp. 9–13. DATeCH2019, Association for Computing Machinery, New York, NY, USA (2019)
24. Simistira, F., Ul-Hassan, A., Papavassiliou, V., Gatos, B., Katsouros, V., Liwicki, M.: Recognition of historical Greek polytonic scripts using LSTM networks. In: 2015 13th International Conference on Document Analysis and Recognition (ICDAR), pp. 766–770 (2015)
25. Teslya, N., Mohammed, S.: Deep learning for handwriting text recognition: existing approaches and challenges. In: 2022 31st Conference of Open Innovations Association (FRUCT), pp. 339–346 (2022)
26. Yang, X., Yan, J.: Arbitrary-oriented object detection with circular smooth label. In: Vedaldi, A., Bischof, H., Brox, T., Frahm, J.-M. (eds.) ECCV 2020. LNCS, vol. 12353, pp. 677–694. Springer, Cham (2020). https://doi.org/10.1007/978-3-030-58598-3_40

A Comprehensive Handwritten Paragraph Text Recognition System: LexiconNet

Lalita Kumari[1(✉)], Sukhdeep Singh[2], Vaibhav Varish Singh Rathore[3], and Anuj Sharma[1]

[1] Department of Computer Science and Applications, Panjab University, Chandigarh, India
{lalita,anujs}@pu.ac.in
[2] D.M. College (Affiliated to Panjab University, Chandigarh), Moga, India
[3] Physical Research Laboratory, Ahmedabad, India
vaibhav@prl.res.in
https://anuj-sharma.in

Abstract. In this study, we have presented an efficient procedure using two state-of-the-art approaches from the literature of handwritten text recognition as Vertical Attention Network and Word Beam Search. The attention module is responsible for internal line segmentation that consequently processes a page in a line-by-line manner. At the decoding step, we have added a connectionist temporal classification-based word beam search decoder as a post-processing step. In this study, an end-to-end paragraph recognition system is presented with a lexicon decoder as a post-processing step. Our procedure reports state-of-the-art results on standard datasets. The reported character error rate is 3.24% on the IAM dataset with 27.19% improvement, 1.13% on RIMES with 40.83% improvement and 2.43% on the READ-16 dataset with 32.31% improvement from existing literature and the word error rate is 8.29% on IAM dataset with 43.02% improvement, 2.94% on RIMES dataset with 56.25% improvement and 7.35% on READ-2016 dataset with 47.27% improvement from the existing results. The character error rate and word error rate reported in this work surpass the results reported in the literature.

Keywords: Paragraph Handwritten Text Recognition · Neural Network · Connectionist Temporal Classification · Word Beam Search · Optical Character Recognition

1 Introduction

A Handwritten Text Recognition (HTR) system understands handwritten text written on digital surfaces, on paper or any other media. The HTR problem is complex enough to be widely studied by computer vision researchers across the community. In the present work, we proposed a tightly coupled HTR system constrained by lexicon and able to recognize the given paragraph-based text in an end-to-end manner without the need for any external segmentation step. The

M. Coustaty and A. Fornés (Eds.): ICDAR 2023 Workshops, LNCS 14194, pp. 226–241, 2023.
https://doi.org/10.1007/978-3-031-41501-2_16

Hidden Markov Model (HMM) is a widely applied technique in an evolutionary period of HTR [28]. The HMM based HTR models lack the use of contextual information because in large sequences of texts the HMM based system only focuses on the current time step as per Markovian assumption.

The Recurrent Neural Network (RNN) based approaches are able to overcome these short coming of HMM. In recent years, due to advancements in computational capacity, deep learning-based models, especially Convolutional Neural Network (CNN), RNN and a combination of both Convolutional Recurrent Neural Network (CRNN) produces state-of-the-art accuracies for HTR task [25].

Although the deep learning based HTR methods give promising results, they require identification of the Region Of Interest (ROI), preprocessing and segmentation as a pre-step to recognize the given document. Hence, these latent errors restrict the system's accuracy and real-world applicability of the systems. In the present work, we aim to propose an end-to-end HTR system that can be as close to real-world applicability as possible. Thus, we are looking forward to developing a segmentation-free HTR system that can be trainable in an end-to-end manner confined by lexicon knowledge. Hence, by developing the system in such a manner, we are free from the latent errors and extra computational costs involved in segmentation followed by recognition approaches. In this study, we focus on offline HTR. In this paper, we presented an efficient and robust end-to-end paragraph based HTR system that takes a paragraph image as an input and generates its features in a line-by-line manner using a series of convolutional and depth-wise separable convolutional operations. These features are further sent to the decoder. The Word Beam Search (WBS) [22] is used as decoder. We trained our NN model on line level first, then transferred these trained weights to the page level model for training the whole page in an end-to-end manner. The following are the major contributions made by this study:

- We proposed an efficient procedure that unified end-to-end trainable HTR system blended with a lexicon decoder.
- The proposed procedure adopts two state-of-the-art algorithms as Vertical Attention Network [7] for training and the WBS [22] in prediction stage.
- We have achieved state-of-the-art results by improving Character Error Rate (CER) and Word Error Rate (WER) as 27.19% and 43.02% respectively on the IAM dataset, 40.83% and 56.25% on the RIMES dataset and 32.31% and 47.27% on READ-2016 dataset from existing results.
- Detailed discussion of each step involved in a generic HTR process. It will be helpful for future readers to understand the whole process in a systematic manner.

The rest of the paper is organized as follows, the key contributions in the literature related to HTR systems are discussed in Sect. 2. Section 3 explains the detailed system design of the HTR system. Section 4 includes the experimental setup and results. In Sect. 5, a detailed discussion is presented on the lexicon decoder including best and worst case analysis. Section 6 draws the conclusion of the present study.

2 Related Works

In this section, the state-of-the-art techniques of the HTR systems have been discussed in an evolutionary manner. We start the section by discussing character-based recognition, followed by word and line level recognition and conclude with a discussion of promising techniques of paragraph text recognition.

2.1 Character/Word Level Recognition

In the early days, methods based on isolated characters were widely used for HTR tasks. In a segmentation-recognition approach, words are segmented into groups of characters using dynamic programming techniques and recognised further using probabilistic models [5]. The isolated characters are obtained using various image processing techniques [23]. Further, the adaptive word recognition method is used with hand-crafted features [17]. While hand-crafted features are used along with HMM based models to recognize the words of large English vocabulary [8]. In a similar study, the input to the system is given by words and these words are segmented into characters for further processing by the HMM [11]. Further, a combination of HMM + RNN is used to recognize at word level [24].

2.2 Line Level Recognition

Initial approaches to solving line level recognition are based on HMM. Statistical language model based techniques are used to recognize complete lines [6]. Further, sliding window-based features are extracted, and a given text line is converted into a sequence of temporal features using HMM [18]. As per Markovian assumption, each observation depends upon the current time step. Hence, context-based information is not fully utilized in HMM based models. In later years, CNN based system is a state-of-the-art method to extract the features of images and RNN is for remembering them for long; hence the combination of both systems, CRNN, emerged as a new state-of-the-art approach. In one such study, line recognition is done using CTC with BLSTM along with a token passing algorithm [9]. Further, the combination of CNN with MDLSTM and BLSTM is used to recognise the handwritten text [15,29]. The CRNN architecture is a combination of deep CNN and RNN layers and is able to produce state-of-the-art results in recognizing sequential objects in images in scene text recognition [25], which is further extended in HTR domain [21]. The total number of trainable parameters can be further reduced using Gated CNN based NN models. In one such study, a system based upon Gated CNN and bidirectional gated recurrent units are able to produce state-of-the-art results [26]. In a similar study, NN based language model is integrated in the decoding step of HTR. In this study, different type of HTR architecture is examined in connection with NNLM at the decoding step on standard datasets [32].

2.3 Paragraph Level Recognition

There are two types of approaches followed in literature. First, the segmentation techniques are used to obtain line level images of paragraph [16]. In one such segmentation technique, there has been a method proposed for word segmentation which is based upon identifying the connecting region [13]. The same algorithm is extended for line segmentation [12]. The statistical methods based segmentation techniques are also used to follow the cursive path of one's handwriting [1]. After obtaining line images, line level optical recognition models are used to obtain the transcriptions of the same. But these segmentation based techniques induced error which resulted in poor recognition results. In the second method, recognition of the paragraph is done without any external segmentation. In these paragraph recognition techniques, either attention based techniques are used to do internal segmentation or the multi-dimensionality of the task is considered to achieve paragraph recognition in a single step. Following the first approach of segmentation free techniques, CNN along with BLSTM is also used to recognise text at paragraph level as an alternative to MDLSTM [19]. In the second approach of paragraph recognition, the two dimensional aspect of the task is considered. It uses a Multi-Dimensional Connectionist Classification (MDCC) technique to process two dimensional features of paragraph images. By using a Conditional Random Field (CRF), a 2D CRF is obtained from the ground truth text. This CRF graph is the control block for line selection among multiple lines [20]. In one similar study, it considered the paragraph in one single large text line. In this work, bi-linear interpolation layers are combined with feature extraction layers of the Fully Convolutional Network (FCN) encoder to obtain a large single line. The need for line breaks in transcription and line level pre-training is not required in this module [31]. We have extended the study of Vertical Attention Network (VAN) [7] by applying the WBS [22] decoder as a post processing step, further improving the accuracies and contributing to the solution of the cursive HTR problem.

3 System Design

In this study, we have performed end-to-end paragraph text recognition on IAM, RIMES and READ-2016 datasets. In this section, we have described the overall architecture in modular form. Module 1 is the recognition module which carries a VAN [7], and module 2 is the decoder module [22]. In Fig. 1, we have shown the functionality of both modules together. In this figure, ConvM denotes a single block of convolution operation of Module 1 and DscbM denotes a single block of depthwise separable operation of Module 1. Initially, in step 1, the input image is preprocessed. In parallel to step 1, the line level model of IAM, RIMES and READ-16 datasets are trained and their weights are used to train the paragraph level model of the same dataset in an end-to-end manner. The attention mechanism is used to obtain lines from paragraphs as step 3. In step 4, the LSTM decoder is used in combination with step 3 to obtain the occurrence of each character. In step 5, depending upon the processing stage, we either backpropagate

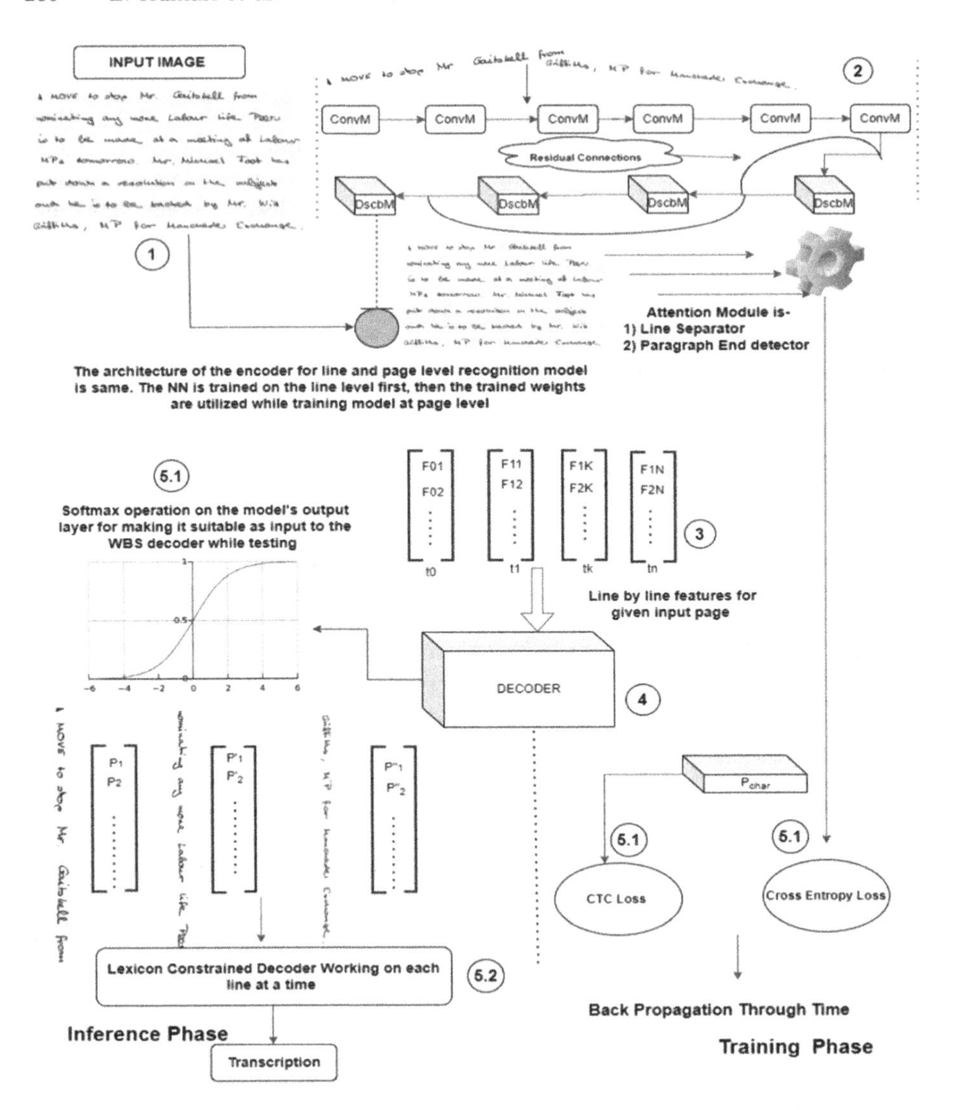

Fig. 1. A Comprehensive Handwritten Paragraph Text Recognition System: LexiconNet

through time or apply the WBS decoding algorithm. We have used all the model hyperparameters of training and testing, the same as mentioned in [7]. Module 1 and Module 2 are explained as follows,

Module-1 consists of a feature extraction module, internal line segmentation and end-of-paragraph detection sub-modules. As the name suggested, this module takes a greyscale or colour input image and produces its feature map. This module consists of Convolution and Depthwise separable Convolutional blocks. These blocks magnify the context. Composite dropout (standard with 0.5 and

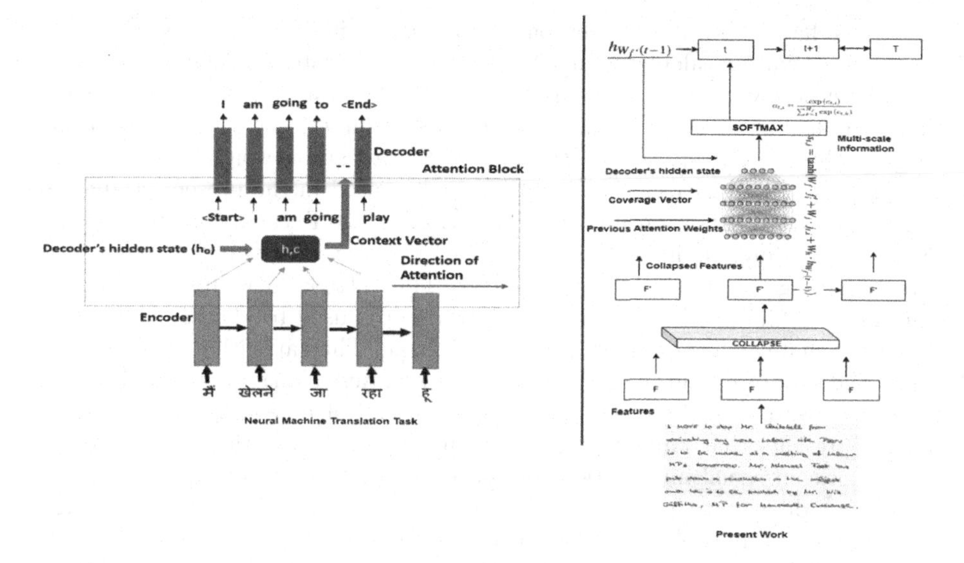

Fig. 2. Comparison of Attention mechanism in Neural Machine Translation (NMT) task (left) and in the present study (right).

2D with 0.25 probability at specified location) is used in this system. The Convolution block has two convolution layers followed by the normalization layer and convolutional layer. The convolution operations are the building block of a convolutional layer. In convolution operation, a kernel slides over the input image and produces feature maps of images. A (3×3) kernel followed by Rectified Linear Unit (ReLU) activation function has been used in this module. A convolution converts all the pixels of the receptive field (961×337) into a single value. Thus, image size is reduced and brings the receptive field's collective information together. In this work, height is reduced by a factor of 32 and width is reduced by a factor of 8 which is controlled by the stride size of the last convolutional layer of each block. The depthwise separable convolution blocks are similar to convolution blocks. In this, instead of using a standard convolutional layer, depthwise separable convolutions are used. The depthwise separable convolutions are computationally less costly than regular convolutions with the same level of performance. The last convolutional layer of the DscbM has a fixed stride of (1, 1) to preserve the input shape. Residual connections are used to address the vanishing gradient problem in deeper NN architectures. In this, residual connections with the element-wise sum operator after the last convolutional layer of the final convolution block and each depthwise separable convolution block are used to propagate the weight updates up to the first layer while training without changing the output. In the Internal line segmentation module, each paragraph image is internally segmented into lines and generates its feature at each time step using Bahdanau soft attention [2] mechanism. The attention is used to identify a particular line of a given page at a specific time step. If we have L lines in a given page so by using the attention mechanism i^{th} line features will be calculated at i^{th} time step. The probabilistic distribution of these attention weights gives the importance of each frame at time t. The attention is applied along the vertical axes such that it only needs one high attention weight value to identify a line, the rest of the attention weights can be near zero. Figure 2 summarizes the attention process used in the present study and compares it with attention used in NMT task. In NMT, each word like 'Main', 'Khelne' has different importance at different timestep of translation and the attention weights are used to signify this. Thus attention is applied horizontally in it. While in the present setup, attention is used to identify features in a linewise manner from a given paragraph image. Thus attention applies vertically. The task of paragraph end detection is separately taken as a classification task with two classes, one class represents the end of the paragraph while the other represents that paragraph has not ended yet. To identify it, we leverage the use of contextual information we obtained from the attention step and the decoder's hidden state information. We have obtained the information about the total number of lines contained by each page while processing the dataset. This serves as one hot encoded ground truth information. The Cross-Entropy (CE) loss function for training purposes, gives us proximity of our predictions to the ground truth and updates the weights accordingly. Thus total loss for training will be the weighted sum of CTC loss and CE loss. Since both the losses are the same equal weights are considered while summation. The LSTM is used as a decoder block. Each decoder's current step is input by the last step's hidden state and current line features. A single LSTM layer is used with 256 cells in the decoder module. The initial hidden state (h0) of LSTM for the first line of the page is initialized with zeros. Hidden states are kept preserved from one line to another line in a paragraph to use contextual information of the paragraph. LSTM's output tensor dimension is further extended to apply convolution with kernel size equal to 1 and input channel size is same as cells in the LSTM layer, and output channel size equal to the total number of character set in dataset +1 (for CTC Blank) [7].

Module-2 is used to predict the final sequence of the text at the testing stage. In this work, we have used the WBS decoding algorithm instead of the best path decoding algorithm. Figure 3 explains the role of the decoder while training and testing. During the training phase, CTC loss is calculated against the predicted, and the actual text and model weights are updated using back propagation accordingly. As in Fig. 3 for each line feature at each time step, we calculate the CTC loss and update the weights. While in testing, the decoder's output is softmax to make it suitable for this module. The decoding algorithm starts with an empty beam. A beam is a notion of a possible transcription candidate at a particular time step. A prefix tree is also built with the available corpus. At $t = 0$, this prefix tree is queried for all possible words from the corpus and beam extended in the same manner. From $t = 1$ to T, where T is the total time steps, beam extension depends upon the state of the beam. A beam can be either in a word state or in a non-word state. A beam is transitioned to

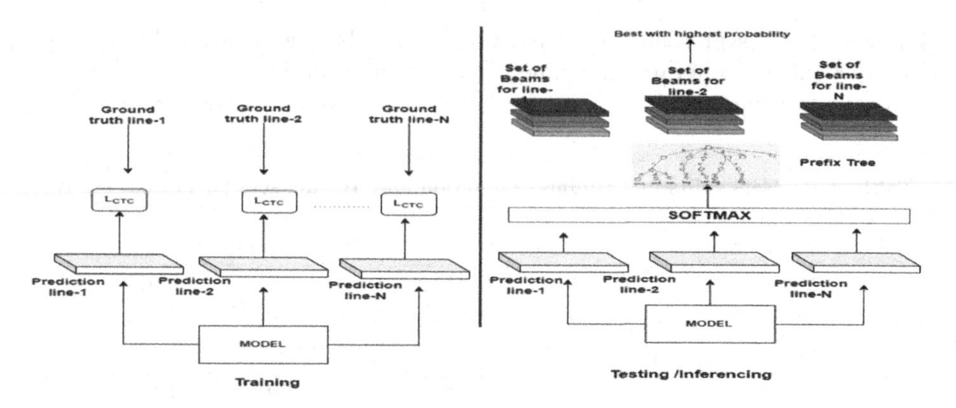

Fig. 3. Various roles of decoder

non-word stage when non-word characters are added to it. Only the 'word' state beam is extended by the characters, that are given by the prefix tree containing the prefix, while in 'non-word' state beam can be extended using all the words as well as non-words characters. We have used the 'Words' operation mode of the decoding algorithm with a beam width of 50.

4 Experimental Setup and Results

In this section, we have covered the experimental setup of the present study. It includes datasets used, preprocessing methods applied, evaluation metrics and training details. We have used the python programming language along with PyTorch framework with the apex to implement the present work. We have implemented CTC WBS decoder[1] with basic architecture of VAN[2]

4.1 Datasets

The architecture presented in the paper is evaluated against IAM [14], RIMES [10] and the READ-2016 [27] datasets which are discussed as follows,

IAM. The IAM handwriting dataset includes the forms of handwritten text, which can be used for training and testing purposes of text recognition experiments. We have used paragraph and line level segmentation in this work with the train, test and validation split are mentioned in Table 1.

RIMES. The RIMES dataset is a collection of handwritten letters sent by individuals to company administrators. Train, validation and test splits are shown in Table 1.

[1] https://github.com/githubharald/CTCWordBeamSearch.
[2] https://github.com/FactoDeepLearning/VerticalAttentionOCR/.

READ-16. It is presented in ICFHR 2016 [27]. It is written in the German language. Page, Paragraph and Line level segmentation is provided by READ-2016. The training, validation and testing splits are shown in Table 1.

Table 1. Datsets splits in Training, Validation and Testing sets for present study

Dataset		Train	Validation	Test
IAM (No. of Characters-79)	Line	6,482	976	2,915
	Page	747	116	336
RIMES (No. of Characters-100)	Line	10,532	801	778
	Page	1400	100	100
READ-2016 (No. of Characters-89)	Line	8,349	1,040	1,138
	Page	1,584	179	197

4.2 Preprocessing and Training Details

Algorithm 1: End-to-End HTR Training and Testing

Input: Paragraph batch images *I*, and ground Truth G_t with lines G_{t1}, G_{t2}...G_{tl} and text corpus $Corp_{text}$

Result: Training using backpropogation and Evaluation Metrics on given Dataset

1 Initialize main();
2 Params=initParams();
3 setDevice(Params);
4 $D_{dataset}$=loadDataset(Params);
5 I=preprocessDataset($D_{dataset}$);
6 Training the model in an end-to-end manner [7];
7 $pred_t$=Testing();
8 $pred_t$=Softmax($pred_t$);
9 Transcript=Concat(Transcript, WBS($pred_t$, $Corp_{text}$)) [22];
10 CER,WER= Accuracy(Transcript,ground truth)

In this work, the methods that have been applied to each dataset for preprocessing are the same. The variable-sized images of datasets have been resized to a fixed size. Images are zero-padded to 800 pixels width and 480 pixels height to ensure minimum feature width of 100 and height of 15. Data augmentation techniques are used to increase variations which is available during training. There were no data augmentation techniques applied during testing. Change in resolution, perspective transformation, elastic distortions, random perspective transformation [31], dilation and erosion, brightness and contrast adjustment are applied with a probability of 0.2 in the given order same as [7]. Algorithm 1, of the HTR system takes input from a set of images, a set of ground truth values and a text corpus. It produces CER and WER of the prediction of handwritten text as output. Algorithm 1 is explained in this section in line by line manner as follows,

Line 1–4:- Device setup in GPU/CPU mode (setDevice()), loading of dataset and set model parameters (loadDataset() and initParams()).
Line 5:- Pre-process the images (processedDataset()) and load train, test and validation batches (loadBatch()).
Line 6:- End-to-end training of the HTR model. CE loss and CTC loss are used to train the model. The attention mechanism helped in internal line segmentation. A pretrained model at line level on the same dataset is also used in the training for quick convergence.
Line 7:- $pred_t$ is the line-by-line character occurrence matrix for a given set of images at the testing stage.
Line 8–9:- The probability matrix is given to WBS algorithm [22]. This returns the recognized text of text line of the paragraph image. By concatenating the WBS decoder's output of each line of the paragraph we have obtained the predicted text of the whole paragraph.
Line 10:-Calculate accuracy of the system.

4.3 Results and Comparison

Table 2. Results and Comparision Table

Reference	CER(%) Valid	WER(%) Valid	CER(%) Test	WER(%) Test
IAM Dataset				
CNN+MDLSTM [4]	–	–	16.2	–
CNN+MDLSTM [3]	4.9	17.1	7.9	24.6
RPN+CNN+BLSTM+LM [30]	–	–	6.4	23.2
VAN [7]	3.02	10.34	4.45	14.55
Ours (LexiconNet)	1.89	5.17	**3.24**	**8.29**
RIMES Dataset				
CNN+MDLSTM [3]	2.5	12.0	2.9	12.6
RPN+CNN+BLSTM+LM [30]	–	–	2.1	9.3
VAN [7]	1.83	6.26	1.91	6.72
Ours (LexiconNet)	1.06	2.91	**1.13**	**2.94**
READ-2016 Dataset				
VAN [7]	3.71	15.47	3.59	13.94
Ours (LexiconNet)	2.29	7.46	**2.43**	**7.35**

In this section, an end-to-end paragraph recognition architecture LexiconNet is evaluated. We compare our results with other works. In decoder [22], we have used the 'Words' mode with beam width 50 and used the corpus present in the test dataset and validation dataset respectively for test and validation error. Table 2 summarizes our findings as we have obtained 3.24% CER and 8.29% WER on the test dataset of IAM with 27.19% improvement in CER and 43.02% improvement in WER, 1.13% CER and 2.94% WER on the test dataset of RIMES with 40.83% improvement in CER and 56.25% improvement in WER and 2.43% CER and 7.35% WER on the test dataset of READ-2016 with 32.31% improvement in CER and 47.27% improvement in WER, respectively.

4.4 Evaluation Metric

State-of-the-art evaluation metrics such as CER and WER are used to evaluate the present study. The CER is based on Levenshtein Distance (L_d) between the ground truth (g) and predicted text (p). The WER is the same as CER, but it is evaluated on word level rather than character level as in Eq. (1) where p_{sub} is the number of substitutions, p_{ins} is the number of insertions and p_{del} is the number of deletions with reference to the predicted text. g_{total} is the total number of characters in the actual string.

$$\text{CER} = (p_{sub} + p_{ins} + p_{del})/g_{total} \tag{1}$$

5 Discussion

In the present study, the proposed end-to-end handwritten paragraph text recognition architecture surpasses the results reported in the literature. An HTR, architecture involves preprocessing, training and post processing modules. Since, various heterogeneous modules are working in abstraction, comparing one deep NN architecture to another is not viable. We have added the WBS decoder [22] as a post processing step and extension to the VAN study [7].

5.1 Improvement in Accuracy upon Applying Lexicon Decoder in HTR System

We have summarized the effective percentage gain in CER and WER in the present study in comparison with the existing one. As shown in Fig. 4 (a) and Fig. 4 (b) comparison of greedy decoder accuracy with lexicon decoder on IAM test dataset and validation dataset respectively. The graph shows that the rate of improvement of WER is more than CER. The decoding algorithm favours WER more. Equation (2) and Eq. (3) are used to calculate the percentage of improvement in CER/WER. y_{cer} is the CER obtained from greedy decoder while $\hat{y}_{cer}$ is CER obtained from WBS decoder and y_{wer} is the WER obtained from greedy decoder while $\hat{y}_{wer}$ is WER obtained from WBS decoder.

$$\%\ \text{Improvement}_{\text{cer}} = \frac{y_{\text{cer}} - \hat{y}_{\text{cer}}}{y_{\text{cer}}} * (100) \tag{2}$$

$$\%\ \text{Improvement}_{\text{wer}} = \frac{y_{\text{wer}} - \hat{y}_{\text{wer}}}{y_{\text{wer}}} * (100) \tag{3}$$

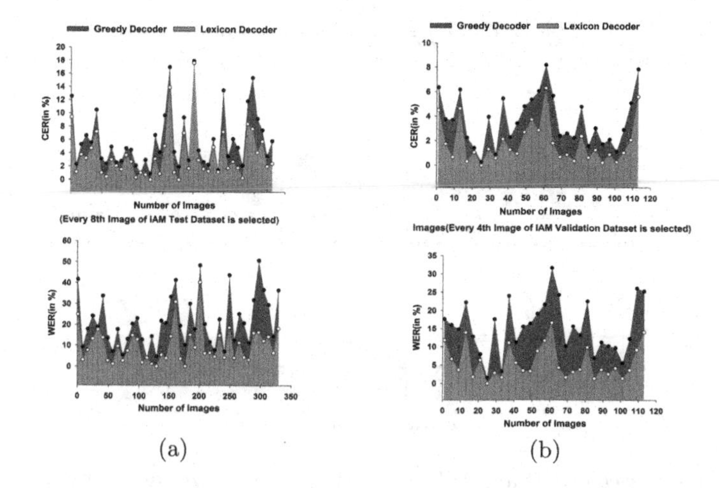

Fig. 4. Comparision of decoder(a) Test dataset (b) Validation dataset.

Table 3. CER and WER value is various corpus sizes of IAM dataset

S.No.	Corpus Size	CER (In %)	WER (in %)
1	Only Test Dataset Corpus with Lexicon Decoder (LD)	3.24	8.29
2	IAM corpus with LD	3.42	9.02
3	IAM corpus + 370K unique words with LD	4.19	12.19

5.2 Effect of Corpus Size

The scenario presented is not always favourable. Internally lexicon decoder creates a prefix tree containing all the possible prefix paths of words of a given dataset corpus for its processing and final transcription containing words of these paths. Hence, the accuracy also depends upon the type of the corpus used. In the present work, we have considered a corpus containing the text of images of the test dataset while testing the NN model and the text of images of the validation dataset while validating the model. As a part of the discussion, we considered different corpus sizes to estimate algorithm performance. In the first case, we consider all the text of the IAM dataset while in the second case we have added 370K unique English language words[3] along with the whole IAM corpus. As per the results shown in Table 3 and Fig. 5 it can be evident that, if we increase the size of the corpus, the total number of paths also increases. Thus the possibility of getting on the wrong path increases. Thus, our CER and WER values increase as we increase the corpus size. Table 3 also shows that there is less increase in error while increasing the corpus size. Hence, the decoding algorithm is robust enough to handle a large corpus.

[3] https://github.com/dwyl/english-words.

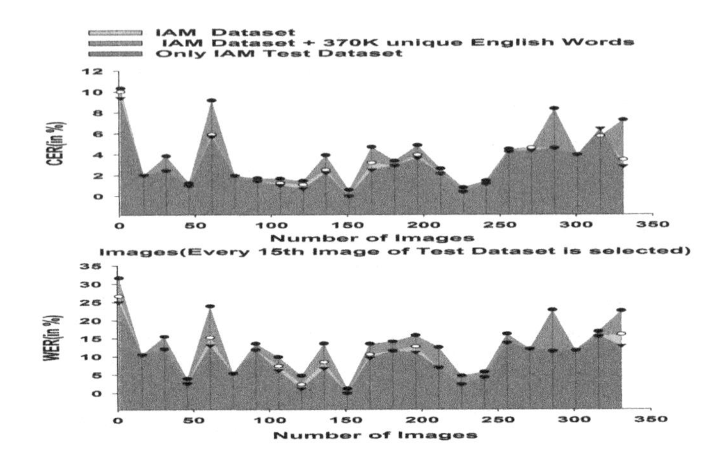

Fig. 5. Graph between CER and WER of various corpus sizes on Test dataset images of IAM dataset. It is also evident from the graph that, the choice of corpus affects the overall accuracy

Ground Truth	Greedy Decoder	Lexicon Decoder
sugya	sagya	sage
quoted	geoted	potentialities
Boaz	Boax	Box
Professor	FroReSsor	Professor
not	mot	not
manuscripts	monuscripts	manuscripts

Fig. 6. Some Positive and Negative outcomes are shown by applying the lexicon decoder as post processing step

5.3 Failure and Best Case Analysis

In this section, failure cases and best case scenarios of the present system have been discussed. The Fig. 6 also depicts the performance of greedy and lexicon decoder on the IAM dataset. In the first three cases, lexicon decoder highly penalizes accuracy by predicting the wrong word. While in the last 3 cases lexicon decoder outperforms the greedy decoder. During the time of testing while applying the CTC WBS decoding algorithm, it internally builds a prefix tree and processes along the edges of the nodes of the prefix tree to predict the handwritten text. In worst cases, the path taken in the prefix tree is totally different from the actual ground truth. This results in a substantial increase in CER due to the prediction of totally a wrong word, in comparison with the wrong prediction of one or two characters in words as in greedy decoder. As shown in Fig. 7, the word 'quoted' is predicted as 'geoted' by the greedy decoder and 'potentialities' by the WBS decoder. While word 'sugya' is predicted as 'seggea' by the greedy decoder and 'sage' by the WBS decoder.

While for best case scenarios, where our greedy decoder already performing with state-of-the-art accuracies (<1% CER), the percentage improvement in such scenario is nearly perfect. For example, in Fig. 7 the word 'Toyohiko' is recognised as 'Toyohits' by the greedy decoder and 'Toyohiko' by the lexicon decoder.

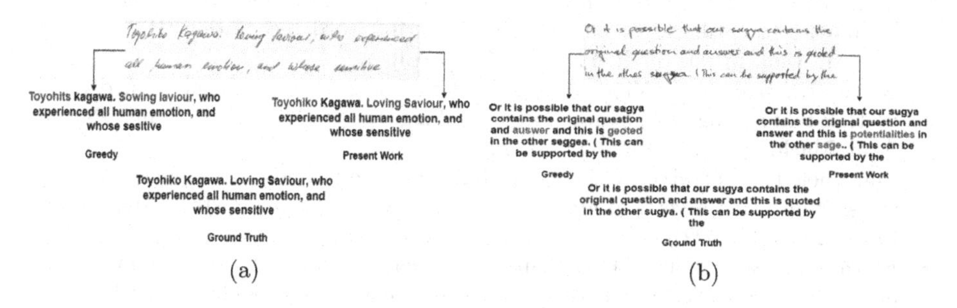

Fig. 7. (a)If Greedy performs better, the WBS decoder made it even better (b)Scenario in which WBS decoder reduces CER and WER in comparison with greedy decoder

6 Conclusion

In this study, we propose a lexicon-based end-to-end paragraph recognition system lexiconNet. It is segmentation free and leverages the current state-of-the-art handwritten text recognition domain techniques. As a post processing step, we have added a WBS decoder to the base model. We have reported substantial improvement in CER and WER on IAM, RIMES and READ-2016 datasets. We have also presented detailed analyses of best-case and worst-case observations due to the addition of a WBS decoder in the system which shows it is not always beneficial for having WBS as a decoder. This system is also able to recognize the complex layout of inclined lines. At present, this system is able to recognize the single-column layout of handwritten text only. To recognize a page in an end-to-end manner is a highly computationally intensive task. One of the future challenges may be making it less computationally extensive.

Acknowledgment. This research is funded by Government of India, University Grant Commission, under Senior Research Fellowship scheme. The authors acknowledge PRL's supercomputing resource PARAM Vikram-1000 made available for conducting the research reported in this paper.

References

1. Arivazhagan, M., Srinivasan, H.: A statistical approach to line segmentation in handwritten documents - art. no. 65000t. In: Proceedings of SPIE - The International Society for Optical Engineering, vol. 6500 (2007)

2. Bahdanau, D., Cho, K., Bengio, Y.: Neural machine translation by jointly learning to align and translate (2014)
3. Bluche, T.: Joint line segmentation and transcription for end-to-end handwritten paragraph recognition. In: Proceedings of the 30th International Conference on Neural Information Processing Systems, pp. 838–846. NIP 2016 (2016)
4. Bluche, T., Louradour, J., Messina, R.: Scan, attend and read: end-to-end handwritten paragraph recognition with mdlstm attention, pp. 1050–1055 (2017)
5. Bozinovic, R., Srihari, S.: Off-line cursive script word recognition. IEEE Trans. Pattern Anal. Mach. Intell. **11**(1), 68–83 (1989)
6. Bunke, H., Bengio, S., Vinciarelli, A.: Offline recognition of unconstrained handwritten texts using HMMs and statistical language models. IEEE Trans. Pattern Anal. Mach. Intell. **26**(6), 709–720 (2004)
7. Coquenet, D., Chatelain, C., Paquet, T.: End-to-end handwritten paragraph text recognition using a vertical attention network. IEEE Trans. Pattern Anal. Mach. Intell. **45**, 1–1 (2022)
8. El-Yacoubi, A., Gilloux, M., Sabourin, R., Suen, C.: An hmm-based approach for off-line unconstrained handwritten word modeling and recognition. IEEE Trans. Pattern Anal. Mach. Intell. **21**(8), 752–760 (1999)
9. Graves, A., Liwicki, M., Fernández, S., Bertolami, R., Bunke, H., Schmidhuber, J.: A novel connectionist system for unconstrained handwriting recognition. IEEE Trans. Pattern Anal. Mach. Intell. **31**(5), 855–868 (2009)
10. Grosicki, E., El-Abed, H.: ICDAR 2011 - French handwriting recognition competition. In: 2011 International Conference on Document Analysis and Recognition, pp. 1459–1463 (2011)
11. Kundu, A., He, Y., Bahl, P.: Recognition of handwritten word: first and second order hidden markov model based approach. In: Proceedings CVPR 1988: The Computer Society Conference on Computer Vision and Pattern Recognition, pp. 457–462 (1988)
12. Manmatha, R., Rothfeder, J.: A scale space approach for automatically segmenting words from historical handwritten documents. IEEE Trans. Pattern Anal. Mach. Intell. **27**(8), 1212–1225 (2005)
13. Manmatha, R., Srimal, N.: Scale space technique for word segmentation in handwritten documents. In: Nielsen, M., Johansen, P., Olsen, O.F., Weickert, J. (eds.) Scale-Space 1999. LNCS, vol. 1682, pp. 22–33. Springer, Heidelberg (1999). https://doi.org/10.1007/3-540-48236-9_3
14. Marti, U.V., Bunke, H.: A full English sentence database for off-line handwriting recognition. In: Proceedings of the Fifth International Conference on Document Analysis and Recognition, p. 705. ICDAR 1999, IEEE Computer Society, USA (1999)
15. Moysset, B., Kermorvant, C., Wolf, C.: Full-page text recognition: Learning where to start and when to stop. In: 2017 14th IAPR International Conference on Document Analysis and Recognition (ICDAR), vol. 01, pp. 871–876 (2017)
16. Papavassiliou, V., Stafylakis, T., Katsouros, V., Carayannis, G.: Handwritten document image segmentation into text lines and words. Pattern Recogn. **43**(1), 369–377 (2010)
17. Park, J.: An adaptive approach to offline handwritten word recognition. IEEE Trans. Pattern Anal. Mach. Intell. **24**, 920–931 (2002)
18. Ploetz, T., Fink, G.: Markov models for offline handwriting recognition: a survey. IJDAR **12**, 269–298 (2009)

19. Puigcerver, J.: Are multidimensional recurrent layers really necessary for handwritten text recognition? In: 2017 14th IAPR International Conference on Document Analysis and Recognition (ICDAR), vol. 01, pp. 67–72 (2017)
20. Schall, M., Schambach, M.P., Franz, M.O.: Multi-dimensional connectionist classification: reading text in one step. In: 2018 13th IAPR International Workshop on Document Analysis Systems (DAS), pp. 405–410 (2018)
21. Scheidl, H.: Handwritten text recognition in historical document. diplom-Ingenieur in Visual Computing, Master's thesis, Technische Universität Wien, Vienna (2018)
22. Scheidl, H., Fiel, S., Sablatnig, R.: Word beam search: a connectionist temporal classification decoding algorithm. In: 2018 16th International Conference on Frontiers in Handwriting Recognition (ICFHR), pp. 253–258 (2018)
23. Seni, G., Cohen, E.: External word segmentation of off-line handwritten text lines. Pattern Recogn. **27**(1), 41–52 (1994)
24. Senior, A., Robinson, A.: An off-line cursive handwriting recognition system. IEEE Trans. Pattern Anal. Mach. Intell. **20**(3), 309–321 (1998)
25. Shi, B., Bai, X., Yao, C.: An end-to-end trainable neural network for image-based sequence recognition and its application to scene text recognition. IEEE Trans. Pattern Anal. Mach. Intell. **39**(11), 2298–2304 (2017)
26. de Sousa Neto, A.F., Bezerra, B.L.D., Toselli, A.H., Lima, E.B.: HTR-Flor: a deep learning system for offline handwritten text recognition. In: 2020 33rd SIBGRAPI Conference on Graphics, Patterns and Images (SIBGRAPI), pp. 54–61 (2020)
27. Sánchez, J.A., Romero, V., Toselli, A.H., Vidal, E.: ICFHR 2016 competition on handwritten text recognition on the read dataset. In: 2016 15th International Conference on Frontiers in Handwriting Recognition (ICFHR), pp. 630–635 (2016)
28. Tay, Y., Khalid, M., Yusof, R., Viard-Gaudin, C.: Offline cursive handwriting recognition system based on hybrid Markov model and neural networks. In: Proceedings 2003 IEEE International Symposium on Computational Intelligence in Robotics and Automation. Computational Intelligence in Robotics and Automation for the New Millennium (Cat. No.03EX694), vol. 3, pp. 1190–1195 (2003)
29. Wigington, C., Stewart, S., Davis, B., Barrett, B., Price, B., Cohen, S.: Data augmentation for recognition of handwritten words and lines using a CNN-LSTM network. In: 2017 14th IAPR International Conference on Document Analysis and Recognition (ICDAR), pp. 639–645 (2017)
30. Wigington, C., Tensmeyer, C., Davis, B., Barrett, W., Price, B., Cohen, S.: Start, follow, read: end-to-end full-page handwriting recognition. In: Ferrari, V., Hebert, M., Sminchisescu, C., Weiss, Y. (eds.) Computer Vision - ECCV 2018, pp. 372–388 (2018)
31. Yousef, M., Bishop, T.E.: OrigamiNet: weakly-supervised, segmentation-free, one-step, full page text recognition by learning to unfold (2020)
32. Zamora-Martínez, F., Frinken, V., España-Boquera, S., Castro-Bleda, M.J., Fischer, A., Bunke, H.: Neural network language models for off-line handwriting recognition. Pattern Recogn. **47**, 1642–1652 (2014)

Local Style Awareness of Font Images

Daichi Haraguchi(✉) and Seiichi Uchida

Kyushu University, Fukuoka, Japan
daichi.haraguchi@human.ait.kyushu-u.ac.jp

Abstract. When we compare fonts, we often pay attention to styles of local parts, such as serifs and curvatures. This paper proposes an attention mechanism to find important local parts. The local parts with larger attention are then considered important. The proposed mechanism can be trained in a quasi-self-supervised manner that requires no manual annotation other than knowing that a set of character images are from the same font, such as Helvetica. After confirming that the trained attention mechanism can find style-relevant local parts, we utilize the resulting attention for local style-aware font generation. Specifically, we design a new reconstruction loss function to put more weight on the local parts with larger attention for generating character images with more accurate style realization. This loss function has the merit of applicability to various font generation models. Our experimental results show that the proposed loss function improves the quality of generated character images by several few-shot font generation models.

Keywords: Font identification · Font generation · Quasi-self-supervised learning · Contrastive learning

1 Introduction

To understand font styles, a reasonable choice is to observe local shapes. Each character has a shape representing font style; however, the whole character shape is unnecessary to understand its style. Assume that a character 'A' is printed with Helvetica (a famous sans-serif font), and we want to understand the style of Helvetica from it. In this case, we must ignore the global shape that makes 'A' as 'A.' In other words, we need to focus on local shapes, such as serifs, corners, stroke width, and local curvatures, which are rather independent of character class 'A.'

Through a *contrastive learning* scheme, this paper tries to determine *local style awareness* representing important local shapes for particular font styles. Imagine a person who has only seen Helvetica in its lifetime — then, the person cannot determine the style of Helvetica. In other words, we can understand the particular style of Helvetica by contrasting (i.e., comparing) it with other fonts, such as Times New Roman and Optima. Moreover, as noted above, we need to ignore the whole letter shape and focus on local shapes during the comparison in some automatic way.

M. Coustaty and A. Fornés (Eds.): ICDAR 2023 Workshops, LNCS 14194, pp. 242–256, 2023.
https://doi.org/10.1007/978-3-031-41501-2_17

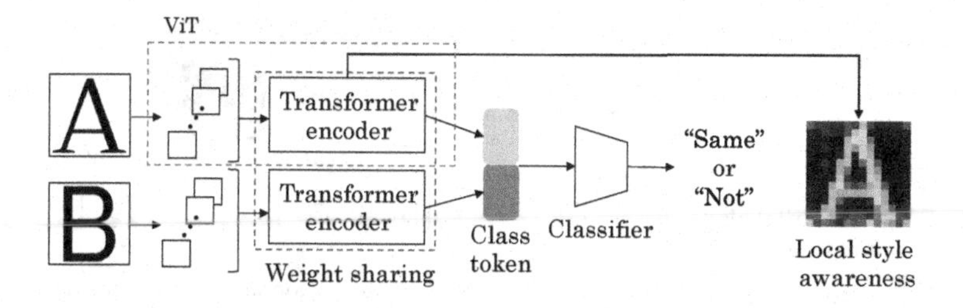

Fig. 1. Overview of the proposed technique to determine *local style awareness*, which indicates important local shapes to describe font styles. The local style awareness is obtained as a fine attention map through a contrastive learning scheme that identifies whether two given character images belong to the same font.

Figure 1 shows the overall structure of the proposed model to determine local style awareness. This figure also shows a heatmap representing local style awareness for the input 'A.' This model is trained via a *font identification* task. This task aims to determine whether two given character images come from the same font[1]. In the case of this figure, a serif-style 'A' and a sans-serif 'B' are given, and therefore, the model must answer "Not." To answer this task correctly, the model needs to ignore the whole shapes of 'A' and 'B' and enhance their local style differences; therefore, the model needs to determine the local style awareness internally. By visualizing this internal representation as a spatial map, we will have local style awareness.

The following two points must be considered for determining local style awareness via the font identification task. First, the task is formulated as a contrastive learning task. As noted above, font style is determined by comparing the target font with other fonts. Therefore, the model of Fig. 1 is trained to enhance local style differences. Second, we need to compare two different alphabets, such as 'A' and 'B,' instead of the same alphabet, such as 'A' and 'A.' If we only compare the same alphabet in the font identification task, it reduces to a trivial task. The model can give perfect identification results by checking whether two inputs are entirely the same or not. In other words, the model cannot learn the local style awareness. By training the model with font pairs including different alphabets, the model can learn local style differences while ignoring the global letter shapes.

The proposed technique uses Vision Transformer (ViT) [4] by expecting the merits from its attention mechanism. In ViT, a character image is decomposed into small patches, and these patches are fed into a transformer encoder. In the encoder, the attention of each patch is calculated by using the mutual relationship between patches. By contrastive learning for the font identification task, ViT will give larger attention to the local patches that are more important for

[1] We can find another type of the font identification task, where a single image is given, and a model chooses its font name, such as `Helvetica`, from the prespecified font classes. In contrast, our font identification task is more general, and its model answers "Same" or "Not" for a given pair of images.

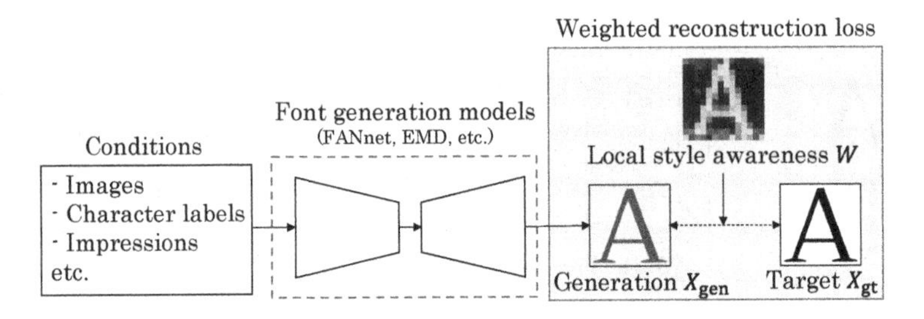

Fig. 2. Training the font generation model using the reconstruction loss weighted by local style awareness.

representing the style. The heatmap of the local style awareness of Fig. 1 is an attention map given by the proposed technique, and each element of the heatmap corresponds to a patch.

It should be emphasized that the ViT-based model of Fig. 1 is trainable very efficiently in *a quasi-self-supervised manner* for the font identification task. The ground truth for our task is whether two character images are from the same font or not. Accordingly, if we prepare the character images from specific font sets, we know the font name of each image and give the ground truth for each image pair without any manual annotation cost. For example, if we prepare font sets of `Helvetica` and `Optima`, the pair 'A' and 'B' from `Helvetica` should have the ground truth of "Same," and the pair 'C' from `Helvetica` and 'D' from `Optima` have "Not." Our experimental results show that the attention learned in this efficient manner becomes larger around important local shapes for individual styles, as expected.

In this paper, we further utilize this attention mechanism to realize local style-aware font generation models. Specifically, as shown in Fig. 2, we utilize the local style awareness representing the importance of individual patches for weighting the reconstruction loss function in various font generation models. This weighting scheme contributes to a more accurate reproduction of the important local parts in the generated images, as proved in our experiments of few-shot font generation in three different font generation models.

Our contributions are as follows.

- To determine local style awareness, we propose an efficient contrastive learning framework to solve the font identification task. Through the solution of the task, our network model can determine the local parts important to describe the font style. Note that the model can be trained without any manual annotation.
- We experimentally proved that the above framework could determine the important local parts.
- We apply the local style awareness to the weight of reconstruction loss in the font generation model. This weighting scheme can be easily introduced in any model trained with a reconstruction loss.
- Our experimental results show that the weighting scheme improves the quality of few-shot font generation.

2 Related Work

2.1 Font Recognition and Identification

Font recognition is a famous font-related task that is also conducted in the DAR community. There are some pioneering works in font recognition [14,30]. Shi *et al.* [14] recognized the font families on documents utilizing page properties and short words in the document images. Zramidin *et al.* [30] used 280 fonts and recognized them using a Bayesian classifier. Following these researches, many font recognition has been studied [2,15,17].

In contrast to the above studies, which address only a small number of fonts and are weak to noise, Chen *et al.* [3] addressed Visual Font Recognition (VFR), which is the task of recognizing font on an image or photo, and they recognized more than 2,000 fonts. Following this, DeepFont, [23] based on deep learning, was proposed for solving VFR. Wang *et al.* [22] conducted VFR for Chinese fonts using deep learning.

Haraguchi *et al.* [7] tackled font identification, identifying whether a pair of fonts are identical. Font recognition deals with fonts registered in advance; in contrast, font identification deals with pairs of arbitrary fonts, even test-time. We utilize font identification to obtain local style awareness of not only training images but also testing images.

2.2 Font Generation

In recent years, many researchers tackled font generation, especially few-shot font generation [6,10,11,16,18,25]. Few-shot font generation is a task that accepts content images or labels (i.g., a character label) and source images for extracting the style and then generating font images with the content and the style. Most of the few-shot font generation approaches inadequately handle aesthetic details in fonts. Fonts have their impression in their details (local structures) [20]; therefore, the aesthetic details in fonts are essential.

Some font generation for Chinese characters studies seeks the local-aware font generation to utilize radical information or structure of characters [10,11, 18]. However, the more detailed font styles in the radicals are not addressed. Additionally, there is no study of local-aware font generation for alphabets.

Some studies address the imbalance between character regions (foreground) and the background or sharpness of characters [16,25]. In our experiment, we use these approaches for comparative methods; therefore, we describe the details of these approaches in Sect. 4.2.

2.3 Fine-Grained Tasks

Fine-grained image recognition and classification focus on learning subtle yet discriminative features. Some studies utilize attention maps to extract such features [5,27–29]. These methods estimate attention maps that localize the discriminative regions through end-to-end training for fine-grained image recognition or classification. And then, they utilize the attention map to emphasize the discriminative features. They do not need extra annotations for the regions; however, they need additional branches to estimate the attention map in each model. Therefore, they need to propose a model for each task to obtain and utilize an attention map.

We obtain the attention map of local style awareness through font identification. In contrast to the above fine-grained recognition and classification, we utilize the attention map for font generation tasks independent of font identification. Therefore, we do not need to prepare the different models to obtain the attention map for each font generation method.

3 Local Style Awareness in Font Images

3.1 Methodology

Font Identification by Contrastive Learning. To determine important local parts for font styles, we realize a font identification model by contrastive learning. As noted in Sect. 1, font styles are defined by comparing various fonts and enhancing their differences. Font identification is the task of determining such differences between two input images by comparing them in a contrastive manner. Therefore, solving the font identification task fits our aim to determine local style awareness.

Figure 1 shows the ViT-based model for the font identification task. A pair of character images are prepared, and each is fed into a ViT, i.e., a transformer encoder, after decomposing into small patches. Each ViT outputs a feature vector called a class token. A pair of class tokens are concatenated and fed to a classifier consisting of fully connected layers to make the binary decision, "Same" or "Not."

Determining Local Style Awareness by Attention. ViT, or transformer encoder, has a patch-wise self-attention mechanism, which evaluates the mutual relationship not only between neighboring patches but also between *distant* patches. This mechanism is useful for acquiring local style features because the style-aware local parts, such as serifs, often exist at distant locations. For example, serifs of 'I' exist at the top and bottom of the vertical stroke. By training the model of Fig. 1 for style identification, this self-attention mechanism is expected to be more sensitive to the local style difference and less sensitive to the global shapes that make, for example, 'A' as 'A.'

Accordingly, if we measure the value of patch-wise self-attention, we can get local style awareness as an attention map. (If an image is decomposed into $M \times N$ patches, the map has $M \times N$ resolution.) Roughly speaking, in the task of font identification, the attention value will become higher (or lower) at patches that are important (or unimportant) for the identification. In our model of Fig. 1, we have two $M \times N$ self-attention maps corresponding to two image inputs. The attention map for each image will show the local style awareness of the image.

To measure the attention values, we use *attention rollout* [1]. Attention rollout is an XAI technique and can visualize the importance of individual patches by using the result of self-attention. For our task of font identification, attention rollout will give higher (or lower) attention to the patches which are important (or unimportant) for the identification.

For local style awareness, it is very important that the patch-wise attention map with attention rollout realizes a higher spatial resolution than other XAI techniques, such as Grad-Cam [13], which is a popular XAI to visualize the regions that contribute to the decision in Convolutional Neural Networks (CNN). It is well-known that the spatial resolution by Grad-Cam is very low because it depends on the size of the deepest convolution layer. In contrast, ours has $M \times N$ resolutions, and theoretically, using smaller patches makes M and N larger. In practice, however, using too small patches is not good to describe the local shape. The current resolution of the local style awareness in Fig. 1 is a good compromise between resolution and descriptive power and still finer than Grad-Cam.

Quasi-self-supervised Learning. To train the model of Fig. 1, we need to give ground truth ("Same" or "Not") for each character image pair. This ground truth information, fortunately, can be given without any manual annotation effort. As noted in Sect. 1, if we can prepare a set of fonts (say, `Helvetica` and `Optima`), they automatically indicate which character images come from `Helvetica` or `Optima`. Such indications are enough to give the ground truth. Since this framework still needs external information (on preparing font sets), it is not fully self-supervised, which does not require any external information. Therefore, we call it *quasi*-self-supervised learning. From a practical viewpoint, however, it is equivalent to self-supervised learning because its annotation cost is zero after font set preparation.

Implementation Details. The transformer encoder in ViT follows the implementation of ViT [4] pretrained by ImageNet-21K. The classifier in Fig. 1 consists of two fully connected layers. The numbers of layers and heads in the transformer are 12. The class token is a 768-dimensional vector. The size of an input image and the patch size are 224×224 and 16×16, respectively. (Therefore, $M = N = 14$.) Batch size and learning rate are set at 64 and 10^{-5}, respectively. We use Adam for the optimizer and cross-entropy loss for the loss function.

3.2 Qualitative Evaluations of Local Style Awareness

Dataset. We used the font dataset from Google Fonts[2] as follows. First, using metadata, we obtained font family name, category name[3], and a character sub-

[2] https://github.com/google/fonts.

[3] We use four categories of fonts included in Google Fonts. In more detail, there are 1,283 Sans-Serif fonts, 630 Serif fonts, 457 Display (i.e., decorative) fonts, and 203 Handwriting fonts.

set, which shows languages included in each font. We chose the fonts with a character subset of "Latin" and discarded the others. We also discarded incomplete fonts. Then, we divided the font into a training, validation, and testing set to 8:1:1. During division, we did our best to avoid very similar fonts in different sets by checking font family names. As a result, we prepared 2,094 fonts, 230 fonts, and 249 fonts for the training, validation, and testing sets, respectively. For simplicity (by avoiding the disturbances of small caps.), we only used 26 capital letters in the following experiments.

The original font data is the vector format (TTF); therefore, we render it to bitmap images of 224×224 pixels with a margin of 5 pixels to use the experiment of font identification. In font generation, we resize these images to 64×64 or 80×80 to adapt to the experimental setting for each baseline of the font generation models.

Comparative Models. Although there is neither similar work nor baseline, we design two comparative models for evaluating local style awareness in font images.

- One is a ViT trained for the font category classification task (instead of font identification). Then, we obtain its attention map by attention rollout. Font category classification is a task that classifies the input font image into one of the four font categories, "Serif," "Sans Serif," "Handwriting" and "Display." These categories are given in Google Fonts. ViT pretrained by ImageNet-21K is fine-tuned for font category classification. The hyper-parameters in the model are the same as the ones in the font identification in Sect. 3.1.
- The other is CNN (instead of ViT) trained for the font identification task. Then, we obtain its Grad-CAM. We employed ResNet-18 [8] as the CNN. The way of making pairs in the training phase is the same as the identification by ViT.

The test accuracy of font identification by ViT, font category classification by ViT, and font identification by CNN are 94.69%, 86.38%, and 94.59%, respectively. Note that font category classification is difficult because of fuzzy class boundaries between four categories. (Especially the boundary between Sans-Serif and Display is often confusing.)

Visualization of Local Style Awareness. Figure 3 (a) visualizes local style awareness obtained by the proposed model. In the first row, strong attention is found in a part of shadows. (Note that the bottom part of these characters are shadows.) We can also see the consistency of attention to serif parts in the second row. In the third row, strong attention is found not only in the serif parts but also curves in 'U' and 'C.' For the sans-serif fonts in the fourth row, attention is found at the bottom, where the stroke thickness and straightness are clearly represented. The fonts in handwriting style in the fifth row show attention around their curvy stroke ends and intersection parts. From observing

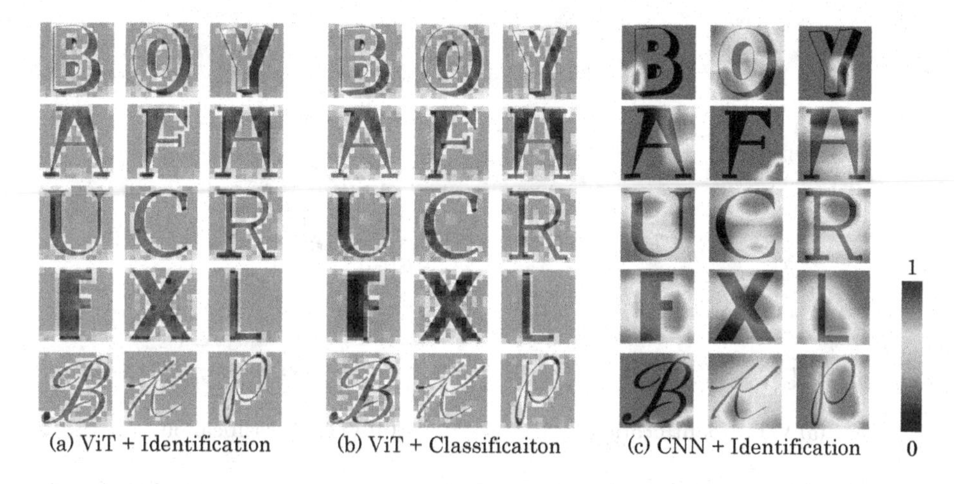

Fig. 3. Visualization of local style awareness. Red regions show strong attention, and blue regions show weak attention. (Color figure online)

those maps, our attention maps, showing local stroke awareness, can find local parts that represent font-specific local structures.

The comparison between (a) and (b) suggests how the font identification task is more suitable for local style awareness than the font category classification task. In the second row of (b), the comparative model of category classification could capture serifs because the model is trained to discriminate serif fonts from others. However, except for the serif parts, the comparable model of (b) often fails to catch representative local parts. For example, for the samples in the third row, this model totally ignores the curves because the curves are not important for the current category classification task. Similarly, in the fourth row, the model also seems to ignore the thickness and straightness — it focuses on the corners to check the existence of serifs. To summarize these observations, this model mainly focuses on the corners to discriminate between serifs and sans-serifs and thus is rather insensitive to other local parts representing the unique structure specific to the font. In the next section, we will see how this comparative model captures different style features from ours.

The differences between ViT (a)(b) and Grad-CAM (c) in their resolution and accuracy are obvious. As expected, the map by Grad-CAM is very coarse and difficult to understand the important local parts for representing font styles. Moreover, the map by CNN shows strong attention not only to the character region but also to the background region — although the background regions are often important for specifying font styles, the Grad-CAM highlight on 'F' and 'B' seems irrelevant to describe the font style.

Distributions of Local Style Features. For a further comparison between the proposed model and the comparative model trained for the category classification, we visualize the distributions of their class tokens, that is, style features,

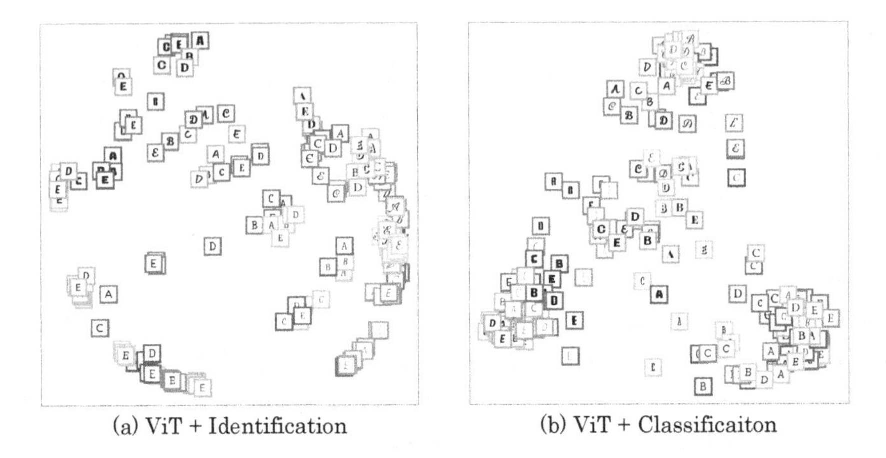

(a) ViT + Identification (b) ViT + Classificaiton

Fig. 4. Distributions of class tokens (i.e., style features) by two-dimensional PCA. The same box color indicates the same font.

by ViT. Figure 4 shows their distributions for the samples of five alphabets from 'A' to 'E' by two-dimensional PCA. The comparative observation of (a) and (b) shows that the characters from the same font are more clustered in (a) than (b). Consequently, our model based on font identification is more sensitive to the style and can ignore the whole letter shapes.

4 Boosting Font Generation Quality by Local Style Awareness

4.1 Methodology

In this section, we utilize local style awareness in font generation tasks for realizing local style-aware font generation. As shown in Fig. 2, local style awareness represents the importance of individual patches. Thus, we can use it for weighting the reconstruction loss function in various font generation models. Adding the weight of local style awareness will contribute to a more accurate reproduction of the important local parts in the generated images.

L1 loss weighted by local style awareness W is as follows:

$$L = ||(W + \alpha) \odot (X_{\text{gt}} - X_{\text{gen}})||_1, \tag{1}$$

where $\odot$ is an element-wise product, and α is a constant value for computing normal L1 loss. We set $\alpha = 0.1$ in the following experiment. Additionally, X_{gt} and X_{gen} indicate a ground truth image and a generated image, respectively.

4.2 Quantitative and Qualitative Evaluation Experiments

Few-Shot Font Generation. We evaluate the usefulness of our proposed loss through few-shot font generation. Few-shot font generation is the task that accepts a few source images and content images (or a content label), and then the style of source images is transferred into the content.

In the training phase of few-shot font generation, almost all of the models optimize a reconstruction loss between generated image X_{gen} and the target image (*e.g.*, ground truth) X_{gt}. Therefore, introducing the proposed reconstruction loss into the font generation model is very straightforward.

In this experiment, we use the same dataset used in Sect. 3.2. We set the image size to 64×64 or 80×80 according to the experimental setting for each baseline of the font generation models. To this end, we resize the attention maps to the same size as each input image.

Three Baseline Models of Few-Shot Font Generation. We picked up three baseline models[4] for few-shot font generation and observed the usefulness of the proposed loss for them.

- **FANnet** [12] is the font generation model that can edit the characters while retaining the font style of a source image. In more detail, FANnet accepts an image of the source font and a one-hot vector corresponding to the character label and then generates the characters with the same style as the source font. In the training phase, it employs L1 loss for reconstruction loss. When we conduct the few-shot font generation, we use average features extracted from source images.
- **EMD** [25,26] is the style transfer model for font style and has often been used as a baseline of the font generation task. EMD accepts content images; therefore, we fix the content images to a simple sans-serif font in the evaluation. EMD employs L1 loss weighted by the character regions to consider the imbalance between background and character regions. We compare the loss with ours in our experiments.
- **Srivatsan *et al.*** [16] proposed a font generation model that disentangles content from style in font images and combines them. They optimized the model by projecting an image onto the frequency using the Discrete Cosine Transform (DCT-II) instead of directly reconstructing an image. Specifically, they impose a Cauchy distribution which is heavy-tailed distribution in the projected space to generate sharper images. We also compare the loss with our proposed one. Note that this model includes a loss function for disentangling the style. Therefore, reconstruction loss is not dominant compared with the other methods.

[4] There are indeed newer font generation methods, and our model can be introduced even to them. Since they have rather complex structures, which might obfuscate the effects of different loss functions (original, L1, and ours), we did not use them.

Table 1. Quantitative evaluation of few-shot font generation. In this experiment, we use five style images. The metrics in magenta are expected to be more sensitive to local differences. The loss of "original" is the loss function proposed in each baseline model.

Baseline	loss	L1 ↓	weighted L1↓	Hausdorff↓	PHD↓	LPIPS↓	IoU ↑	SSIM↑
FANnet [12]	L1	0.0855	0.0347	7.172	1083	0.1542	0.6503	0.6810
	ours	**0.0843**	**0.0319**	**5.602**	**803**	**0.1294**	**0.6764**	**0.6896**
EMD [25,26]	original	**0.0916**	0.0364	8.997	2049	**0.1523**	0.6404	**0.6829**
	L1	0.0938	0.0376	9.267	2139	0.1564	0.6306	0.6794
	ours	0.0988	**0.0358**	**8.600**	**2013**	0.1543	**0.6434**	0.6666
Srivatsan *et al.* [16]	original	0.1219	0.0429	6.0439	1026	0.2182	**0.6486**	0.6182
	L1	**0.0901**	**0.0362**	6.866	990	0.1339	0.6335	**0.6735**
	ours	0.1007	0.0378	**5.699**	**888**	**0.1130**	0.6307	0.6417

Evaluation Metrics. We evaluate the quality of font generation with various evaluation metrics[5]. L1, LPIPS (Learned Perceptual Image Patch Similarity), and SSIM (Structural Similarity) are evaluation metrics commonly used for font generation. Hausdorff distance and IoU are used for the quantitative evaluation of several font generations [9,21,24]. When we calculate the Hausdorff distance and IoU, we binarize the image by Otsu's method. In the Hausdorff distance, we conduct canny edge detection as preprocessing. Additionally, we use Pseudo Hamming Distance (PHD) [19], an evaluation metric, to calculate the similarity between fonts. PHD might be the most appropriate way to evaluate font styles among the above metrics because PHD can directly evaluate the difference between two shapes. Hausdorff distance also directly evaluates the difference. Roughly speaking, PHD evaluates an average difference over all shape contours, whereas Hausdorff distance evaluates the maximum difference. Consequently, it is sensitive to slight font shape differences and has been used for evaluating the similarity between font images. We also evaluate the quality of font generation using our loss function Eq. 1 to set $\alpha = 0$ and call it weighted L1.

Quantitative Evaluation. As shown in Table 1, our loss could improve all evaluation metrics for FANnet. FANnet is a simple model; therefore, our loss directly improves the font generation quality. In EMD, ours is better than the others in more than half of the metrics. Especially, ours is best in Hausdorff distance and PHD. These two metrics are more sensitive to the little difference between the images than L1 loss. This indicates that our loss contributes to generating fonts keeping detailed styles more than the others. Original loss takes into the imbalance between character regions and background regions. However, for sustaining the font style, using local style awareness for font generation is more effective than the original one.

In Srivatsan *et al.*, ours is better than the other model in several metrics. In particular, ours is much better than the original loss function in almost all metrics. Ours is worse than the L1 loss in several evaluation metrics (such as

[5] Some metrics might take infinite value when the generated image becomes empty. Therefore, we exclude such images.

weighted L1 and SSIM). This model includes not only reconstruction loss but also a loss function for disentangling the style. The balance between the loss function and reconstruction loss is crucial. Therefore, the order difference between ours and the L1 loss might be one of the reasons for the lower results than the simple L1 loss. It is a limitation of our loss to tune the hyper-parameter (e.g., weight between loss).

Through the experiment of all three baseline models, the results by our loss tend to be better in Hausdorff distance and PHD, as shown in Table 1. This indicates that our loss contributes to improving font generation quality in detailed font styles because these losses are sensitive to differences between images, and especially, PHD is a metric to evaluate the similarity between fonts. Note that our loss aims to sustain the detailed local styles in font generation; therefore, seeing a clear improvement in font generation by using our loss might be difficult in the other evaluation metrics.

However, in some metrics, degradations are caused by two reasons. The first reason is the characteristics of evaluation metrics. For example, we sometimes have better (low) L1 scores when the font generation model generates empty images than generating deformed font images. The second reason is the limitation of our loss function. Our loss function focuses on local shapes representing the style; this implies that some parts unimportant for the style sometimes become noisy. A typical case is 'J' in the first example of Fig. 5 (c), whose stroke width is not constant.

Qualitative Evaluation. Figure 5 shows the font generation examples by each baseline model and loss. To generate fonts, we use five source images marked by orange boxes. The source images are chosen randomly.

In the first example in FANnet (a), the L1 loss tends to defect to thin strokes. Especially, 'A,' 'J,' and 'M' are likely to defect their strokes. However, ours is effective in fonts with thin strokes. This is because our loss is correctly weighting to the local style awareness of the font with thin strokes. The second example in (a) shows that ours has serif parts more clearly than L1, especially 'E,' 'F,' and 'T.' From this example, our loss effectively generates font with keeping the local style.

In the first example of EMD (b), the original can not generate serif parts precisely, and some images defect the strokes. L1 can not capture the stress of stroke width (e.g., 'C' and 'G'). In contrast, ours can clearly generate fonts with keeping its serif style. In the second example, there is not much difference between the generated fonts. However, we emphasize that only ours can generate 'J' correctly. This result comes from our loss functions advantage that can pay attention to local style awareness.

In Srivatsan *et al.* (c), the first examples show that ours can generate fonts sustaining the detailed serif parts, especially the top serif of 'A.' This trend can be seen in ours. Local style awareness contributes to generating the serif parts like the above 'J.' The second example shows that ours can generate thin fonts than L1. This trend is the same as in (a). The original method also generates the images; however, several images have a blurry noise.

Fig. 5. Results of few-shot font generation by each baseline model and loss. Orange boxes show the source, which is used for extracting font styles. "Target" shows the ground truth. (Color figure online)

5 Conclusion

This paper proposed local style awareness, which represents important shapes for particular font styles. Local style awareness is acquired by solving a font identification task in a contrastive learning scheme. This task is solved very efficiently in a quasi-self-supervised learning manner where no manual annotation is necessary. In other words, we can get local style awareness without human efforts. Our model is based on ViT instead of CNN because ViT and its attention mechanism help us to have finer local style awareness that can catch a small style structure such as serifs.

As an application task, we utilized local style awareness in few-shot font generation to generate font images whose local structures are realized more accurately. In this application, we simply use the local style awareness as the weight for the reconstruction loss function; this simplicity allows us to use the local style awareness in various state-of-the-art font generation models. In our experiments, we prove quantitatively and qualitatively that our loss could improve the performance of three baseline models.

In future work, we will obtain local style awareness of other languages and apply them to a font generation task in the language. Additionally, we will apply the awareness to other tasks, such as style analysis, style transfer, and style domain adaptation.

Acknowledgment. This work was supported in part by JST, the establishment of university fellowships towards the creation of science technology innovation, Grant Number JPMJFS2132, JSPS KAKENHI Grant Number JP22H00540, and JST ACT-X Grant Number JPMJAX22AD.

References

1. Abnar, S., Zuidema, W.: Quantifying attention flow in transformers. In: ACL, pp. 4190–4197 (2020)
2. Avilés-Cruz, C., Rangel-Kuoppa, R., Reyes-Ayala, M., Andrade-Gonzalez, A., Escarela-Perez, R.: High-order statistical texture analysis–font recognition applied. Pattern Recognit. Lett. **26**(2), 135–145 (2005)
3. Chen, G., et al.: Large-scale visual font recognition. In: CVPR, pp. 3598–3605 (2014)
4. Dosovitskiy, A., et al.: An image is worth 16 × 16 words: transformers for image recognition at scale. arXiv preprint arXiv:2010.11929 (2020)
5. Fu, J., Zheng, H., Mei, T.: Look closer to see better: Recurrent attention convolutional neural network for fine-grained image recognition. In: CVPR, pp. 4438–4446 (2017)
6. Gao, Y., Guo, Y., Lian, Z., Tang, Y., Xiao, J.: Artistic glyph image synthesis via one-stage few-shot learning. TOG **38**(6), 1–12 (2019)
7. Haraguchi, D., Harada, S., Iwana, B.K., Shinahara, Y., Uchida, S.: Character-independent font identification. In: DAS, pp. 497–511 (2020)
8. He, K., Zhang, X., Ren, S., Sun, J.: Deep residual learning for image recognition. In: CVPR, pp. 770–778 (2016)
9. Kang, J., Haraguchi, D., Matsuda, S., Kimura, A., Uchida, S.: Shared latent space of font shapes and their noisy impressions. In: MMM, pp. 146–157 (2022)
10. Kong, Y., et al.: Look closer to supervise better: one-shot font generation via component-based discriminator. In: CVPR, pp. 13482–13491 (2022)
11. Liu, W., Liu, F., Ding, F., He, Q., Yi, Z.: XMP-Font: self-supervised cross-modality pre-training for few-shot font generation. In: CVPR, pp. 7905–7914 (2022)
12. Roy, P., Bhattacharya, S., Ghosh, S., Pal, U.: STEFANN: scene text editor using font adaptive neural network. In: CVPR, pp. 13228–13237 (2020)
13. Selvaraju, R.R., Cogswell, M., Das, A., Vedantam, R., Parikh, D., Batra, D.: Grad-CAM: visual explanations from deep networks via gradient-based localization. In: ICCV, pp. 618–626 (2017)
14. Shi, H., Pavlidis, T.: Font recognition and contextual processing for more accurate text recognition. In: ICDAR, pp. 39–39 (1997)

15. Solli, M., Lenz, R.: FyFont: find-your-font in large font databases. In: SCIA, pp. 432–441 (2007)
16. Srivatsan, N., Barron, J., Klein, D., Berg-Kirkpatrick, T.: A deep factorization of style and structure in fonts. In: EMNLP-IJCNLP, pp. 2195–2205 (2019)
17. Sun, H.M.: Multi-linguistic optical font recognition using stroke templates. In: ICPR, vol. 2, pp. 889–892 (2006)
18. Tang, L., et al.: Few-shot font generation by learning fine-grained local styles. In: CVPR, pp. 7895–7904 (2022)
19. Uchida, S., Egashira, Y., Sato, K.: Exploring the world of fonts for discovering the most standard fonts and the missing fonts. In: ICDAR, pp. 441–445 (2015)
20. Ueda, M., Kimura, A., Uchida, S.: Which parts determine the impression of the font? In: ICDAR, pp. 723–738 (2021)
21. Wang, Y., Gao, Y., Lian, Z.: Attribute2Font: creating fonts you want from attributes. TOG **39**(4), 69–1 (2020)
22. Wang, Y., Lian, Z., Tang, Y., Xiao, J.: Font recognition in natural images via transfer learning. In: MMM, pp. 229–240 (2018)
23. Wang, Z., et al.: DeepFont: identify your font from an image. In: ACMMM, pp. 451–459 (2015)
24. Wen, Q., Li, S., Han, B., Yuan, Y.: ZiGAN: fine-grained Chinese calligraphy font generation via a few-shot style transfer approach. In: ACMMM, pp. 621–629 (2021)
25. Zhang, Y., Zhang, Y., Cai, W.: Separating style and content for generalized style transfer. In: CVPR, pp. 8447–8455 (2018)
26. Zhang, Y., Zhang, Y., Cai, W.: A unified framework for generalizable style transfer: style and content separation. TIP **29**, 4085–4098 (2020)
27. Zheng, H., Fu, J., Mei, T., Luo, J.: Learning multi-attention convolutional neural network for fine-grained image recognition. In: ICCV, pp. 5209–5217 (2017)
28. Zheng, H., Fu, J., Zha, Z.J., Luo, J.: Looking for the devil in the details: learning trilinear attention sampling network for fine-grained image recognition. In: CVPR, pp. 5012–5021 (2019)
29. Zhuang, P., Wang, Y., Qiao, Y.: Learning attentive pairwise interaction for fine-grained classification. In: AAAI, pp. 13130–13137 (2020)
30. Zramdini, A., Ingold, R.: Optical font recognition using typographical features. TPAMI **20**(8), 877–882 (1998)

Fourier Feature-based CBAM and Vision Transformer for Text Detection in Drone Images

Ayush Roy[1](✉), Palaiahnakote Shivakumara[2], Umapada Pal[1], Hamam Mokayed[3], and Marcus Liwicki[3]

[1] Computer Vision and Pattern Recognition Unit, Indian Statistical Institute, Kolkata, India
aroy80321@gmail.com, umapada@isical.ac.in

[2] Faculty of Computer Science and Information Technology, University of Malaya, Kuala Lumpur, Malaysia

[3] Department of Computer Science, Electrical and Space Engineering, Lulea University of Technology, Luleå, Sweden
{hamam.mokayed,marcus.liwicki}@ltu.se

Abstract. The use of drones for several real-world applications is increasing exponentially, especially for the purpose of monitoring, surveillance, security, etc. Most existing scene text detection methods were developed for normal scene images. This work aims to develop a model for detecting text in drone as well as scene images. To reduce the adverse effects of drone images, we explore the combination of Fourier transform and Convolutional Block Attention Module (CBAM) to enhance the degraded information in the images without affecting high-contrast images. This is because the above combination helps us to extract prominent features which represent text irrespective of degradations. Therefore, the refined features extracted from the Fourier Contouring Network (FCN) are supplied to Vision Transformer, which uses the ResNet50 as a backbone and encoder-decoder for text detection in both drone and scene images. Hence, the model is called Fourier Transform based Transformer. Experimental results on drone datasets and benchmark datasets, namely, Total-Text and ICDAR 2015 of natural scene text detection show the proposed model is effective and outperforms the state-of-the-art models.

Keywords: Scene text detection · Drone images · Deep learning · Transformer · Detection transformer

1 Introduction

For protecting sensitive areas, monitoring several situations, such as exhibitions, and possessions, tracking the person in the crowd and vehicles in traffic, use of drones and unmanned aerial vehicles have been used [1]. However, due to height, angle variations, and weather conditions during capturing scenes from drones, one can expect images affected by multiple adverse factors, such as poor quality, degradations, loss of vital information, and occlusion. Therefore, detecting text in drone images is challenging compared to text detection in scene images. There are models developed in the past

M. Coustaty and A. Fornés (Eds.): ICDAR 2023 Workshops, LNCS 14194, pp. 257–271, 2023.
https://doi.org/10.1007/978-3-031-41501-2_18

for addressing the challenges of scene text detection. However, most existing models consider scene images but not drone images, and hence the scope of the state-of-the-art methods are limited to scene images [1, 2]. As a result, the existing models may not be effective for drone images. It is illustrated in Fig. 1, where for the drone and scene images, the existing method [2], which was developed based on deep learning, fails to fix proper bounding boxes for the text in the drone image while the same method detects text in the scene image. This shows that the fine-tuned pre-defined model [2] does not have the ability to cope with the challenges of drone images. However, on the other hand, the proposed model performs well for both drone and scene images. Therefore, detecting text in drone and scene images remains a challenge for the document analysis community.

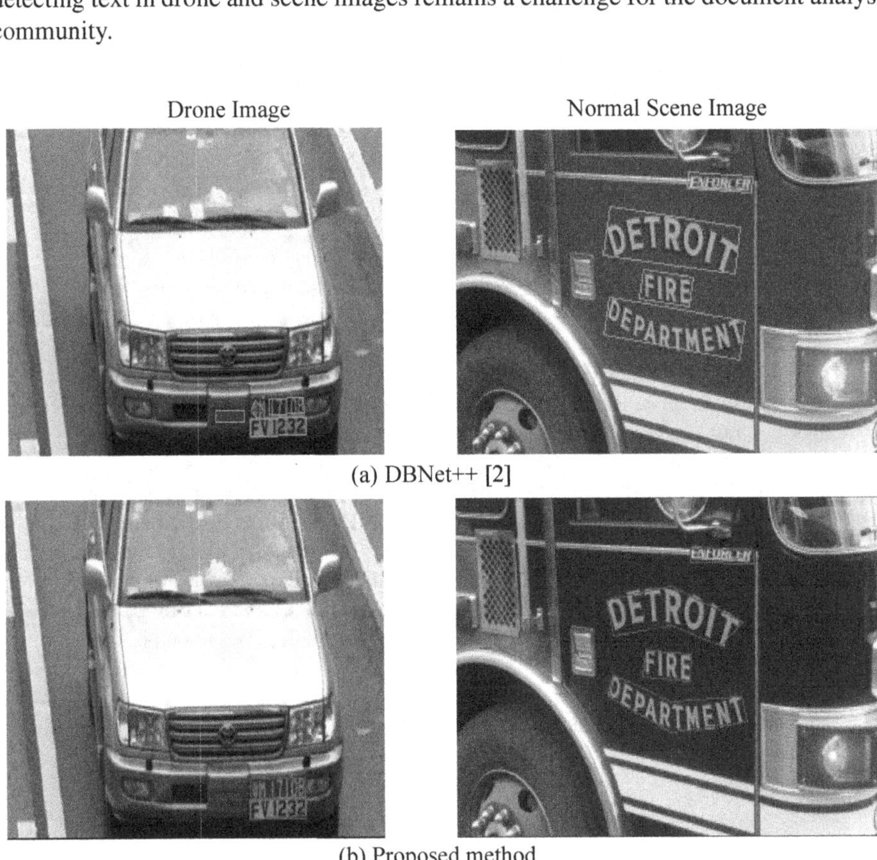

Fig. 1. Illustrating challenges of text detection in drone images. Results of (a) DBNet++ and (b) Proposed method and (B)

To address the challenges of drone and scene images, motivated by the Fourier transform, which can extract the features that represent vital edge information in the images irrespective of contrast variations [3], we explore the Fourier transform for feature extraction in this work. Similarly, inspired by the Convolutional Block Attention Module (CBAM) that integrates prominent features extracted through the channel and spatial attention to represent dominant information in the images [4], this work supplies features maps obtained from Fourier convolutional layers to CBAM for extracting accurate and efficient features that represent text in the drone/scene images. In the same way, inspired by the success of vision transformers for object detection and classification in complex situations [5], the proposed work explores the vision transformer in this work. Overall, the combination of Fourier transforms, CBAM, and Vision Transformer is the key contribution to tackling the challenges of drone as well as scene images in contrast to existing methods.

The rest of the paper is organized as follows. Since scene text detection methods are related to text detection in drone images, we review scene text detection methods in Sect. 2. In Sect. 3, the details of the Fourier transform, and vision transformer are presented. Section 4 discusses experiments on drones and benchmark scene text detection datasets to validate the performance of the proposed and existing methods. Section 5 summarizes the proposed work based on experimental results.

2 Related Work

We review the existing methods of scene text detection and text detection in drone images. Liao et al. [2] developed a model based on a differential binarization network for addressing the challenges of arbitrarily shaped text in scene images. Zhang et al., [6] proposed a network utilizing the Graph Convolutional Network (GCN). Every text instance is divided into small rectangular boxes (estimated by the text proposal network). The local graph construction model creates the linkage between the components and GCN is applied to these graphs followed by link merging. FCENet [3] uses the feature pyramid network (FPN) along with Fourier Transform to detect the text regions. The image is represented as a complex vector and the points are reconstructed from the Fourier domain to the spatial domain using the Inverse Fourier Transform (IFT) along with non-maximal suppression (NMS). PANet [7] is an efficient model extracting the features using the FPN via the FPEM network. The multi-scale features of the FPEM networks are added elementwise, upsampled, and concatenated into one feature map by the FFM network for text detection. PSENet [8] generates different scales of kernels for each instance and generally expands the kernel with the minimum scale to the text instance.

Zheng et al. [9] noted that most existing methods assume that the training and testing data are chosen from the same domain (distribution). This is the reason to get poor results from the deep learning-based model. To overcome this problem, the authors proposed a domain adaptation model which combines low-level and high-level features for achieving the best results. Zhao et al. [10] pointed out that most methods are expensive due to heavy labeling of a huge sample. To reduce time, the approach explores mixed supervised learning which uses weakly supervised data and a small number of labeled data. In contrast to other existing methods which consider scene images, Banerjee et al. [11] focused on addressing the challenges of watermark text and scene text images. The work uses UNet and Fourier contour embedding networks for detecting both watermarks and scene text in video frames. Bagi et al. [12] proposed a model for detecting traffic signs and scene text in adverse weather conditions. For this, the work proposes an end-to-end multi-oriented spotting approach based on hierarchical spatial context. Dai et al. [13] observed that due to limited training samples, there are chances of causing overfitting problems and hence the performance of the method degrades. To solve this problem, the authors proposed an idea of scale-aware data augmentation to achieve better generalization ability. Keserwani et al. [14] noted that to achieve the best results, the method should be fully supervised. In the case of partially annotated instances, the performance of the methods degrades. To address this challenge, the approach uses pseudo labels to refine text regions. To improve the performance of the method for arbitrarily oriented and shaped text detection, Deng et al. [15] proposed an enhancement network based on cross-domain, which explores low- and high-level features extracted from different domains. Unlike the above methods, Zeng et al. [5] explored the Swin transformer with a feature pyramid network for detecting text in circuit cabinet wiring. The approach shows that the Swin transformer is capable of handling the effect of occlusion, wire bending, and lighting.

From the above review, one can notice that most of the methods consider scene images for text detection. None of the methods consider drone images for text detection. Jain et al. [16] proposed Yolov5 architecture for number plate detection in drone images. In the same way, Mokayed et al. [1] extended their conventional edge-based method of license plate number detection [17] to drone images. The approach explores the combination of the Discrete Cosine Transform and Phase Congruency Model for addressing the challenges of drone images. Although the models [1, 16] found the solution to drone images, the scope of the method is limited to license plate numbers but not both license plate numbers and scene images. Recently, Pal et al. [18] introduced adaptive Swin Transformer-based RCNN architecture for text detection in drone as well as scene images. The approach considers the text components detected by MSER and RPN (Regional Proposal Network) as input for text detection. Therefore, the main weakness of the approach is that text detection performance depends on the success of text component detection. Since MSER is sensitive to degradations, the text component detection step is not effective for drone images. Thus, detecting text in drone and scene images is considered a challenging problem. This is because both images pose different threats to detection.

3 Proposed Model

The proposed work aims to detect text in drone and scene images. It is noted that drone and scene images pose different challenges, such as degradations due to poor quality, distortion, occlusion, weather conditions, and arbitrarily oriented text and irregularly shaped text, respectively. To reduce the effect of degradations, motivated by the ability of Fourier transform which can extract features that represent vital text information regardless of contrast variations [3], we explore the same for feature extraction from drone and scene images. Similarly, it is noted that the Convolutional Block Attention Module (CBAM) has the ability to extract features that represent dominant information in the images irrespective of distortions [4]. We explore CBAM for feature extraction that represents text in distorted drone images without affecting text in scene images. Overall, the combination of Fourier and CBAM (F-CBAM), is a new combination proposed in this work to tackle the challenges of drone images. In the same way, for text detection, inspired by the success of the transformer for object detection in a complex situation [5], we explore the vision transformer. For successful text detection in drone and scene images, the proposed work combines the F-CBAM and vision transformer in this work as shown in Fig. 3 (Fig. 2).

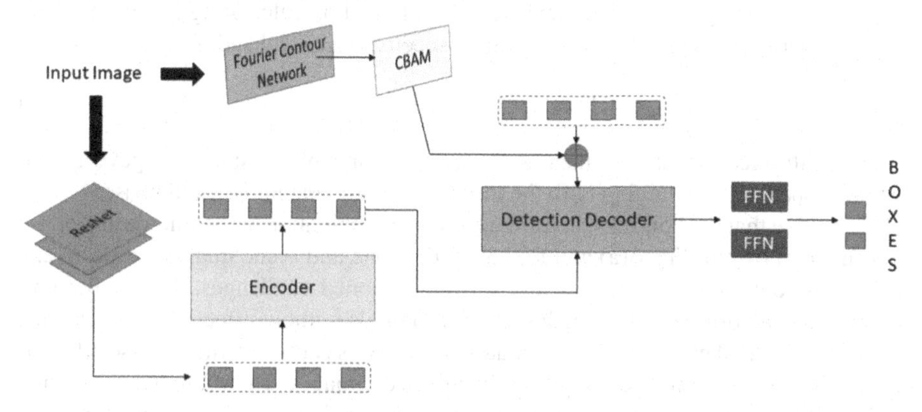

Fig. 2. Block diagram of the proposed work

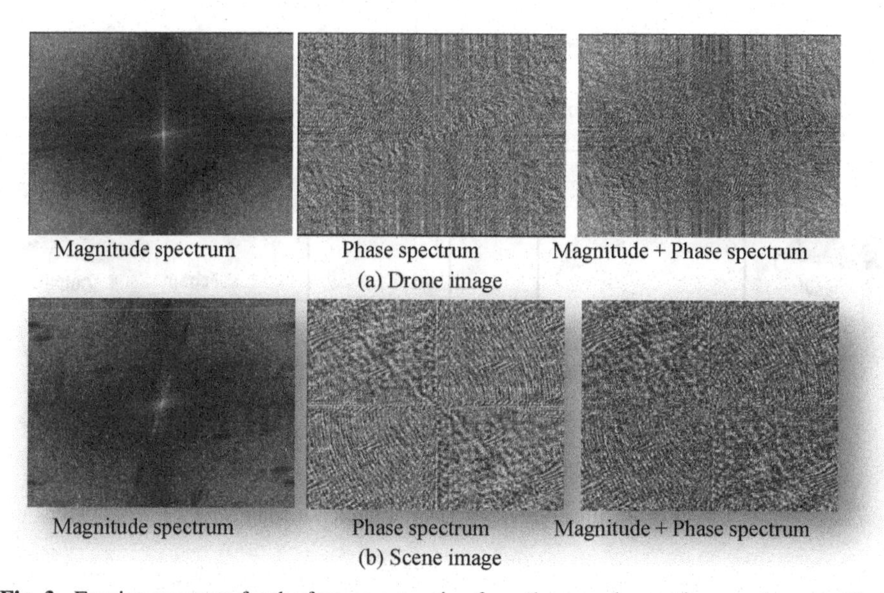

Fig. 3. Fourier spectrum for the feature extraction from drone and scene images shown in Fig. 1.

3.1 Fourier Transform Spectral Features

To cope with the challenges of drone and scene images, we explore Fourier transformation to obtain Fourier spectrum with the following steps. The Fourier coefficients are calculated using Eq. (1).

$$F_{(u,v)} = \frac{1}{MN\sum_{m=0}^{M-1}\sum_{n=0}^{N-1} e^{-2\pi i\left(\frac{mu}{M}+\frac{nv}{N}\right)} f_{(m,n)}} \tag{1}$$

where, $F_{(u,v)}$ are the coefficients of DFT, N, and M are the height and width of the image, and $f_{(m,n)}$ is the image. The phase spectrum is obtained by calculating the angle of the complex Fourier coefficients. The magnitude spectrum is calculated using Eq. (2).

$$\mathrm{Fmag} = 20\log(|\mathrm{F}_{(u,v)}|) \tag{2}$$

The multiplication of the magnitude spectrum and phase spectrum gave us the combined spectrum containing both the information, i.e., magnitude and phase.

We believe that the Fourier spectrum which is a combination of magnitude and phase spectrum as shown in Fig. 3(a) and Fig. 3(b) for drone and scene images, respectively, provide fine details irrespective of the above-mentioned challenges. This is because the Fourier transform can enhance low-contrast details without affecting high-contrast details [3]. This makes sense as it eliminates noisy pixels caused by the adverse effect of different threats in the images, resulting in enhanced images. Therefore, the combined Fourier spectrum is fed to Fourier convolutional pyramid layers for feature extraction as shown in Fig. 4.

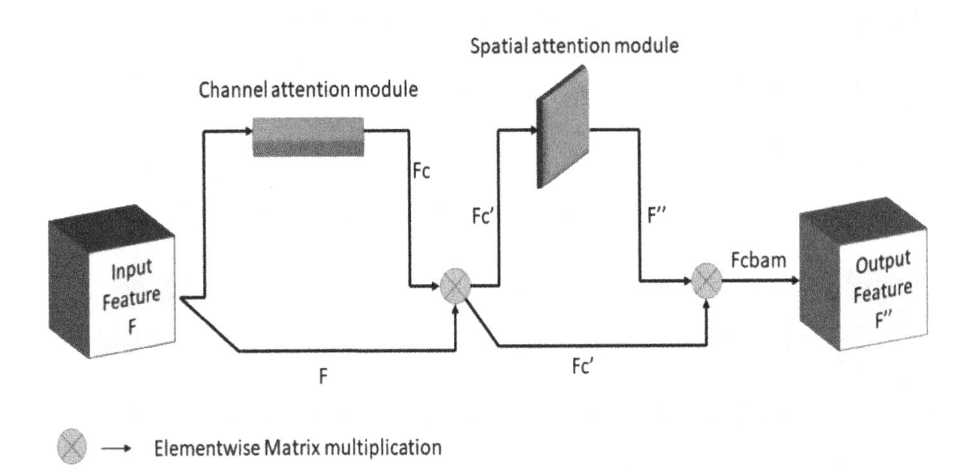

Fig. 4. Illustration of CBAM attention module

3.2 CBAM Attention Module

Since the complexity of drone images is unpredictable, to strengthen features extracted by Fourier convolutional layers, this work proposes CBAM (Convolutional Block Attention Module), which integrates the features extracted through the channel and spatial attention [4]. As a result, the CBAM helps us to extract prominent features that represent text in the images irrespective of the type of images. Overall, the combination of Fourier and CBAM (FCBAM) extracts accurate and efficient features to tackle the challenges of drone images. The CBAM uses a two-stage attention mechanism that comprises a 1D Channel Attention Module (CAM) and a 2D Spatial Attention Module (SAM). The CAM assigns weights to channels of the feature maps, emphasizing the more informative channels that contribute to better performance. The SAM weights the spatial dimensions of the feature maps, allowing the network to attend to the most relevant regions of an image. The CBAM attention mechanism is typically applied to the last feature map of dimension C × H × W generated from any CNN architecture. Here, C, H, and W represent the number of channels, height, and width of the feature map, respectively. The 1D channel attention network outputs a feature map (say, F_c) of dimension C × 1 × 1, which can be defined using Eq. (3).

$$F_c = \sigma(MLP(GAP(F)) + MLP(GMP(F))) \tag{3}$$

In Eq. (3), '+' denotes the element-wise addition operation, F is the input feature map, and σ represents the sigmoid activation function. Now, $F_{c'} = F_c * F$ is fed to the SAM ('*' denotes element-wise matrix multiplication), which is the domain space encapsulation attention mask applied to enhance the feature representation $F_{c'}$. It outputs a feature map (say, F'') of dimension C × H × W. F'' can be formulated using Eq. (4).

$$F'' = f^{7\times7}[DL(GAP(F_{c'}))\,;\; DL(GMP(F_{c'}))] \tag{4}$$

In Eq. (4), ';' denotes the concatenation of the two features. $f^{7\times7}$ is the convolutional layer of kernel size 7 × 7 with a dilation of 4, and DL represents Dense Layers. The Dense Layers (DL) consist of two dense layers that use the ReLU activation function. The first dense layer takes C dimensional input and outputs C/r (r is the reduction ratio) dimensional vector. This output is fed to the second dense layer which returns an output feature of dimension C. DL is shared by the Global Average Pooling (GAP) layer and Global Max Pooling (GMP) layer. Thus, Fcbam (see Eq. (5)) is the output of the CBAM attention module having dimensions C × H × W.

$$Fcbam = F'' * F_{c'} \tag{5}$$

3.3 Fourier Transform and Convolutional Block Attention Module for Feature Extraction

The architecture described in Fig. 4 involves a convolutional layer that resembles a feature pyramid network with 128 filters of kernel size 3 × 3. The stride of the first convolutional layer is 1, and subsequent layers have a stride of 2, resulting in a fourth layer feature map that is 1/8th the size of the first layer. Kaiming Normal initialization is used for all convolutional layers. The feature maps are then upsampled to match the size of the first feature map and concatenated. The concatenated feature map is passed on to the CBAM attention module. The attention-aided feature map is then flattened and passed through a dense layer of 32 units with a relu activation function. The resulting output is then used as input to two parallel sets of dense layers, each with 16 units and

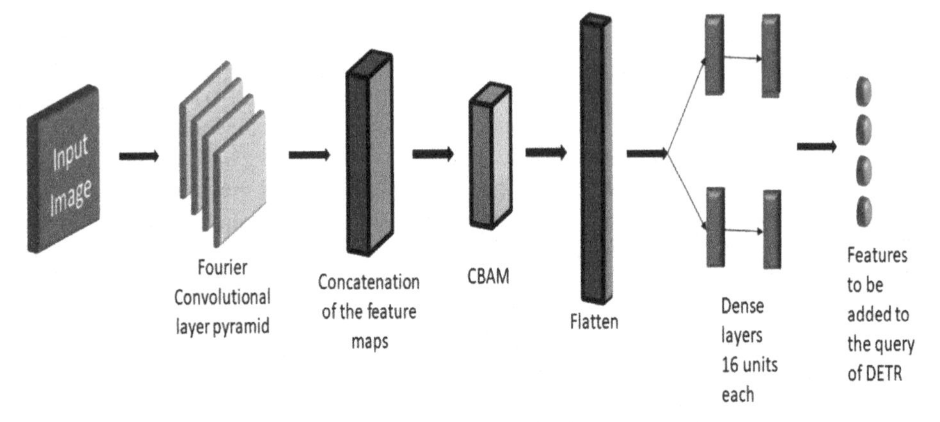

Fig. 5. Fourier based architecture for feature extraction.

a relu activation function. The features from both branches are combined and passed on to the final dense layer, which has 15 units and a tanh activation function. These 15 features represent the object queries, which are then used as input to DETR for frequency feature-guided detection. The overall block diagram is shown in Fig. 5.

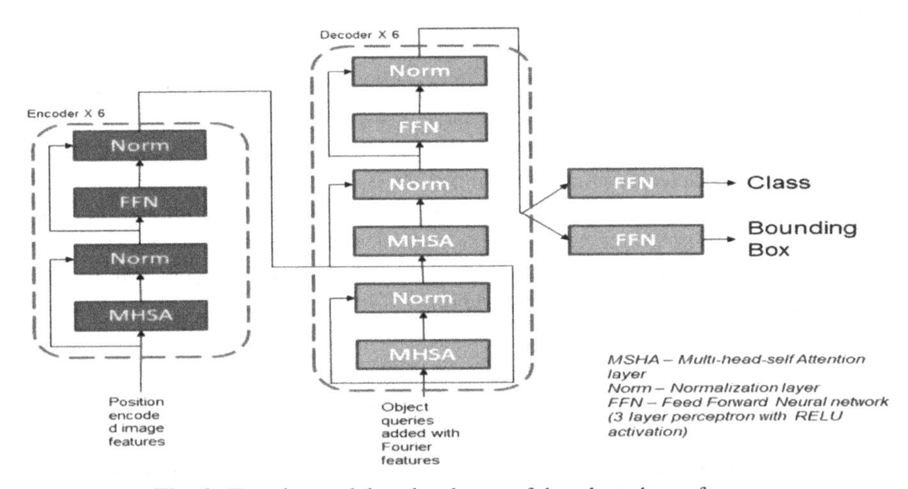

Fig. 6. Encoders and decoders layers of the adapted transformer.

3.4 Transformer for Text Detection

It is noted that the DETR is an established powerful spotting and detection transformer in the field of object detection [19]. Therefore, we explore the same to adapt for text detection in drone and scene images. The details of adaptation are shown in Fig. 6, where it can be seen how the work changes the layers of the encoder and decoder. The combined features extracted from the FCN and the input images are fed a conventional

CNN model for a better feature representation. This 2D map is flattened and positionally encoded before passing to the encoding architecture (due to permutation invariance). The encoder layer performs a higher dimensional feature matrix to a lower dimensional feature matrix. Each encoder unit is made of a Multi-Head-Self attention layer, normalizer, and feed-forward neural modules (FFN). The decoder follows a similar architecture to the encoder and is used for parallelly decoding the embeddings (N embeddings are parallelly decoder in each decoder layer). Since the decoder is also permutation invariant, the input embeddings i.e., the object queries are positionally encoder and fed to each of the attention layers and are then individually decoded into bounding boxes and their class labels by the FFN network which comprises of 3-layer perceptron with Relu activation layer having dimensions of the encodings and a projection layer. The linear projection layer predicts the class of the bounding box (text or background) whereas the perceptron is responsible for the prediction of the normalized center coordinates, height, and width of the box w.r.t. the input image.

After the bounding boxes and their class labels are predicted using the FFN network, they are compared with the ground truth using bipartite matching. The matching algorithm uses a permutation-based algorithm to search for the lowest loss between the ground truth and a predicted bounding box. The bounding box with the lowest loss is then assigned to the corresponding ground truth. The loss function used for this matching is the Hungarian loss function, which is a linear combination of two loss functions: negative log-likelihood for class prediction and box loss. The box loss is a weighted addition of IoU loss, which is used for scale invariance, and l1 loss. For more information about this loss function, refer to [19].

4 Experimental Results

We construct our own drone dataset which contains 1142 images (combining the Kaggle dataset[1] and our drone images) for experimentation and evaluating the proposed model. In our dataset, the majority of images are mainly traffic signals or parking images. As a result, our dataset includes low-resolution, degraded, good-quality, poor-quality, partially occluded license plate number images, and images with tiny text. For evaluating the ability of the proposed model to text detection in natural scene images, we consider the ICDAR 2015 and Total-Text datasets in this work [20, 21]. The ICDAR 2015 dataset consists of challenging arbitrarily oriented rectangular texts on natural images of malls, stations, etc. The images consist of blurry images which are difficult for text detection purposes and consist of varying intensities of texts. On the other hand, the Total-Text dataset consists of curved as well as arbitrarily oriented texts. Since these two datasets are popular and benchmark datasets for evaluating text detection methods on horizontal, multi-oriented, arbitrarily oriented, and arbitrarily shaped text, we use the same datasets for judging the performance of the proposed method on the aforementioned challenges. Sample images shown in Fig. 7 illustrate the challenges of text detection in drone and scene images.

[1] https://www.kaggle.com/datasets/andrewmvd/car-plate-detection.

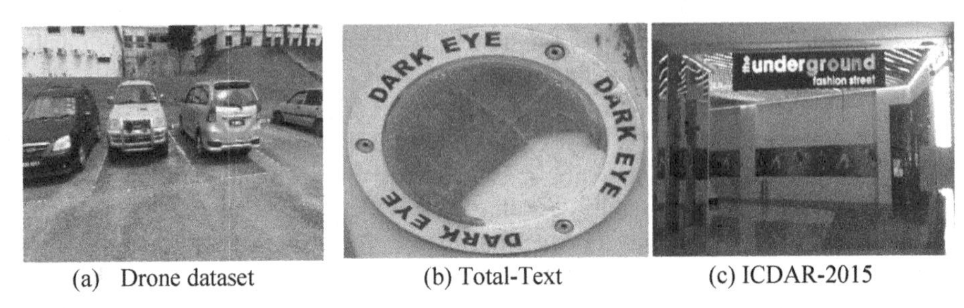

(a) Drone dataset (b) Total-Text (c) ICDAR-2015

Fig. 7. Sample images of our dataset and standard datasets of scene text detection

For measuring the performance of the proposed model, we consider the standard metrics as defined in Eq. (6)–Eq. (8). Similar to the traditional Precision, Recall, and F1-score, the measures are defined as follows for evaluating the performance of the methods. The pixels that are inside the ground truth bounding box are defined as True Positives. The pixels that are inside the predicted box but outside the ground truth bounding box are defined as False Positives. The pixels that are outside the ground truth bounding box are defined as True Negatives. The pixels that are outside the predicted bounding box but inside the ground truth bounding box are defined as False Negatives.

Precision – It is the ratio of the total area of intersection between the ground truth and the predicted bounding box and the area of the predicted bounding box. It is represented in Eq. (6)

$$P = \frac{area(ground\ truth) \cap area(predicted\ box)}{area(predicted\ box)} \tag{6}$$

Recall - It is the ratio of the total area of intersection between the ground truth and the predicted bounding box and the area of the predicted bounding box as shown in Eq. (7).

$$R = \frac{area(ground\ truth) \cap area(predicted\ box)}{area(ground\ truth)} \tag{7}$$

F1 score – It is the harmonic mean of recall and precision calculated using Eq. (8).

$$F = 2 \times \frac{(P \times R)}{(P + R)} \tag{8}$$

For calculating measures, we manually annotate the text in drone images of our dataset.

Implementation Details: The proposed Fourier contouring network was initially trained on the Total text and Drone datasets for 30 epochs to predict the number of detections to be made in an image. During this training, the number of ground truths for a particular image was used to train the network. The model achieved a validation accuracy of 94.59% with a validation loss of 0.1987 (mean squared error loss). Next, the last Dense layer, which predicts the number of detections, was retrained while freezing all other layers of the Fourier Contouring Network along with the DETR network for text detection datasets. The overall model was trained for 250 epochs using ResNet-50 as the backbone CNN model. The number of object queries was kept at 15, and 6 encoder and decoder blocks were used along with 15 FFNs for 15 decoded embeddings. Popular Python libraries such as TensorFlow, Keras, OpenCV, and NumPy were used for experimentation. A 70:30 split was used for training and testing.

4.1 Ablation Study

To evaluate the effectiveness of the proposed method for text detection in drone images, the authors conducted the following ablation experiments and reported the results in Table 1. (i) Using baseline DETR with original images as input. This experiment is to show the baseline architecture alone is not sufficient to detect text in distorted drone images successfully. At the same time, the results of the baseline architecture are considered benchmark results to compare with the results of the proposed method. (ii) Feeding the image to a conventional FCN for feature extraction and then adapted DETR is used for detection. This is to validate the contribution of the combination of FCN and DETER for text detection. (iii) Feeding the magnitude spectrum obtained by Fourier transform to FCN for feature extraction and adapted DETR for detection. This is to validate the effectiveness of the Fourier spectrum and DETR for text detection. (iv) Feeding the phase spectrum obtained by Fourier transform to FCN for feature extraction and adapted DETR for detection. This is to study the effectiveness of Phase spectrum and DETER for text detection. (v) Using the combined Fourier spectrum for feature extraction with FCN and adapted DETR is used for text detection. This is to show the combination of Fourier spectrums, FCN and DETR is effective for text detection in drone images. (vi) Feeding images to CBAM for feature extraction and adapted DETR for text detection. This experiment assesses the contribution of CBAM to text detection. (vii) Applying CBAM attention to the combined Fourier spectrum for feature extraction with FCN and adapted DETR is used for text detection. This is nothing but the proposed method.

It is observed from the results reported in Table 1 that the text detection performance gradually improves compared to the results of baseline architecture when the method includes new modules, namely, FCN, the magnitude, phase spectrums, and the combined Fourier spectrum, respectively. The baseline DETR was found to be not effective in addressing the challenges of drone images. The combined Fourier spectrum provided the highest result compared to the other two, indicating that both magnitude and phase spectrums contribute to text detection. Similarly, it is noted from the results of FCN, CBAM, and the proposed method that FCN and CBAM alone are not capable of achieving the best results for drone images as the proposed method. However, when the FCN and CBAM are combined with Fourier spectrums and DETR, the steps are effective for achieving the best results.

4.2 Experiments on Detection

The results of our and existing methods are presented in Fig. 8(a) and (b), respectively on sample images of drone and scene datasets. It is observed from the results in Fig. 7 that the text in different images is detected accurately by the proposed method. However, the existing method [2], which is a state-of-the-art model, uses a deep learning model for detecting arbitrarily oriented and arbitrarily shaped text, performs well for scene image,

Table 1. Studying the effectiveness of the key steps using the drone dataset

Exp.	Steps	P	R	F
(i)	Baseline DETR (original images as input)	0.746	0.699	0.722
(ii)	FCN + DETR	0.753	0.711	0.731
(iii)	Magnitude spectrum-FCN + DETR	0.809	0.776	0.792
(iv)	Phase spectrum-FCN + DETR	0.816	0.784	0.799
(v)	Magnitude + Phase spectrum-FCN + DETR	0.849	0.772	0.811
(vi)	CBAM + DETR	0.766	0.720	0.742
(vii)	Proposed model	0.864	0.821	0.842

and misses text in the drone images. At the same time, the proposed method works well for both drone and scene images. Therefore, we can infer that the existing SOTA of scene text detection methods are not effective for drone images. In the same way, the proposed method is domain independent.

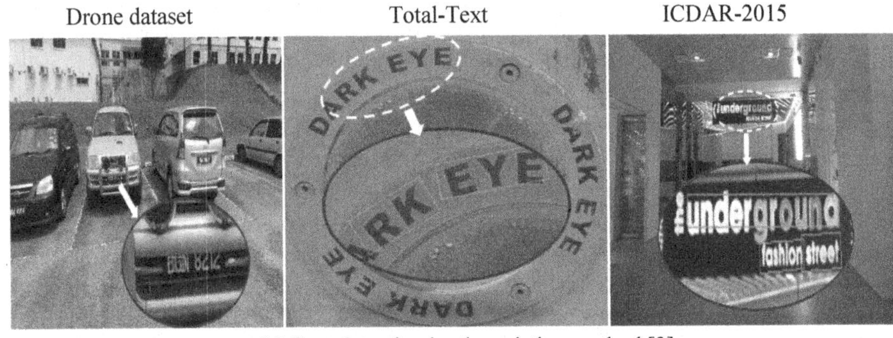

(a) Text detection by the existing method [2]

(b) Text detection by the proposed model

Fig. 8. Illustrating text detection performance of our and existing methods on different datasets.

To show the superiority of the proposed method for text detection in drone and scene images, we compare the performance with the SOTA of scene text detection methods [2,

3, 6–8] and text detection in drone images [1, 18]. The results are reported in Table 2. It is observed from Table 2 that the F-score of the proposed method is higher than the existing methods for all three datasets including the drone dataset. However, the precision is lower than the existing methods [2, 3, 7] for the three datasets. This is because the methods lack an end-to-end approach, which can reduce the effect of false positives for improving the results. At the same time, since the key objective of existing scene text detection methods [2, 3, 6–8] is to obtain the best results for scene images, the existing methods perform well for scene images, and they are promising for the drone dataset in terms of precision. For the tiny text and occluded text, the existing methods do not perform well and hence recall and F-measure are low compared to the proposed method. The same conclusion can be drawn from the results of existing drone text detection methods [1, 18] and the proposed method. Although these two methods were developed for drone images, the detection performance is not high compared to the proposed approach, especially for the drone images dataset. This is because the performance of these two methods depends on the success of preprocessing steps and hence not capable of handling the unpredictable behavior of drone images. On the other hand, overall, when we consider the text detection performance of the proposed model on all three datasets, our model is better than existing methods and is capable of addressing the challenges of different types of images. Therefore, one can infer that the proposed combination of FCN, CBAM, and DETR is effective.

Table 2. Evaluating the proposed and existing methods for text detection on drone and standard scene text datasets

Models	Drone Dataset			Total Text			ICDAR 2015		
	P	R	F	P	R	F	P	R	F
DBNet++ [2]	**0.91**	0.63	0.74	0.87	0.79	0.83	**0.90**	0.77	0.831
DRRG [6]	0.68	0.62	0.65	0.86	0.84	0.85	0.88	**0.84**	0.865
FCENet [3]	**0.91**	0.66	0.76	**0.89**	0.82	0.85	**0.90**	0.82	0.862
PANet [7]	0.70	**0.91**	0.79	**0.89**	0.81	0.85	0.84	0.81	0.829
PSENet [8]	0.73	0.89	0.80	0.84	0.78	0.80	0.86	**0.84**	0.857
Mokayed et al. [1]	0.862	0.832	0.839	0.80	**0.85**	0.82	0.81	**0.84**	0.825
Pal et al. [18]	79.86	77.99	78.91	0.76	0.79	0.77	0.89	0.80	0.84
Proposed method	0.864	0.821	**0.842**	0.87	**0.85**	**0.86**	0.89	**0.84**	**0.868**

Sometimes, when the image is hazy and is of extremely bad quality (very low pixels or highly noisy data) as shown in samples in Fig. 9, the model fails to detect the texts properly. When the images are affected by multiple adverse factors mentioned above, the FCE may miss vital features that represent text in the images. Therefore, there is a scope for improvement in addressing the challenges of text detection in different domains, such as drone and scene images. The results can be improved by exploring the end-to-end model at the feature level, which will be our future work.

Fig. 9. Limitation of the proposed method

5 Conclusion and Future Work

This work proposes a new combination of Fourier spectrum, FCN network, CBAM, and DETR for text detection in drones as well as scene images. Since the work targets two different domains (drone and scene images), each domain poses a different challenge and complexities. To overcome this problem, the proposed method obtains the Fourier spectrum, and the same is fed to FCN to extract features. In addition, to strengthen the features, the proposed work adapts CBAM for feature extraction. We believe these features are invariant to domain. The features and input images are fed to DETR for text detection in the images of both the domains. The results on drone and scene image datasets show that the proposed method outperforms the existing methods in terms of F-measure. However, the proposed method does not perform well for the images affected by multiple adverse factors such as severe degradations and tiny text in the images. This can be solved by exploring an end-to-end transformer, which includes one module for smoothing the images without affecting high-contrast information and one more module for text detection.

Acknowledgement. This work is partly supported by Technology Innovation Hub (TIH), Indian Statistical Institute. Kolkata.

References

1. Mokayed, H., Shivakumara, P., Woon, H.H., Kankanhali, M., Lu, T., Pal, U.: A new DCT-PCM method for license plate number detection in drone images. Pattern Recognit. Lett. **148**, 45–53 (2021)
2. Liao, M., Zou, Z., Wan, Z., Yao, C., Bai, X.: Real-time scene text detection with differentiable binarization and adaptive scale fusion. IEEE Trans. Pattern Anal. Mach. Intell. **45**, 919–931 (2023)
3. Zhu, Y., Chen, J., Liang, L., Kuang, Z., Jin, L., Zhang, W.: Fourier contour embedding for arbitrary-shaped text detection. In: Proceedings of the CVPR, pp. 3122–3130 (2021)
4. Woo, S., Park, J., Lee, J.-Y., Kweon, I.S.: CBAM: convolutional block attention module. In: Ferrari, V., Hebert, M., Sminchisescu, C., Weiss, Y. (eds.) ECCV 2018. LNCS, vol. 11211, pp. 3–19. Springer, Cham (2018). https://doi.org/10.1007/978-3-030-01234-2_1
5. Zeng, C., Song, C.: Swin transformer with feature pyramid networks for scene text detection of the secondary circuit cabinet wiring. In: Proceedings of the ICPICS, pp. 255–258 (2022)

6. Zhang, S., et al.: Deep relational reasoning graph network for arbitrary shape text detection. In: Proceedings of the CVPR, pp. 9699–9708 (2020)
7. Wang, W., et al.: Efficient and accurate arbitrary-shaped text detection with pixel aggregation network. In: Proceedings of the ICCV, pp. 8440–8449 (2019)
8. Wang, W., et al.: Shape robust text detection with progressive scale expansion network. In: Proceedings of the CVPR, pp. 9336–9345 (2019)
9. Zheng, J.: Multi-level alignment for cross-domain scene text detection. In: Proceedings of the ICCECE, pp. 671–675 (2022)
10. Zhao, M., Feng, F., Yin, F., Zhang, X.Y., Liu, C.L.: Mixed-supervised scene text detection with expectation-maximization algorithm. IEEE Trans. Image Process. **31**, 5513–5528 (2022)
11. Banerjee, A., Shivakumara, P., Acharya, P., Pal, U., Canet, J.L.: TWD: a new deep E2E model for text watermark/caption and scene text detection in video. In: Proceedings of the ICPR, pp. 1492–1498 (2022)
12. Bagi, R., Dutta, T., Nigam, N., Verma, D., Gupta, H.P.: Met-MLTS: leveraging smartphones for end-to-end spotting of multilingual oriented scene texts and traffic signs in adverse meteorological conditions. IEEE Trans. Intell. Transp. Syst. **23**, 12801–12810 (2021)
13. Dai, P., Li, Y., Zhang, H., Li, J., Cao, X.: Accurate scene text detection via scale-aware data augmentation and shape similarity constraint. IEEE Trans Multimed. **24**, 1883–1885 (2021)
14. Keserawani, P., Saini, R., Liwicki, M., Roy, P.P.: Robust scene text detection for partially annotated training data. IEEE Trans. Circuits Syst. Video Technol. **32**, 8635–86745 (2022)
15. Deng, J., Luo, X., Zheng, J., Dang, W., Li, W.: Text enhancement network for cross-domain scene text detection. IEEE Signal Process. Lett. **29**, 2203–2207 (2022)
16. Jain, S., Patel, S., Mehat, A., Verma, J.P.: Number plate detection using drone surveillance. In: Proceedings of the UPCON, pp. 1–6 (2022). https://doi.org/10.1109/UPCON56432.2022.9986360
17. Mokayed, H., Meng, L.K., Woon, H.H., Sin, N.H.: Car plate detection engine based on conventional edge detection technique. In: Proceedings of the International Conference on Computer Graphics, Multimedia and Image Processing (CGMIP) (2014)
18. Pal, S., Roy, A., Shivakumara, P., Pal, U.: Adapting a Swin transformer for license plate number/text detection in drone images. In: Artificial Intelligence and Applications (2023). https://doi.org/10.47852/bonviewAIA3202549
19. Carion, N., Massa, F., Synnaeve, G., Usunier, N., Kirillov, A., Zagoruyko, S.: End-to-end object detection with transformers. In: Vedaldi, A., Bischof, H., Brox, T., Frahm, J.-M. (eds.) ECCV 2020. LNCS, vol. 12346, pp. 213–229. Springer, Cham (2020). https://doi.org/10.1007/978-3-030-58452-8_13
20. Karatzas, D., et al.: ICDAR 2015 competition on robust reading. In: Proceedings of the ICDAR, pp. 1156–1160 (2015)
21. Ch'ng, C.K., Chan, C.S., Liu, C.: Total-text: towards orientation robustness in scene text detection. Int. J. Doc. Anal. Recognit. (IJDAR) 31–52 (2020)

Document Binarization with Quaternionic Double Discriminator Generative Adversarial Network

Giorgos Sfikas[1(✉)], George Retsinas[2], and Basilis Gatos[3]

[1] Department of Surveying and Geoinformatics Engineering, School of Engineering, University of West Attica, Athens, Greece
gsfikas@uniwa.gr

[2] School of Electrical and Computer Engineering, National Technical University of Athens, Athens, Greece
gretsinas@central.ntua.gr

[3] Computational Intelligence Laboratory, Institute of Informatics and Telecommunications, National Center for Scientific Research "Demokritos", Athens, Greece
bgat@iit.demokritos.gr

Abstract. Quaternionic networks have emerged as a lightweight alternative to standard neural networks. We propose using a Quaternionic conditional Generalized Adversarial Network adapted to document image binarization. A double discriminator ensures that the output is consistent over a coarse and a finer level of resolution, while the generator is tasked with producing the binarized document. We achieve excellent binarization results, while our network is significantly smaller (4x smaller) than its real-valued counterpart.

Keywords: Binarization · Quaternions · Generative Adversarial Nets

1 Introduction and Related Work

Document image binarization is the task of converting a scanned document image into a binary segmentation. One segment corresponds to the content of the document, while the other segment must cover non-textual components. Binarization is usually desired in order to reduce storage space for large document collections, save communication bandwidth, prepare the document for a subsequent document processing step, or simply enhance the document for better readability. Binarization can be especially challenging when it comes to handling scanned historical manuscripts, as its efficiency is affected by various types of degradation. Poor preservation, bleed-through, faded ink, stains, are some the most common "culprits" that make successful binarization, as well as other processes of document analysis, even more difficult. When it comes to modern documents, another set of problems is related to contemporary picture capturing conditions, e.g. non-uniform lighting due to use of strobe flash [13].

M. Coustaty and A. Fornés (Eds.): ICDAR 2023 Workshops, LNCS 14194, pp. 272–284, 2023.
https://doi.org/10.1007/978-3-031-41501-2_19

Classical binarization techniques were relying on treating the problem as that of choosing a single, global threshold of intensity that would be used to separate background from foreground segments. A threshold is selected automatically according to optimizing an objective that is based on the image intensity statistics [19]. Global thresholding is definitely suboptimal, as its underlying premise is false: No single threshold is applicable for all pixels in an image, and image statistics are in general non-stationary. The immediate extension of global thresholding has been adaptive thresholding, or choosing a different threshold for each different area of patch of the image [8,20,25]. Another group of methods has focused on fitting the image to a more elaborate model of image formation. A Total Variation framework is used in [11], where a data fidelity is traded off a regularization term, which encodes a prior belief that both segments must be spatially consistent, all the while preserving separating edges. Postprocessing is not uncommon, and can be used to rectify the output of an initial processing phase. Such tools may include morphological image processing operations (opening, closing, erosion, dilation, etc.) or non-local filtering [11].

After the starting gun for the comeback of neural networks and deep learning has been fired with Alexnet and other developments around 2014, it has not been long before document image processing techniques were also flooded with deep model solutions. Regarding binarization in particular, neural networks are essentially treated as complex, non-linear filters that are to be learned from data. To this end, fully convolutional networks (FCNs) have been the go-to solution in most cases [2,4,31,33,34]; exceptions include the recurrent architecture proposed in [34], where Grid Long-Short Term Memory units are used. In general however, convolutional architectures provide a very useful inductive bias when processing images. Although more recent competitors such as transformers [23] claim to have rendered convolutions and recurrent units next to unnecessary [32], convolutions especially seem to remain always ubiquitous in cutting-edge vision models in general. In [17], a "morphological" neural network learns a set of morphological operations over an input image; these operations include the weights of a layer that combines dilation and erosion operations, as well as the parameters of the morphological structuring element.

Recently, authors have set forward considerations other than processing accuracy [13]. Model size is another factor that can be very important, especially when it comes to applying vision techniques on embedded systems, or in general machines that are heavily resource-constrained. Methods that involve pruning weights or neurons, or whole blocks of model components is one strategy that may result in a neural network that is as efficent, or almost as efficient, as the non-pruned network [5,15,24] The advantage in this case is that due to the smaller size of total parameters, we have in this sense a smaller network and less storage requirements, as well as potentially speedier inference. Recent work in network pruning methods shows that it is not rare to attain good model compression rates at little loss of efficiency (e.g. [5]).

Hypercomplex architectures constitute a line of work within the research direction that aims for smaller models [21]. This is a type of neural network architectures that is based on challenging the *de facto* choice of using real-valued representations in all aspects of neural networks. In hypercomplex networks, we have inputs, intermediate outputs, weights, biases and activations that are defined on hypercomplex number domains. Hypercomplex numbers in a sense generalize the notion of complex numbers, where each number is defined as a composition of a real and an imaginary part, with an imaginary part that can itself comprise multiple, independent dimensions and multiple, orthogonal imaginary axes. The most well-studied hypercomplex algebra is that of quaternions. Aside from the major application of quaternions involving representations of spatial rotations [10], in computer science quaternions have been used in the domains of signal processing, digital image processing and computer vision [1,6,30]. An application, for which the theoretical prerequisite has been laid down in the 90s [18], is that of quaternion neural networks. It may seem that superficially there is no gain in representing groups of 4 as quaternions on the basis of a trivial bijection between quaternions $\mathbb{H}$ and four-dimensional vectors in $\mathbb{R}^4$; however, due to using quaternion operations – and specifically, quaternion multiplication – the network under the hood uses significantly less parameters than a real-valued model of corresponding size [21].

In this work, we propose a lightweight, quaternion-based architecture over a state-of-the-art backbone that is based on a conditional Generative Adversarial Network geared for the task of document image binarization. To this end, the network includes two key components: namely, i. a double discriminator that is intended to check that the generated binarization over a coarse and a fine resolution. ii. focal loss, which acts to mitigate the detrimentary effect of having imbalanced background and foreground classes. All components are defined in terms of quaternionic operations, including fully-connected layers, convolutions, and activations. We achieve results with little or minimal loss in accuracy over tests in the DIBCO 2017 dataset [22], all the while using a network with a total size that is 4 times smaller than the real-valued model.

The remainder of this paper is structured as follows. In Sect. 2 we briefly review the theoretical requirements concerning quaternions. In Sect. 3 we present the proposed model, and in Sect. 4 we present numerical evaluation results. We close with our concluding remarks in Sect. 5.

2 Quaternion Neural Networks

2.1 Preliminaries

In this section, we shall provide a brief introduction in the theoretical preliminaries regarding quaternions and hypercomplex numbers in general. Sets of hypercomplex numbers are forms of mathematical structures that are comprised of numbers of "higher-dimensionality". Dimensionality in this sense can be understood as the dimension of the linear space with which each corresponding set is isomorphic. For example, the set $\mathbb{H}$ of quaternions is isomorphic to the set of 4-dimensional vectors of $\mathbb{R}^4$ (disregarding for the time being other forms of algebraic structure), as we can trivially define a bijection between the two sets.

This is straightforward from the definition of a quaternion, which is as follows. Any $q \in \mathbb{H}$ can be written as:

$$q = a + b\boldsymbol{i} + c\boldsymbol{j} + d\boldsymbol{k}, \tag{1}$$

where a, b, c, d are real numbers, and $\boldsymbol{i}, \boldsymbol{j}, \boldsymbol{k}$ are so-called imaginary units which correspond to an equal number of imaginary axes. The "a" component corresponds to the real axis. From this definition, we can see that quaternions are a generalization of the set of real numbers $\mathbb{R}$ and the set of complex numbers $\mathbb{C}$. Indeed, for $b = c = d = 0$ we obtain a quaternion that is also a real number; for $c = d = 0$ we obtain a complex number.

With respect to algebraic structure, we can endow $\mathbb{H}$ with summation and product operations so that they form a division algebra or skew-field [7]. This means quaternions $\mathbb{H}$ adhere to all the properties of an algebraic field, such as e.g. the field $\mathbb{R}$, with the exception of being commutative with respect to multiplication. So, in general we have $pq \neq qp$ for $p, q \in \mathbb{H}$. Regarding the definitions of addition and multiplication, the former is simply a sum of corresponding real or imaginary components. In particular,

$$p + q = (a_p + a_q) + (b_p + b_q)\boldsymbol{i} + (c_p + c_q)\boldsymbol{j} + (d_p + d_q)\boldsymbol{k}, \tag{2}$$

where we use $p = a_p + b_p\boldsymbol{i} + c_p\boldsymbol{j} + d_p\boldsymbol{k}$ and $q = a_q + b_q\boldsymbol{i} + c_q\boldsymbol{j} + d_q\boldsymbol{k}$. Multiplication over $\mathbb{H}$ is more complex; we can break down its definition by defining first a multiplication rule between pairs of real and imaginary units. The identity 1 of course leaves any q intact by definition, $1q = q$, and for the imaginary units we have:

$$i^2 = j^2 = k^2 = -1, ij = -ji = k, jk = -kj = i, ki = -ik = j, ijk = -1. \tag{3}$$

From the above relation we immediately can see that the square root of -1 does not correspond only to $\pm\boldsymbol{i}$ as in the set of complex numbers $\mathbb{C}$, but to the further two imaginary units as well; furthermore, the equation $\mu^2 + 1 = 0$ in fact possesses an infinite number of solutions in $\mathbb{H}$. By combining the Eqs. 2 and 3 with the distributive property, we readily obtain the multiplication rule for quaternions p, q:

$$\begin{aligned} pq =&(a_p a_q - b_p b_q - c_p c_q - d_p d_q)+ \\ &(a_p b_q + b_p a_q + c_p d_q - d_p c_q)\boldsymbol{i}+ \\ &(a_p c_q - b_p d_q + c_p a_q + d_p b_q)\boldsymbol{j}+ \\ &(a_p d_q + b_p c_q - c_p b_q + d_p a_q)\boldsymbol{k}. \end{aligned} \tag{4}$$

This rule is also know as a *Hamilton* product [21]. It is especially important regarding the application of quaternions to neural networks, as one of the major differences between "standard" neural network layers and quaternionic layers is that multiplication in all cases is performed according to Eq. 4.

Another way of representing quaternions, is by writing them as a sum of two components. In turn, this can be done in (at least) two ways. One way is to consider the real part to be a single component $S(q)$, and the rest as another component $V(q)$. In this manner, one part corresponds to the real axis, and the other part corresponds to the imaginary axes collectively.

$$q = S(q) + V(q), \tag{5}$$

where $S(q) = a$ and $V(q) = b\boldsymbol{i} + c\boldsymbol{j} + d\boldsymbol{k}$. Another way to represent quaternions is by the Caley-Dickson construction. This amounts to writing $q \in \mathbb{H}$ as a sum of a real and imaginary part, like members of $\mathbb{C}$:

$$q = \alpha + \beta\boldsymbol{k}, \tag{6}$$

where $\boldsymbol{k}$ is the imaginary unit we defined previously, and $\alpha, \beta \in \mathbb{C}$ (generalizing the analogous construction for $\mathbb{C}$, where $\alpha, \beta \in \mathbb{R}$). Assuming $\alpha = \gamma + \delta\boldsymbol{i}$ and $\beta = \epsilon + \zeta\boldsymbol{i}$, it is straightforward again to combine with the imaginary unit multiplication rule of Eq. 3 and the distributive property in $\mathbb{H}$, to obtain our initial definition in Eq. 1.

Furthermore, many properties and notions that are well-known from $\mathbb{R}$ or $\mathbb{C}$ are inherited to, or are generalized gracefully to elements of $\mathbb{H}$. For example, we define a length or magnitude of a quaternion as: $|q| = \sqrt{q\overline{q}} = \sqrt{\overline{q}q} = \sqrt{a^2 + b^2 + c^2 + d^2}$, where $\overline{q}$ is the conjugate of q, defined as $\overline{q} = a - b\boldsymbol{i} - c\boldsymbol{j} - d\boldsymbol{k}$. Quaternions with a zero real part are called pure quaternions, and quaternions with unitary length $|q| = 1$ are called unit quaternions. The Taylor series is a very useful tool that generalizes to $\mathbb{H}$, namely as $e^p = \sum_{n=0}^{\infty} \frac{p^n}{n!}$. Given a quaternion p that is both unit and pure, we also obtain a generalization of Euler's identity, $e^{p\theta} = cos\theta + p \sin\theta$. It is trivial to see that for $p = \boldsymbol{i}$ we get Euler's identity for complex numbers. Quaternions can also be written in polar form: $q = |q|e^{\mu\theta}$, with $\theta \in \mathbb{R}$, and $p \in \mathbb{H}$ again unit and pure. Quaternion p and real angle θ are referred to as the eigenaxis and eigenangle [1]. The eigenaxis and eigenangle can be computed as: $\mu = V(q)/|V(q)|, \theta = \tan^{-1}(|V(q)|/S(q))$. For pure q, hence $S(q) = 0$, we have $\theta = \pi/2$.

Extensions of convolution are also of special importance to applications to quaternionic neural networks. As quaternion multiplication is non-commutative, it is perhaps unsurprising that convolution of signals f and g is also non-commutative, $f * g \neq g * f$. Hence, different variants of convolution can be employed, depending on whether the kernel multiplies the signal from the left or the right; in $2D$ especially, a bi-convolution operation can also be defined, which has one part of the kernel multiplying the input from the left, and the other part from the right. For all intents and purposes within the scope of the current application, we treat these variants as equivalent; we shall use a left-side convolution by convention:

$$(w * f)(k, l) = \int_{\Omega_y} \int_{\Omega_x} w(x, y) f(k - x, l - y) dx dy \tag{7}$$

As a sidenote, a set of other interesting properties of quaternions emerge once we consider matrices with quaternionic values. Properties that are well-known from complex matrix algebra may generalize naturally to quaternions, while others may diverge immensely from our experience on $\mathbb{R}$ or even $\mathbb{C}$. For example, square quaternionic matrices do have sets of eigenvalues and eigenvectors, in the sense of vectors that solve the equation $Ax = \lambda x$, for $A \in \mathbb{H}^{N \times N}$, $x \in \mathbb{H}^N$, $\lambda \in \mathbb{H}$. However, since in general $\lambda x \neq x\lambda$ due to non-commutativity of multiplication in $\mathbb{H}$, the problems $Ax = \lambda x$ are $Ax = x\lambda$ are different. The two sets of eigenvalues are distinct, and are referred to as *left eigenvalues* and *right eigenvalues* respectively. Very little is known regarding connections between the two sets [16,36].

2.2 Quaternion Layers

Conversion of "standard" linear transformations (in the sense of assuming real values on inputs, outputs and transformation parameters) is the "workhorse" behind converting real-valued networks to quaternionic ones. Recall that in general a feed-forward neural network can be written as a composition of ℓ layers, which are in turn defined as compositions of a linear and a non-linear part. Formally we write:

$$f[x, \theta] = f^{\ell} \circ f^{\ell-1} \circ \cdots \circ f^{1}[x, \theta], \tag{8}$$

where $f[x, \theta]$ represents the NN, which takes an input x conditioned on parameters θ and outputs a vector y. All components are made up of quaternionic variates, $x \in \mathbb{H}^{d_x}$, $\theta \in \mathbb{H}^{d_\theta}$, $y \in \mathbb{H}^{d_y}$, where d_x, d_θ, d_y represent input, parameter and output dimensionalities. Note that in practical terms, assuming that we have a network consisting of d_θ quaternionic parameters, they will take up as much space as $4d_\theta$, since each quaternion is intrinsically 4-dimensional. Layers are represented as $f^1, \cdots, f^\ell$ in the above formulation. In general, each layer consists of a linearity or linear component $f_L(\dot{})$ and a non-linearity or activation $f_{NL}(\dot{})$.

Quaternion Linearities. The linear component is written as:

$$g = Wf + b, \tag{9}$$

where $g \in \mathbb{H}^M$, $f \in \mathbb{H}^N$ represent layer outputs and inputs, and W, b are the weights and biases of the layer, which are of course part of the full set of parameters θ. W is a quaternionic matrix $\mathbb{H}^{M \times N}$ and b is a quaternionic vector $\mathbb{H}^M$. We can rewrite this relation as:

$$g^i = \sum_{j=1}^{N} w^{ij} f^j + b^i, \tag{10}$$

where additions and multiplications follow the rules for quaternions (cf. Sect. 2). Multiplication between w^{ij} and f^j is effectively a Hamilton product (Eq. 4). We then use the fact that we can rewrite Eq. 4 as a matrix-vector product, where one of the quaternions (here, the weight component w_{ij}) is rewritten as a 4×4 (real) matrix:

$$\begin{bmatrix} g_a \\ g_b \\ g_c \\ g_d \end{bmatrix} = \begin{bmatrix} w_a & -w_b & -w_c & -w_d \\ w_b & w_a & -w_d & w_c \\ w_c & w_d & w_a & -w_b \\ w_d & -w_c & w_b & w_a \end{bmatrix} \begin{bmatrix} f_a \\ f_b \\ f_c \\ f_d \end{bmatrix}, \tag{11}$$

The key observation now is that we can combine Eqs. 9 and 11 and use the trivial bijection between $\mathbb{R}^4$ and $\mathbb{H}$, i.e. $(a, b, c, d)^T \rightarrow (a + b\boldsymbol{i} + c\boldsymbol{j} + d\boldsymbol{k})$, in order to write Eq. 9 in a block-matrix form as follows:

$$\begin{bmatrix} \boldsymbol{g}^1 \\ \boldsymbol{g}^2 \\ \cdots \\ \boldsymbol{g}^M \end{bmatrix} = \begin{bmatrix} \boldsymbol{w}^{11} & \ldots & \boldsymbol{w}^{1N} \\ \vdots & \ddots & \vdots \\ \boldsymbol{w}^{M1} & \ldots & \boldsymbol{w}^{MN} \end{bmatrix} \begin{bmatrix} \boldsymbol{f}^1 \\ \boldsymbol{f}^2 \\ \cdots \\ \boldsymbol{g}^N \end{bmatrix} + \begin{bmatrix} \boldsymbol{b}^1 \\ \boldsymbol{b}^2 \\ \cdots \\ \boldsymbol{b}^N \end{bmatrix}. \tag{12}$$

In the above equation each boldface element represents a 4×1 vector (on the vectors) or a 4×4 submatrix (on the matrix). Hence, we are dealing with dimensions equal to $4d_g$, $4d_f$ for the input and output vectors. However, due to the bijective relation between the two equation forms Eq. 11 and 12, *the multiplying matrix only has* $4d_g d_f$ *independent parameters.* A real matrix construction would have $4 * 4d_g d_f$ independent parameters, i.e. equal to the number of all matrix elements. Thus, we have four-fold saving in number of parameters (similar discussions concerning why quaternion layers lead to extensive parameter savings can be found in e.g. [21] or [28, Section 3])

Convolutions can be interpreted as a constrained version of the fully connected layer, where extensive parameter sharing is employed in the form of the convolution kernel. It is well-known that convolutions, as linear operations, can be written in a matrix-vector form as Töplitz matrices [9]. Hence, the entirety of the aformentioned analysis also applies in their case. Similar considerations hold for transpose convolutions or deconvolutions.

Quaternion non-linearities. Regarding activation functions, we formally require a mapping from $\mathbb{H}$ to $\mathbb{H}$. In practice, so-called split activation functions are usually employed, where simply real-valued activations are used over each of the quaternionic (real/imaginary) componenets separately.

3 Proposed Model

The proposed architecture is based on a Quaternionic conditional Generative Adversarial Network, comprised of a total of three composing networks: A generator network, tasked with producing a binarization given the original input image; a global discriminator network, tasked with discerning between produced binarization that are unlikely to be artificial and those that are; a local discriminator network, which acts similarly as the global discriminator but on the level of small-sized patches (32×32 pixels). In this manner, the binarization estimate can be evaluated by the networks in two resolution scales: a coarse one

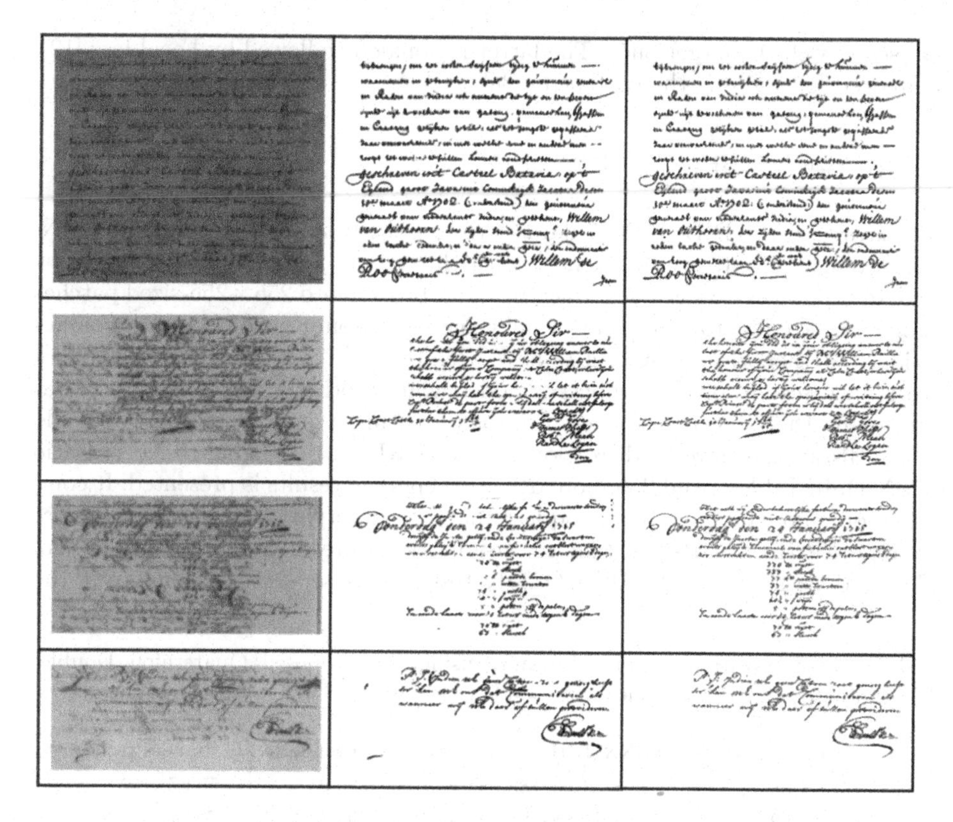

Fig. 1. Visual results using the proposed Quaternionic Double-Discriminator GAN model. Lines correspond to selected images from the DIBCO 2017 test set. Columns correspond to: original image, binarization result, ground truth.

that corresponds to the global discriminator, and a finer one that corresponds to the local discriminator. The generator is structured as a U-Net, with skip connections between corresponding layers of the same size. Global and local discriminators are defined as in [4], with the exception of adding 1×1 convolution layers when it is necessary to change input or output channel depth to a multiple of 4. Training loss is defined as a weighted average of Binary cross-entropy losses for the three composing networks. Formally we have:

$$L = \mu(L_{global} + \sigma L_{local}) + \lambda L_{gen}. \tag{13}$$

We set loss aggregation hyperparameters to $\mu = 0.5, \sigma = 5, \lambda = 75$, as suggested in [4]. Concerning imbalances in the number of background and foreground elements, one major strategy is using resampling in order to prioritize classes that are initially under-represented, by sampling more augmented samples from low-volume classes. The other major strategy is to tweak the loss function, so that under-represented classes are artifically assigned a larger loss. In this manner, training can be implicity manipulated towards an optimum that classifies small classes as well the larger ones. The latter rationale is followed by Focal loss [12], employed for training the current model.

4 Experiments

We have used the DIBCO 2017 dataset to test the proposed quaternionic GAN. All grayscale inputs are augmented with three extra zero channels so as to be able to be cast as quaternions. Each image is broken into 256×256-sized patches with a stride equal to 128. As in [4], all the images from previous DIBCO competitions together with DIBCO 2018 have been used to train the model. In Table 3 we review results of the proposed model against other state-of-the-art methods, using the following metrics: F-measure (F-m), pseudo F-measure (pseudo F-m), distance reciprocal distortion metric (DRD), peak Signal-to-Noise Ratio (PSNR). In Tables 1 and 2, a more detailed report of results is presented. It compares our model versus its non-quaternionic counterpart [4], and furthermore we report extra metrics – Recall, Precision, PseudoRecall and PseudoPrecision. This is done for all of the DIBCO 2017 test images separately. Comparing the

Table 1. Binarization numerical results using the proposed Quaternion Double-Discriminator GAN. Each line corresponds to a different DIBCO 2017 test image.

	F-m	Pseudo F-m	PSNR	DRD	Recall	Precision	PseudoRecall	PseudoPrecision
1	71.85	68.02	12.08	10.52	83.82	62.87	85.05	56.67
2	84.81	82.54	15.60	6.81	94.22	77.59	96.77	74.21
3	75.56	71.83	13.46	9.77	89.28	65.5	91.09	59.3
4	71.88	68.32	14.11	0.87	83.28	63.23	84.89	57.16
5	74.84	70.68	16.08	11.54	95.94	61.35	96.78	55.67
6	94.4	95.33	15.96	2.22	96.1	92.76	99.7	91.33
7	94.12	94.4	15.99	2.35	96.83	91.55	99.56	89.74
8	92.79	95.1	19.43	2.64	92.97	92.61	99.09	91.42
9	86.55	84.1	14.95	4.52	96.35	78.57	96.13	74.74
10	88.6	86.19	14.38	4.03	99.16	80.08	99.62	75.95
11	92.84	91.72	16.79	3.73	99.57	86.96	99.76	84.87
12	84.59	83.93	14.86	7.69	96.83	75.09	99.1	72.79
13	66.2	64.55	11.5	25.38	97.65	50.07	99.65	47.74
14	87.51	87.29	16.78	5.75	96.98	79.73	99.26	77.9
15	93.31	93.42	16.77	2.72	97.04	89.86	98.99	88.45
16	87.49	84.86	18.09	5.23	99.44	78.11	99.77	73.83
17	77.4	72.64	15.37	7.68	95.3	65.16	95.72	58.53
18	84.4	87.75	14.95	7.05	88.18	80.93	98.21	79.3
19	91.67	92.57	18.98	3.1	93.98.	89.46	99.55	86.51
20	91.42	93.92	16.95	3.29	92.82	90.06	99.73	88.75
Avg	84.61	83.45	15.60	6.81	94.22	77.59	96.77	74.21

Table 2. Binarization numerical results using a non-quaternionic Double-Discriminator GAN (baseline model, cf. [4]). Each line corresponds to a different DIBCO 2017 test image. Compared to the proposed model, similar results are achieved, albeit at a much heavier computational burden.

	F-m	Pseudo F-m	PSNR	DRD	Recall	Precision	PseudoRecall	PseudoPrecision
1	72.25	68.25	12.12	10.36	84.68	63	85.77	56.68
2	85.79	83.29	14.86	5.92	96.46	77.24	97.26	72.83
3	76.28	72.16	13.45	9.8	93.2	64.56	94.89	58.22
4	72.25	68.34	13.94	11.44	87.96	61.3	89.57	55.24
5	73.55	69.74	15.82	12.56	95.27	59.89	96.28	54.67
6	93.98	94.75	15.62	2.52	96.29	91.78	99.7	90.26
7	93.79	94.03	15.74	2.63	96.95	90.83	99.68	9.05
8	93.23	95.05	19.68	2.51	94.06	92.42	99.19	91.24
9	86.43	83.96	14.86	4.65	97.17	77.82	97.31	73.83
10	88.6	86.23	14.38	4.04	99.08	80.12	99.6	76.03
11	92.9	91.85	16.84	3.68	99.49	87.13	99.75	85.11
12	85.13	84.48	15.04	7.3	97.01	75.84	99.33	73.5
13	66.6	64.94	11.59	24.7	997.48	50.58	99.63	48.17
14	88.7	88.5	17.26	5.07	97.18	81.57	99.44	79.7
15	93.2	693.3	16.73	2.7	97.27	89.57	99.21	88.05
16	88.8	86.39	18.64	4.42	99.27	80.33	99.72	76.2
17	75.76	71.46	15.28	7.86	88.87	66.03	89.49	59.48
18	85.37	88.75	15.25	6.49	88.79	82.2	98.91	80.48
19	92.68	93.92	19.59	2.5	93.87	91.53	99.47	88.95
20	92.65	95.17	17.67	2.6	92.92	92.37	99.71	91.03
Avg	84.9	83.72	15.71	6.69	94.66	77.80	97.19	74.43

Table 3. Comparison of state-of-the-art document binarization methods, using DIBCO 2017 test set accuracy as a benchmark. Lines denoted as "Comp #x" refer to winners of the corresponding competition [22]. DD-GAN-x refers to a baseline, real-valued double discriminator model. The results under DD-GAN-1 are the figures reported in the original publication, [4], while DD-GAN-2 corresponds to the figures we computed after running the authors' implementation [3], without any parameter finetuning. Quaternion DD-GAN refers to the proposed method, which achieves good results while being very economical in terms of network size (4× smaller compared to DD-GAN).

	F-measure	Pseudo F-measure	PSNR	DRD
Comp #1 (U-Net)	91.04	92.86	18.28	3.4
Comp #1 (FCN-VGG)	89.67	91.03	17.58	4.35
Comp #3 (Deep SN)	89.42	91.52	17.61	3.56
Otsu	77.73	77.89	13.85	15.54
Sauvola	77.1	77.89	14.25	8.85
DD-GAN-1	90.98	92.85	17.6	3.34
DD-GAN-2	84.9	83.72	15.71	6.69
Quaternion DD-GAN	84.61	83.45	15.6	6.81

two tables, we can conclude that in may cases we can observe a slight loss in performance, though for the most part losses are insignificant. On the other hand, the proposed model is 4× smaller (cf. Subsect. 2.2) than the real-valued state-of-the-art GAN of [4].

5 Conclusion

We have presented a model for document image binarization that encompasses two key components: i. a Generative Adversarial architecture that is comprised of two discriminators, aimed to capture data interdependencies on a coarse as well as a finer scale of the input document image; ii. use of quaternionic layers, that replace real-valued fully connected, convolution and deconvolution layers. The end-result is a model that attains state-of-the-art performance in a multitude of binarization metrics, all the while being several times (4×) more compact than its real-valued counterpart (Fig. 1). For future work, we envisage exploring uses of more recent developments in hypercomplex architectures for binarization [35], or ways to fusion quaternion networks with other methodological paradigms, like probabilistic approaches on inference [26,27]. Also, more extensive tests are to be conducted, including other DIBCO datasets, or more recently published datasets [14] and other binarization methods [29].

Acknowledgments. This research has been partially co - financed by the EU and Greek national funds through the Operational Program Competitiveness, Entrepreneurship and Innovation, under the calls : "RESEARCH - CREATE - INNOVATE", project *Culdile* (code T1EΔK - 03785) and "OPEN INNOVATION IN CULTURE", project *Bessarion* (T6YBΠ - 00214).

References

1. Alexiadis, D.S., Daras, P.: Quaternionic signal processing techniques for automatic evaluation of dance performances from mocap data. IEEE Trans. Multimedia **16**(5), 1391–1406 (2014)
2. Ayyalasomayajula, K.R., Malmberg, F., Brun, A.: PDNet: semantic segmentation integrated with a primal-dual network for document binarization. Pattern Recogn. Lett. **121**, 52–60 (2019)
3. Chakraborty, A.: Implementation of binarization with dual discriminator GAN (2023). https://github.com/anuran-Chakraborty/BinarizationDualDiscriminatorGAN. Accessed Jan 2023
4. De, R., Chakraborty, A., Sarkar, R.: Document image binarization using dual discriminator generative adversarial networks. IEEE Signal Process. Lett. **27**, 1090–1094 (2020)
5. Dimitrakopoulos, P., Sfikas, G., Nikou, C.: Variational feature pyramid networks. In: International Conference on Machine Learning, pp. 5142–5152. PMLR (2022)
6. Ell, T.A., Sangwine, S.J.: Hypercomplex Fourier transforms of color images. IEEE Trans. Image Process. **16**(1), 22–35 (2007)
7. Fraleigh, J.B.: A First Course in Abstract Algebra, 7th (2002)

8. Gatos, B., Pratikakis, I., Perantonis, S.J.: Adaptive degraded document image binarization. Pattern Recogn. **39**(3), 317–327 (2006)
9. Jain, A.K.: Fundamentals of Digital Image Processing. Prentice-Hall Inc., Upper Saddle River (1989)
10. Kuipers, J.B.: Quaternions and Rotation Sequences: A Primer with Application to Orbits, Aerospace and Virtual Reality. Princeton University Press, Princeton (1999)
11. Likforman-Sulem, L., Darbon, J., Smith, E.H.B.: Enhancement of historical printed document images by combining total variation regularization and non-local means filtering. Image Vis. Comput. **29**(5), 351–363 (2011)
12. Lin, T.Y., Goyal, P., Girshick, R., He, K., Dollár, P.: Focal loss for dense object detection. In: Proceedings of the IEEE International Conference on Computer Vision, pp. 2980–2988 (2017)
13. Lins, R.D., Bernardino, R.B., Barboza, R., Oliveira, R.: The winner takes it all: choosing the "best" binarization algorithm for photographed documents. In: Uchida, S., Barney, E., Eglin, V. (eds.) Document Analysis Systems. DAS 2022. Lecture Notes in Computer Science, vol. 13237, pp. 48–64. Springer, Cham (2022). https://doi.org/10.1007/978-3-031-06555-2_4
14. Lins, R.D., Bernardino, R.B., Smith, E.B., Kavallieratou, E.: ICDAR 2021 competition on time-quality document image binarization. In: Lladós, J., Lopresti, D., Uchida, S. (eds.) ICDAR 2021. LNCS, vol. 12824, pp. 708–722. Springer, Cham (2021). https://doi.org/10.1007/978-3-030-86337-1_47
15. Louizos, C., Welling, M., Kingma, D.P.: Learning sparse neural networks through l_0 regularization. arXiv preprint: arXiv:1712.01312 (2017)
16. Macías-Virgós, E., Pereira-Sáez, M., Tarrío-Tobar, A.D.: Rayleigh quotient and left eigenvalues of quaternionic matrices. Linear Multilinear Algebra, 1–17 (2022)
17. Mondal, R., Chakraborty, D., Chanda, B.: Learning 2D morphological network for old document image binarization. In: 2019 International Conference on Document Analysis and Recognition (ICDAR), pp. 65–70. IEEE (2019)
18. Nitta, T.: A quaternary version of the backpropagation algorithm. In: Proceedings of ICNN'95 - International Conference on Neural Networks, pp. 2753–2756 (1995)
19. Otsu, N.: A threshold selection method from gray-level histograms. IEEE Trans. Syst. Man Cybern. **9**(1), 62–66 (1979)
20. Papamarkos, N., Gatos, B.: A new approach for multilevel threshold selection. CVGIP: Graph. Models Image Process. **56**(5), 357–370 (1994)
21. Parcollet, T., Morchid, M., Linarès, G.: A survey of quaternion neural networks. Artif. Intell. Rev. **53**(4), 2957–2982 (2020)
22. Pratikakis, I., Zagoris, K., Barlas, G., Gatos, B.: ICDAR2017 competition on document image binarization (DIBCO 2017). In: 2017 14th IAPR International Conference on Document Analysis and Recognition (ICDAR), vol. 1, pp. 1395–1403. IEEE (2017)
23. Prince, S.J.: Understanding Deep Learning. MIT Press, Cambridge (2023). https://udlbook.github.io/udlbook/
24. Retsinas, G., Elafrou, A., Goumas, G., Maragos, P.: Online weight pruning via adaptive sparsity loss. In: 2021 IEEE International Conference on Image Processing (ICIP), pp. 3517–3521. IEEE (2021)
25. Sauvola, J., Pietikäinen, M.: Adaptive document image binarization. Pattern Recogn. **33**(2), 225–236 (2000)
26. Sfikas, G., Nikou, C., Galatsanos, N., Heinrich, C.: MR brain tissue classification using an edge-preserving spatially variant Bayesian mixture model. In: Metaxas,

D., Axel, L., Fichtinger, G., Székely, G. (eds.) MICCAI 2008. LNCS, vol. 5241, pp. 43–50. Springer, Heidelberg (2008). https://doi.org/10.1007/978-3-540-85988-8_6
27. Sfikas, G., Nikou, C., Galatsanos, N., Heinrich, C.: Majorization-minimization mixture model determination in image segmentation. In: CVPR 2011, pp. 2169–2176. IEEE (2011)
28. Sfikas, G., Retsinas, G., Giotis, A.P., Gatos, B., Nikou, C.: Keyword spotting with quaternionic ResNet: application to spotting in Greek manuscripts. In: Uchida, S., Barney, E., Eglin, V. (eds.) Document Analysis Systems. DAS 2022. Lecture Notes in Computer Science, vol. 13237, pp. 382–396. Springer, Cham (2022). https://doi.org/10.1007/978-3-031-06555-2_26
29. Souibgui, M.A., Biswas, S., Jemni, S.K., Kessentini, Y., Fornés, A., Lladós, J., Pal, U.: DocEnTr: an end-to-end document image enhancement Transformer. In: 2022 26th International Conference on Pattern Recognition (ICPR), pp. 1699–1705. IEEE (2022)
30. Subakan, Ö.N., Vemuri, B.C.: A quaternion framework for color image smoothing and segmentation. Int. J. Comput. Vision **91**(3), 233–250 (2011)
31. Tensmeyer, C., Martinez, T.: Document image binarization with fully convolutional neural networks. In: 2017 14th IAPR International Conference on Document Analysis and Recognition (ICDAR), vol. 1, pp. 99–104. IEEE (2017)
32. Vaswani, A., et al.: Attention is all you need. In: Advances in Neural Information Processing Systems, vol. 30 (2017)
33. Vo, Q.N., Kim, S.H., Yang, H.J., Lee, G.: Binarization of degraded document images based on hierarchical deep supervised network. Pattern Recogn. **74**, 568–586 (2018)
34. Westphal, F., Lavesson, N., Grahn, H.: Document image binarization using recurrent neural networks. In: 2018 13th IAPR International Workshop on Document Analysis Systems (DAS), pp. 263–268. IEEE (2018)
35. Zhang, A., et al.: Beyond fully-connected layers with quaternions: parameterization of hypercomplex multiplications with $1/n$ parameters. In: International Conference on Learning Representations (ICLR 2021) (2021). arXiv:2102.08597
36. Zhang, F.: Quaternions and matrices of quaternions. Linear Algebra Appl. **251**, 21–57 (1997)

Crosslingual Handwritten Text Generation Using GANs

Chun Chieh Chang[1,2](✉), Leibny Paola Garcia Perera[1,2], and Sanjeev Khudanpur[1,2]

[1] Center for Language and Speech Processing, Johns Hopkins University, Baltimore, USA
{cchunch1,khudanpur}@jhu.edu

[2] Human Language Technology Center of Excellence, Johns Hopkins University, Baltimore, USA

Abstract. Generative Adversarial Networks, also called GANs, have been able to produce realistic looking handwritten images. However, training a GAN to generate usable data for improving handwriting recognition is a chicken and egg problem. In a low-resource language scenario, it would be beneficial to have a method that can generate more labeled data. But training such a GAN requires an amount of data that would not be available in a low-resource setting.

In this paper, we present our work in data augmentation with a GAN that is independent of language and can be used to generate handwritten images by learning a mapping between printed and handwritten text. Our method is able to leverage training data from a source language and generate handwriting in a different target language. We show that in scenarios with adequate amounts of target language data, similar improvements in WER can be made by augmenting with either synthetic handwritten or printed text. However, in low resource scenarios, our GAN generated handwriting improves recognition results by 5–10% absolute over the baseline and 3–5% absolute over adding rendered printed text.

Keywords: Generative Adversarial Networks · Crosslingual Handwriting Generation · Data Augmentation

1 Introduction

Optical Character Recognition (OCR) has achieved impressive results in recognizing lines of printed text and the field has largely moved towards the localization and recognition of 'in scene' text [23,29]. However, recognition rates for Handwritten Text Recognition (HTR) systems are not as good as those for OCR. Recognizing handwriting is a more challenging task, but the lack of large amounts of training data also plays a part as well. Recognition rates for OCR and HTR models improve with increasing amounts of data and it is easy to synthetically render printed images to augment training data for an OCR model [6,33]. Unfortunately, it is less clear how to synthetically generate handwritten images for the purposes of training an HTR model.

M. Coustaty and A. Fornés (Eds.): ICDAR 2023 Workshops, LNCS 14194, pp. 285–301, 2023.
https://doi.org/10.1007/978-3-031-41501-2_20

Generative Adversarial Networks (GANs) are able to generate handwritten images that, when used to augment training data, can improve recognition rates of HTR systems [1,7]. However, these methods often still require large amounts of labeled data for the training of the handwritten text generator. Where in order to create more labeled handwritten data, some amount of data must first be obtained. For low resource languages, there may not be enough labeled images to properly train a GAN that can generate handwritten text images.

The following paper will address the problems faced with data augmentation for low resource languages. More specifically, the training of a GAN to generate handwritten text in a low resource language. We will focus on the zero-shot scenario where there is no labeled text in the target, low resource language. Our model is a crosslingual GAN where the source language used to train the GAN is different from the target language used to train the HTR system. Comparisons will be made to the naive scenario where data augmentation is done by rendering printed text via a font.

2 Previous Work

Generative Adversarial Networks have provided exciting ways to artificially generate realistic looking data [8]. A paradigm was introduced in which data can be generated through a minimax competition between a generator G and discriminator D.

$$\min_{G} \max_{D} V(D, G) = \mathbb{E}_{x\ p_{\text{data}}(x)}[\log D(x)] + \mathbb{E}_{z\ p_z(z)}[\log(1 - D(G(z)))], \quad (1)$$

where x is a sample from the distribution of real data p_{data} and x is a sample from noise. This competition between the generator and the discriminator leads to the training of a generator that can generate data with a distribution similar to that of the real data. The next sections will cover advancements in GANs to generate handwritten text and fonts.

2.1 Text Generation

There has been some work done with directly generating handwritten images from text [1,7]. These techniques require converting the text into text embeddings before feeding them into the standard GAN paradigm. In addition to providing the Adversarial loss from the Discriminator, feedback in the form of text recognition loss from a Recognizer is propagated up to the Generator as well. Pooling the Discriminator feedback on individual patches allows for a variable length input and output in the ScrabbleGAN paper [7]. Another paper introduces a model to perform image-to-image translation between printed and handwritten text [21]. In addition, this paper adds an author style bank to create additional variability in handwriting style.

Our work is largely inspired by these text generation methods. However, most of these methods have a limitation in that it is unclear how to generate characters that do not appear in the training set. In that regard, image-to-image translation methods may be better suited to generate handwritten images in a crosslingual manner. Furthermore, none of these papers address a crosslingual scenario where a GAN may be trained on a source language that is different than the target language. An image-to-image approach has the potential to generalize the differences between printed and handwritten text in order to generate handwritten text in a variety of different languages.

2.2 Image to Image Translation

Some of the first techniques for image-to-image translation between two domains were papers such as pix2pix [14] and CycleGAN [34]. Later innovations involved applying image-to-image translation techniques to printed text images [15,20,28,32] for the purposes of font generation. Work has been done with font generation in a crosslingual manner [19,25] as well. However, many of these methods are for individual characters in character based languages such as Chinese or Korean. These papers focus primarily on font generation and not handwriting generation.

There has been work done with image-to-image translation for text lengths beyond single characters. These works are meant for data augmentation in their respective text domains. CycleGAN has been shown to be useful in converting machine print Mongolian to woodblock print Mongolian [31]. Other applications involve removing artifacts from handwritten text such as noise or strike-through [11,27].

There has been prior work applying image-to-image translation techniques in a multilingual setting [13]. In this paper, the authors train a GAN that does image-to-image translation between printed and handwritten images. However, their model is pre-trained on the source language and further trained on data in the target domain. We believe that further work can be done by generating handwritten text in a crosslingual fashion, specifically without any labeled text from the target domain.

2.3 Contributions

In this paper, we present our work with the generation of a target language using only labeled data from a different source language. Our model learns an image-to-image translation between printed images and handwritten images using the standard GAN paradigm [8]. The generator takes printed images, created through rendering corpus text, as input and outputs realistic looking handwritten images. The trained generator is independent of language and learns an image-to-image mapping between the printed and handwritten domains. The generated images are of sufficient quality to improve text recognition results when used for data augmentation. The advantage of training a crosslingual GAN is that we are able to leverage the data from a 'nearby' script. Our method is 'zero-shot' and is able to generate handwritten images of a target language without seeing any labeled handwritten data of that language. To the best of our knowledge, this has not been studied before as previous literature have focused on the 'supervised' scenario, where labeled handwritten data of the target language has been provided to the model.

The results of two sets of experiments will be provided. The first will demonstrate the ability of a generator trained on English images to generate handwritten images of a variety of other languages and scripts. An ablation study is also performed. We will demonstrate how the performance of the crosslingual GAN changes as the script of the source language gets 'further' away in visual appearance from the target language. A GAN is trained on either Vietnamese, English, or Arabic and then used to generate handwritten Vietnamese.

Since the method used is an image-to-image translation, the handwriting of the output images is controllable by applying standard image augmentation techniques on the input printed images.

3 Method

Our GAN model is comprised of a generator, discriminator, and recognizer; similar to previous text generation architectures [1,7]. Adversarial loss from the discriminator and CTC [9] loss from the recognizer are combined with a parameter λ to provide feedback for the generator.

$$L_{total} = L_{adversarial} + \lambda * L_{CTC} \tag{2}$$

Our model is different from prior work because we use printed images as input instead of text embeddings. Our system renders printed images from text and uses them as input to the generator. This model is shown in Fig. 1. After such a GAN is trained, the generator can be used to generate handwritten images of any low-resource language. Text of a target language can be used to render printed images. These printed images are fed into the generator to generate handwritten images. The output handwritten images are then used for data augmentation when training a text recognition system. Examples of the target language can be optionally added during training of the GAN in a semisupervised manner. An unpaired dataset can be created where random corpus text of the target language is rendered as a printed image to be used alongside unlabeled handwritten images of the target language. Adding training images in a semisupervised manner is also described in the ScrabbleGAN paper [7].

3.1 GAN Model

In order to generate handwritten images in a crosslingual manner, our generator needs to learn a mapping between the printed and handwritten image domains which is independent of language. An analogous scenario might be how a human is able to produce a handwritten version of a given printed document in a foreign language without necessarily understanding the words themselves. Since the

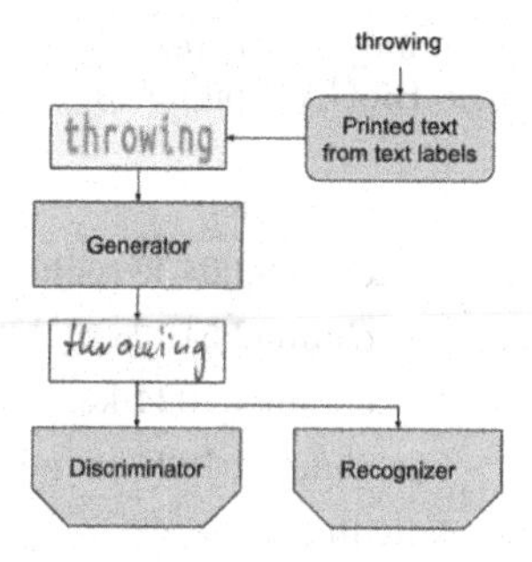

Fig. 1. Generation of handwritten text with discriminator and recognition loss. Text labels are first fed into a text renderer in order to generate printed images as input.

input to the GAN is printed images, there is no need for text embeddings as is done in [1] or filterbanks as is done in [7]. Control over the appearance of the handwriting output can be done by varying the style of the input printed image instead of a noise vector. Without needing text embeddings, our model is an image-to-image translation between the printed and handwritten text domain. The generator consists of 'GBlocks' described by the BigGAN paper [3]. The full model parameters are shown in Table 1.

Table 1. Generator used for handwritten text generation. GBlocks are blocks of convolution layers described in the BigGAN paper [3]

Type	Configuration
SNConv2d	filter_size:1024 kernel:3×3
GBlock1	out_channel:1024 kernel:3×3
GBlock2	out_channel:512 kernel:3×3
GBlock3	out_channel:512 kernel:3×3
GBlock4	out_channel:256 kernel:3×3
Attention	out_channel:256
GBlock5	out_channel:128 kernel:3×3
GBlock6	out_channel:64 kernel:3×3
GBlock7	out_channel:64 kernel:3×3
SNConv2d	filter_size:1 kernel:3×3

The discriminator and recognizer are the same as in ScrabbleGAN [7] but will be described here for completeness. The discriminator consists of 'DBlocks' from the BigGAN paper [3]. The full architecture can be seen in Table 2. The recognizer is comprised of 7 CNN layers with max-pooling layers in between; a final linear layer maps to the character dictionary.

Table 2. Discriminator used for handwritten text generation. DBlocks are blocks of convolution layers described in the BigGAN paper [3]

Type	Configuration
DBlock1	out_channel:64 kernel:3 × 3
DBlock2	out_channel:512 kernel:3 × 3
DBlock3	out_channel:1024 kernel:3 × 3
DBlock4	out_channel:1024 kernel:3 × 3
SNLinear	out_dim:1

3.2 Handwritten Text Recognition (HTR) Model

The goal of our method is to generate handwritten text in a low-resource setting usable for data augmentation. To this end, the HTR model used is a standard Encoder-Decoder model in the Espresso ASR toolkit [30]. The encoder consists of 4 layers of CNN with kernel size 3×3, and subsampling on layers 2 and 4. After the CNN layers, there are 3 Bi-LSTM layers [10]. The decoder is 3 Bi-LSTM layers with attention [2]. LSTM-based language model using characters trained on the training text was applied.

3.3 Datasets

Several handwriting datasets of various different languages are needed to evaluate crosslingual handwritten text generation. The IAM handwritten English [22] dataset was used as the primary source language. The VNonDB handwritten Vietnamese [24], Madcat handwritten Arabic [16–18], and Kaggle handwritten Cyrillic [5] datasets were used as the target languages. These different datasets were chosen to get a broad range of different scripts. For example, English is a Latin alphabet-based script written from left to right. Whereas, Arabic is cursive and written right to left. This broad range of scripts was chosen to test the robustness of our crosslingual GAN. Table 3 shows some comparisons between the different scripts.

Table 3. General Comparison Between Scripts

	Latin-based	Direction
English	Yes	left-to-right
Vietnamese	Yes	left-to-right
Cyrillic	No	left-to-right
Arabic	No	right-to-left

IAM English Dataset. The IAM dataset [22] is an offline handwritten English dataset that is largely black text on a white background. There are 9 k lines total in the dataset with 6 k lines in the train set. Word level segmentation of this dataset was used as the source language for training our GAN. The example image is shown as Fig. 2.

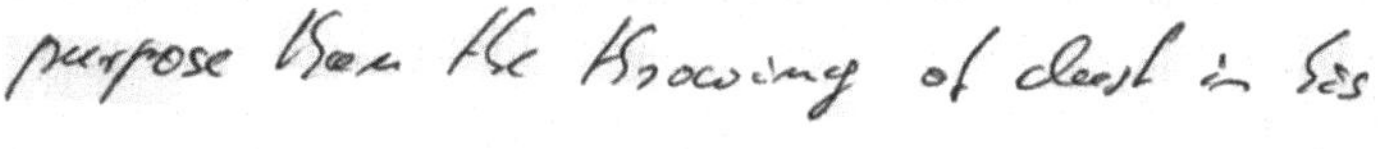

Fig. 2. Example image for IAM

VNonDB Vietnamese Dataset. The VNonDB dataset [24] is an online handwritten Vietnamese dataset released as a challenge for ICFHR2018. This online dataset was made into an offline dataset using a github tool[1]. The parameters used was a line width of 3 and dpi of 300. This dataset includes 67 k train, 18 k validation, 25 k test images. These images are word-level segmentations that are entirely black text on a white background as a result of the conversion code. The example image is shown as Fig. 3.

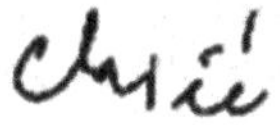

Fig. 3. Example image for VNonDB

Kaggle Cyrillic Dataset. The Kaggle Cyrillic [5] dataset is an offline handwritten Russian dataset. The data is of handwritten Russian phrases with a variety of backgrounds in color. For our experiments, the images were all converted to grayscale. There are 72 k train and 1.5 k test examples. The other datasets used are primarily black text on a white background. However, this dataset has a more varied background. In the results section we will discuss the effects that this difference had on the generated images. Another characteristic of this dataset is that the handwriting is written in cursive. The example image is shown as Fig. 4.

MADCAT Arabic Dataset. The MADCAT dataset [16–18] is an offline handwritten Arabic dataset. This data is also black text on a white background. The official test set has not been released so unofficial train, dev, and test data splits were used[2]. The dataset includes line-level segmentation with 600 k train,

[1] https://github.com/VinhLoiIT/vnondb_convert.
[2] https://openslr.org/48.

Fig. 4. Example image for Kaggle Cyrillic

74 k dev, and 72 k test examples. However, for the sake of training speed, only the first 100 k of the training data was actually used. The example image is shown as Fig. 5.

Fig. 5. Example image for MADCAT Arabic

Printed Dataset. Printed versions of every dataset were created with Synthtiger [33] using the ground-truth text labels. These images are also created to be dark text on a light background in order to be visually similar to the ground truth images. Various distortions such as stretch, skew, and rotation were also applied but mostly kept at a minimum. These images were used as input to our GAN. Example generated images are shown as Fig. 6.

Fig. 6. Example rendered printed images

3.4 Implementation Details

For the GANs model, all the printed and ground truth images were scaled to a height of 32 pixels. Since the generator is an image to image translation there is no need to scale the image width according to the text labels as is done with ScrabbleGAN [7]. The generated handwritten images were then scaled to 60 pixels for the HTR model. The evaluation was done by comparing the Frèchet Inception Distance (FID) [12,26] between the generated and ground truth images. An additional metric used was word error rate (WER). The generated images were used as data augmentation for training a HTR model and WER was reported on the ground truth test set.

Three sets of images were compared against the ground truth images; two sets of GAN generated handwritten images and the rendered printed images. The first set of handwritten images was generated by our model trained only on paired English data where the ground truth images are paired with the corresponding rendered printed text. The second set of handwritten images was generated by our model trained on the paired English data and unpaired target domain data. This unpaired target domain data consists of the ground truth images of the unrelated language and printed images rendered from text corpora. Words from the target language text corpora were selected randomly, so the content of the printed images is unrelated to that of the ground truth images. To ensure that our proposed method improves the input printed images, the results of using the printed images directly for text recognition were also considered. In order to assess the robustness of our model, three target languages were used: Russian, Vietnamese, and Arabic.

The low resource scenario was simulated by randomly sampling a subset of data when training the HTR model. Data augmentation was done by taking the ground truth labels of the sampled data and folding in the images generated from those selected ground truth labels. The augmentation images were either the printed images or the output of our GAN after feeding the printed images into the generator.

Table 4. Output from the model for various different languages: Vietnamese, Russian, and Arabic.

Ground Truth	Printed	Crosslingual	Semisupervised
	Chiếc		
	tốc		
	độ		
	mặt		
	Даже работа		
	людей).		
	богу		
	جدا.إذا لم تجر 43862 السنة 131 -العد ر الى25 في المئة .		

4 Results

In this section, we refer to the GAN trained only on labeled English as 'crosslingual' and the GAN trained on both labeled English and unpaired, unlabeled target domain data as 'semisupervised'. Table 4 shows the printed images fed into the generator, the output images from Crosslingual generator, the output images from the semisupervised generator, and the ground truth images for comparison.

4.1 Qualitative Observations

Some qualitative observations will be made on the example images given in Table 4. Overall, the handwritten text generated for Vietnamese is the most visually impressive. This makes intuitive sense given that Vietnamese and English, the source language, share the same script.

Vietnamese. For Vietnamese, the crosslingual model is able to generate readable text. However, the diacritics do not seem to be properly generated. For example, the word 'Chiếc' has two diacritics on top of 'ế' that are not readable for output of the crosslingual model. The two diacritics, 'circumflex' and 'acute accent' have been combined into a single mark. The inability to generate such an images is understandable as English does not have as many diacritics and so the character 'ế' is an unseen symbol. There is also background noise that is visible around each character which is particular to the IAM dataset.

However, the same GAN model architecture is able to produce diacritics correctly when unpaired Vietnamese data is included in the training. The two diacritics mentioned previously are visually distinct for the semisupervised model. The background noise from the IAM dataset is also gone. Though the semisupervised model also hallucinates extra diacritics below the word 'Chiếc'.

Russian. The biggest limitation on Russian is that the ground truth images are written in cursive while our rendered printed image is in print. Subsequently, our generated handwritten Russian is also in print handwriting. The crosslingual model is able to faithfully reproduce the characters of the rendered printed text but is unable to generate cursive Russian.

The semisupervised model trained with unpaired Cyrillic does not produce images that are as clear as the crosslingual images. However, the addition of unpaired Cyrillic does cause the generated images to have a background closer to that of the ground truth images. The content of the images also seem to be closer to cursive handwriting than in the crosslingual case. For example, compare the first letter in 'Даже работа' between the crosslingual, semisupervised, and ground truth images.

Arabic. The overall observations between the crosslingual and semisupervised images are also apparent with Arabic. The crosslingual model generates handwritten images that have the background noise inherent to the IAM dataset. The observation that the semisupervised model generates worse results for Russian and Arabic suggests that the addition of unpaired data is detrimental if the script of the source language, English, is too different from the script of the target language, Russian or Arabic.

4.2 FID Results

The two generated handwritten images and the input printed images were compared against the ground truth images. The FID results were as expected, with the printed images having the highest computed FID. The crosslingual GAN trained on handwritten English is able to generate handwritten images in the target language somewhat convincingly and so has a lower computed FID when compared to the printed images. Finally, the addition of extra unlabeled target language data allows the semisupervised GAN to generate images even closer to the target domain and has the lowest computed FID. The improvements in FID score mirror the qualitative assessments in visual similarity between the generated images and the ground truth images. Though it is hard to say if the semisupervised images are better than the crosslingual images for Russian and Arabic. Table 5 lists the computed FID scores between the ground truth dataset and the generated images.

Table 5. FID Between Ground Truth and Generated Images

	Printed	Crosslingual GAN	Semisupervised GAN
Kaggle Cyrillic	118.97	101.93	83.90
VNonDB Vietnamese	142.08	77.10	27.46
MADCAT Arabic	111.74	70.56	23.28

4.3 WER Results

We trained an HTR on increasing amounts of data to observe how much the generated images improve word error rate. The WER results were a little unexpected in that data augmentation with the semisupervised images did not yield the best WER despite having the lowest FID score. Table 6 shows the full WER results for the different datasets.

For Vietnamese, we randomly sampled a range of training data from 1 k to 30 k ground truth images. For data augmentation, a handwritten image was created for each ground truth image selected. At 5 k ground truth images, the baseline model achieves 39.83% WER. The 2x data augmentation with generated handwritten text improves the results to 34.34% WER. If the rendered printed images are used as data augmentation directly without using the Crosslingual

Table 6. WER results on various datasets with 2x Data Augmentation. Results on varying amounts of ground truth data.

	Baseline	Printed	Crosslingual GAN	SemiSupervised GAN
VNonDB				
1 k	77.86	75.72	72.35	**71.06**
5 k	39.83	37.76	**34.34**	35.95
10 k	27.95	24.18	**22.21**	23.88
20 k	19.47	15.98	**15.87**	17.07
30 k	16.18	**12.83**	13.23	14.49
Kaggle				
5 k	99.44	65.18	**64.81**	78.86
10 k	59.39	52.02	**49.05**	54.24
20 k	44.74	44.18	**41.63**	43.81
30 k	41.91	39.96	**39.17**	40.01
MADCAT				
1 k	110.8	**75.73**	101.69	95.96
5 k	46.32	42.13	41.9	**41.61**

generator, the text recognition system has 37.76% WER. The generated handwritten images improve the WER from a low-resource scenario all the way up to around 20 k ground truth images. This indicates that in low-resource scenarios, the HTR performs best when the images used for data augmentation are visually similar to the training data. At higher amounts of training data, there is enough data to train a reasonable text recognition system. So the variety that the rendered printed images provides adds to the robustness of the model. And so the rendered printed images perform best at 12.84% WER for data augmentation at 30 k ground truth images. Our overall baseline with all 67 k ground truth training images is 12.16%. These results are comparable to the 13.2% in literature [4]. Though the results in literature are on the online version of the dataset.

The trend for Cyrillic is similar to that of Vietnamese. The improvements over the baseline for the crosslingual images outperform those of the rendered printed images until around 30 k ground truth images. At 10 k images, the baseline result is 59.39% WER, whereas the results for the crosslingual and rendered printed images are 49.05% WER and 52.02% WER, respectively. Our overall baseline with all 72 k ground truth images is 37.09% WER. Even with the full dataset, augmenting with extra data still gives improvements with rendered printed images improving results to 35.14% WER and crosslingual images improving to 35.28% WER.

The GAN-generated handwritten images for Arabic are not of sufficient quality to use as data augmentation. Even though the crosslingual images still improve recognition results over the baseline, better performance can be achieved by using the rendered printed images directly as data augmentation. This was not expected, considering the crosslingual and semisupervised images had much lower FID scores than the printed images.

4.4 Ablation on Training Script

An ablation study was performed to assess the effects of using different scripts when training a crosslingual GAN. The baseline model is a standard GAN trained on all available handwritten Vietnamese data and used to generate Vietnamese. The crosslingual GANs were trained on English and Arabic handwritten data respectively. These models were then used to generate handwritten Vietnamese for the purposes of data augmentation when training a Vietnamese HTR system. In the scenario with large amounts of data, data augmentation with printed text outperforms the GANs generated handwritten text. However, when there are fewer than 10 k words of ground truth training data, the use of a crosslingual GAN is beneficial when performing data augmentation. The amount of improvement in WER depends on the script used in training. English and Vietnamese both use Latin scripts while Arabic does not. This explains why the crosslingual GAN for English outperforms that of Arabic. Table 7 shows the full WER results.

Table 7. WER results on the VNonDB test data with 2x Data Augmentation. Results on varying amounts of ground truth data.

	Baseline	Monolingual GAN VN	Crosslingual GAN EN	Crosslingual GAN AR	Printed
1 k	77.86	68.60	72.35	74.60	75.72
5 k	39.83	29.67	34.34	36.23	37.76
10 k	27.95	20.51	22.21	24.43	24.18
30 k	16.18	13.03	13.23	14.01	12.83

4.5 Image Augmentation

As mentioned before, one advantage of using rendered printed images as input is that the input noise vector that other text generation models have [1,7] are unnecessary. Different handwritten images can be created by augmenting the input printed image itself. Table 8 shows that image rotations and shear augmentations on the input printed side yield the same augmentations on the output handwritten side. This provides for much finer control over the desired output image. Table 9 shows the different effects of a combination of image augmentation techniques on the resulting output handwritten text. Unfortunately, the

Table 8. Different input images of the same label text yield different handwriting text. Image augmentation techniques such as rotation and shear on printed input causes output handwritten images to be augmented as well.

Printed Input	Semisupervised GAN Output
that Labour should not take any steps	that labour should not take any steps
that Labour should not take any steps	that labour should not take any steps
that Labour should not take any steps	that labour should not take any steps
that Labour should not take any steps	that labour should not take any steps
that Labour should not take any steps	that labour should not take any steps
that Labour should not take any steps	that labour should not take any steps
that Labour should not take any steps	that Labour should not take any steps

style of the output handwriting seems arbitrary, and there is no reliable control over the handwriting style.

One qualitative observation between the images from [7] and the images generated from our method is that our images do not have as large of a variation in handwriting styles. However, the Vietnamese handwriting dataset that we used is converted from an online dataset, so the ground truth images themselves have a fixed line width and is not as varied. More variety in author styles can be achieved by including author id as input in the generator [21].

Table 9. Output from the model for English with each word generated individually from our model.

GAN model trained on English
He believes that the House of Lords should be abolished
He believes that the House of Lords should be abolished
and that Labour should not take any steps which would
and that Labour should not take any steps which would
appear to prop an out-dated institution
appear to prop an out-dated institution

5 Conclusion

We present our work with the crosslingual generation of handwritten text. We demonstrate that when there is an adequate amount of training data available, data augmentation with printed text is sufficient to achieve improvements in WER. The improvement is comparable to that of GAN generated handwritten images. However, in a low resource scenario, it is beneficial to perform data augmentation using a crosslingual GAN. And this improvement is dependent on how visually 'similar' the source language is to the target language.

Our GAN model, trained on only labeled English text, is able to generalize and produce handwritten text in Vietnamese, Russian, and Arabic. In all cases, our model generates handwritten images that are closer to the ground truth images than printed images when measured by FID. We also show that the generated handwritten images are of sufficient quality to improve recognition results over the baseline when used for data augmentation. However, the generated handwritten images only outperform printed images for Vietnamese and Russian but not for Arabic. This suggests that visual similarity between the scripts of the source and target language is important when generating handwritten text in a crosslingual manner. The importance of visual similarity is highlighted in the ablation study. The crosslingual GAN trained on English outperforms the one trained on Arabic in generating Vietnamese handwritten text.

Further work can be done to expand this area of crosslingual handwritten text generation. In the future, applying our model to multiple source languages could train a large multilingual handwritten text generator. This may allow our model to better generalize a mapping between the printed and handwritten text-domain. In this way, there is potential to generate handwritten images of any script supported in Unicode.

References

1. Alonso, E., Moysset, B., Messina, R.: Adversarial generation of handwritten text images conditioned on sequences. In: 2019 International Conference on Document Analysis and Recognition (ICDAR), pp. 481–486. IEEE Computer Society, Los Alamitos (2019). https://doi.org/10.1109/ICDAR.2019.00083, https://doi.ieeecomputersociety.org/10.1109/ICDAR.2019.00083
2. Bahdanau, D., Cho, K., Bengio, Y.: Neural machine translation by jointly learning to align and translate. CoRR abs/1409.0473 (2014)
3. Brock, A., Donahue, J., Simonyan, K.: Large scale GAN training for high fidelity natural image synthesis. In: International Conference on Learning Representations (2019). https://openreview.net/forum?id=B1xsqj09Fm
4. Carbune, V., et al.: Fast multi-language LSTM-based online handwriting recognition. Int. J. Doc. Anal. Recogn. (IJDAR) **23**(2), 89–102 (2020). https://doi.org/10.1007/s10032-020-00350-4
5. CFT SL: Cyrillic handwriting dataset (2022). https://www.kaggle.com/datasets/constantinwerner/cyrillic-handwriting-dataset
6. Etter, D., Rawls, S., Carpenter, C., Sell, G.: A synthetic recipe for OCR. In: 2019 International Conference on Document Analysis and Recognition (ICDAR), pp. 864–869 (2019). https://doi.org/10.1109/ICDAR.2019.00143
7. Fogel, S., Averbuch-Elor, H., Cohen, S., Mazor, S., Litman, R.: ScrabbleGan: semi-supervised varying length handwritten text generation. In: IEEE/CVF Conference on Computer Vision and Pattern Recognition (CVPR) (2020)

8. Goodfellow, I., et al.: Generative adversarial networks. In: Advances in Neural Information Processing Systems, vol. 3 (2014). https://doi.org/10.1145/3422622
9. Graves, A., Fernandez, S., Gomez, F., Schmidhuber, J.: Connectionist temporal classification: labelling unsegmented sequence data with recurrent neural nets. In: ICML '06: Proceedings of the International Conference on Machine Learning (2006)
10. Graves, A., Fernández, S., Schmidhuber, J.: Bidirectional LSTM networks for improved phoneme classification and recognition. In: Duch, W., Kacprzyk, J., Oja, E., Zadrożny, S. (eds.) ICANN 2005. LNCS, vol. 3697, pp. 799–804. Springer, Heidelberg (2005). https://doi.org/10.1007/11550907_126
11. Heil, R., Vats, E., Hast, A.: Strikethrough removal from handwritten words using CycleGANs. In: Lladós, J., Lopresti, D., Uchida, S. (eds.) ICDAR 2021. LNCS, vol. 12824, pp. 572–586. Springer, Cham (2021). https://doi.org/10.1007/978-3-030-86337-1_38
12. Heusel, M., Ramsauer, H., Unterthiner, T., Nessler, B., Hochreiter, S.: GANs trained by a two time-scale update rule converge to a local Nash equilibrium. In: Guyon, I., et al. (eds.) Advances in Neural Information Processing Systems, vol. 30. Curran Associates, Inc. (2017). https://proceedings.neurips.cc/paper_files/paper/2017/file/8a1d694707eb0fefe65871369074926d-Paper.pdf
13. Huu, M.K.N., Ho, S.T., Nguyen, V.T., Ngo, T.D.: Multilingual-GAN: a multilingual GAN-based approach for handwritten generation. In: 2021 International Conference on Multimedia Analysis and Pattern Recognition (MAPR), pp. 1–6 (2021). https://doi.org/10.1109/MAPR53640.2021.9585285
14. Isola, P., Zhu, J.Y., Zhou, T., Efros, A.A.: Image-to-image translation with conditional adversarial networks. In: Proceedings of the IEEE Conference on Computer Vision and Pattern Recognition (CVPR) (2017)
15. Kong, Y., et al.: Look closer to supervise better: One-shot font generation via component-based discriminator. In: Proceedings of the IEEE/CVF Conference on Computer Vision and Pattern Recognition (CVPR), pp. 13482–13491 (2022)
16. Lee, D., Ismael, S., Grimes, S., Doermann, D., Strassel, S., Chen, S.: MADCAT phase 1 training set (2012). https://doi.org/10.35111/9bm5-nz55
17. Lee, D., Ismael, S., Grimes, S., Doermann, D., Strassel, S., Chen, S.: MADCAT phase 2 training set (2013). https://doi.org/10.35111/044b-ah68
18. Lee, D., Ismael, S., Grimes, S., Doermann, D., Strassel, S., Chen, S.: MADCAT phase 3 training set (2013). https://doi.org/10.35111/w1px-d922
19. Li, C., Taniguchi, Y., Lu, M., Konomi, S.: Few-shot font style transfer between different languages. In: Proceedings of the IEEE/CVF Winter Conference on Applications of Computer Vision (WACV), pp. 433–442 (2021)
20. Liu, W., Liu, F., Ding, F., He, Q., Yi, Z.: XMP-FONT: self-supervised cross-modality pre-training for few-shot font generation. In: Proceedings of the IEEE/CVF Conference on Computer Vision and Pattern Recognition (CVPR), pp. 7905–7914 (2022)
21. Luo, C., Zhu, Y., Jin, L., Li, Z., Peng, D.: SLOGAN: handwriting style synthesis for arbitrary-length and out-of-vocabulary text. IEEE Trans. Neural Netw. Learn. Syst. (2022)
22. Marti, U.V., Bunke, H.: The IAM-database: an English sentence database for offline handwriting recognition. Int. J. Doc. Anal. Recogn. **5**, 39–46 (2002)
23. Nayef, N., et al.: ICDAR 2019 robust reading challenge on multi-lingual scene text detection and recognition - RRC-MLT-2019 (2019). https://doi.org/10.48550/ARXIV.1907.00945, https://arxiv.org/abs/1907.00945

24. Nguyen, H.T., Nguyen, C.T., Nakagawa, M.: ICFHR 2018 - competition on vietnamese online handwritten text recognition using HANDS-VNonDB (VOHTR2018). In: 2018 16th International Conference on Frontiers in Handwriting Recognition (ICFHR), pp. 494–499 (2018). https://doi.org/10.1109/ICFHR-2018.2018.00092
25. Park, S., Chun, S., Cha, J., Lee, B., Shim, H.: Multiple heads are better than one: few-shot font generation with multiple localized experts. In: Proceedings of the IEEE/CVF International Conference on Computer Vision (ICCV), pp. 13900–13909 (2021)
26. Seitzer, M.: PyTorch-FID: FID Score for PyTorch. https://github.com/mseitzer/pytorch-fid (2020). version 0.2.1
27. Sharma, M., Verma, A., Vig, L.: Learning to clean: a GAN perspective. In: Carneiro, G., You, S. (eds.) ACCV 2018. LNCS, vol. 11367, pp. 174–185. Springer, Cham (2019). https://doi.org/10.1007/978-3-030-21074-8_14
28. Tang, L., et al.: Few-shot font generation by learning fine-grained local styles. In: Proceedings of the IEEE/CVF Conference on Computer Vision and Pattern Recognition (CVPR), pp. 7895–7904 (2022)
29. Veit, A., Matera, T., Neumann, L., Matas, J., Belongie, S.: Coco-text: dataset and benchmark for text detection and recognition in natural images. arXiv preprint: arXiv:1601.07140 (2016). http://vision.cornell.edu/se3/wp-content/uploads/2016/01/1601.07140v1.pdf
30. Wang, Y., et al.: Espresso: a fast end-to-end neural speech recognition toolkit. In: 2019 IEEE Automatic Speech Recognition and Understanding Workshop (ASRU), pp. 136–143 (2019). https://doi.org/10.1109/ASRU46091.2019.9003968
31. Wei, H., Liu, K., Zhang, J., Fan, D.: Data augmentation based on CycleGAN for improving woodblock-printing Mongolian words recognition. In: Lladós, J., Lopresti, D., Uchida, S. (eds.) ICDAR 2021. LNCS, vol. 12824, pp. 526–537. Springer, Cham (2021). https://doi.org/10.1007/978-3-030-86337-1_35
32. Wen, C., et al.: Handwritten Chinese font generation with collaborative stroke refinement. In: Proceedings of the IEEE/CVF Winter Conference on Applications of Computer Vision (WACV), pp. 3882–3891 (2021)
33. Yim, M., Kim, Y., Cho, H.-C., Park, S.: SynthTIGER: synthetic text image GEneratoR towards better text recognition models. In: Lladós, J., Lopresti, D., Uchida, S. (eds.) ICDAR 2021. LNCS, vol. 12824, pp. 109–124. Springer, Cham (2021). https://doi.org/10.1007/978-3-030-86337-1_8
34. Zhu, J.Y., Park, T., Isola, P., Efros, A.A.: Unpaired image-to-image translation using cycle-consistent adversarial networks. In: Computer Vision (ICCV), 2017 IEEE International Conference on (2017)

Knowledge Integration Inside Multitask Network for Analysis of Unseen ID Types

Timothée Neitthoffer[1,3](✉), Aurélie Lemaitre[1], Bertrand Coüasnon[1], Yann Soullard[1,2], and Ahmad Montaser Awal[3]

[1] Univ Rennes, CNRS, IRISA, Rennes, France
timothee.neitthoffer@irisa.fr
[2] Université Rennes 2, CNRS, LETG, IRISA, Rennes, France
[3] IDNow, AI & ML Center of Excellence, Cesson-Sévigné, France

Abstract. Identity Document recognition is a key step in Know Your Customer applications where identity documents (IDs) are verified. IDs belonging to the same type share the same field structure called template. Traditional ID pipelines leverage this template to guide the localisation of the fields and then the text recognition. However, they have to be tuned to the different templates to correctly perform on those. Thus, such pipelines can not be directly used on new types of IDs. In this work, we address the task of text localisation and recognition in the context of new document types, where only the template is available with no labeled samples from the new ID type. To that end, we propose the use of Context Blocks (CB) performing template self-attention to guide the features of the network by the template. We propose three ways to leverage CB in a multitask architecture. To evaluate our approach, we design a new public task for the MIDV2020 database from rectified in-the-wild photos. Our method achieves the best results for two datasets including an industrial one composed of real examples.

Keywords: Knowledge Integration · Multitask Learning · OCR · Text localisation · Identity Documents · Deep Network

1 Introduction

Optical Character Recognition (OCR) is a major research topic in computer vision where text is recognized from captured images. Although it first appears in the 80's, text recognition remains a challenging problem with highly variable documents and capture conditions. In this work, we are interested in Know Your Customer applications where a service provider extracts user's information from their identity documents (IDs). Countries issue IDs for different uses (id cards, driving licenses, etc) while following the international norms (such as ICAO).

Each kind of ID contains specific information and thus present different template, background and security elements. IDs belonging to the same type and the same version (for example "2015 French ID card") share the same template as shown in Fig. 1. Although the template possesses strong structural knowledge,

M. Coustaty and A. Fornés (Eds.): ICDAR 2023 Workshops, LNCS 14194, pp. 302–317, 2023.
https://doi.org/10.1007/978-3-031-41501-2_21

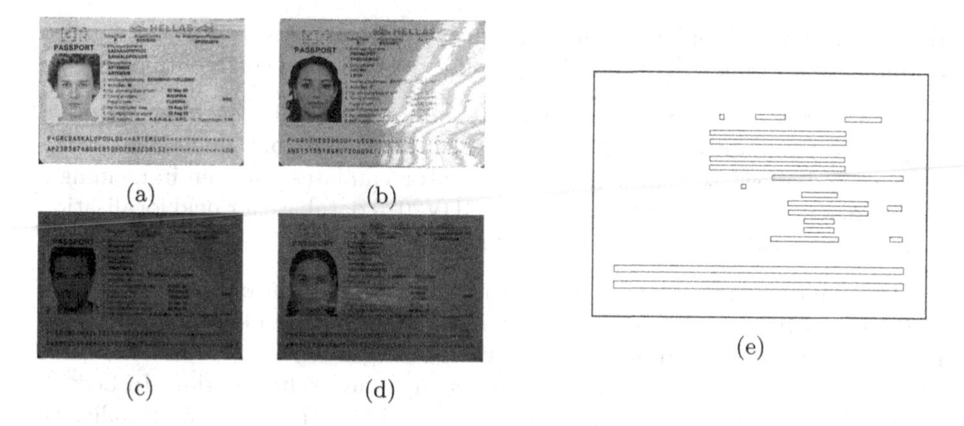

Fig. 1. a,b,c,d) Greek passports from MIDV2020 e) Greek template

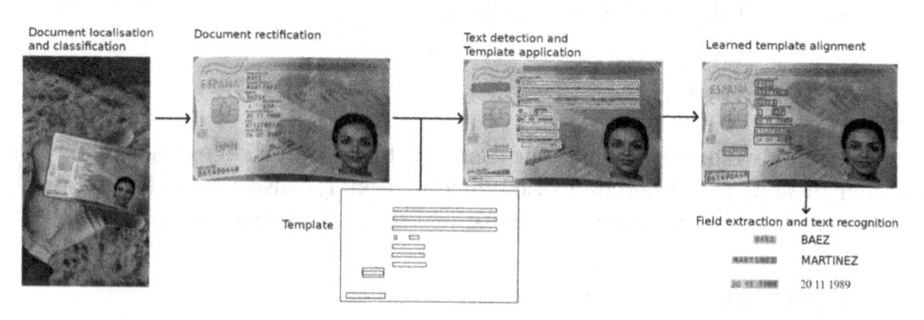

Fig. 2. Pipeline of information extraction in IDs.

it is not enough to localize the fields. Indeed, as shown in Fig. 2, the images are captured in real conditions, mainly with smartphones, and then rectified from their detected quadrilateral to a plane document. However, the uncontrolled environment and the variable capture conditions with projections, introduce shifts and deformations: the fields are distorted and different in size from the template. Thus, it is not possible to apply the template directly onto an image.

Traditional pipelines (Fig. 2) deal with in-the-wild ID images captured from various devices. The document image is first localised and classified in the input image and then rectified. The fields are detected by aligning the result of text localisation and the known template. The text fields are extracted from the rectified image and then recognized by a generic OCR. The quality of the predictions highly depends on the alignment process, that may propagate errors and must be trained for each kind of document.

In this work, we consider as input a rectified document image. We focus on field localisation and recognition. We especially focus on dealing with new ID types from new countries or newer versions. To correctly handle new types, the traditional methods need to be hand-tuned in an *ad hoc* manner on annotated data. However, it is difficult to obtain labeled data for those new ID types, due to the sensitive information they contain. In that context, we only benefit from the known template as an *a priori* knowledge, without the need of any labeled example from the new ID types. We propose the following contributions:

- define core concepts for knowledge integration in neural networks;
- propose template self-attention using a Context Block that guides the predictions and allows to deal with new ID types (templates not seen in training);
- design a new dataset and task on the MIDV2020 database for field localisation and recognition, available to the community[1].

This paper is organized as follows. In Sect. 2, we introduce the works related to ID recognition, joint text localisation and recognition and knowledge integration. Then, we present proposals on knowledge integration inside neural networks through the use of an auxiliary branch and template self-attention. In Sect. 4, we describe the new dataset and task we designed for MIDV2020 [5]. Finally, we present the result of our experiments on the new MIDV2020 dataset and on a private industrial dataset before concluding.

2 Related Works

In this section, we first introduce the methods for information extraction on IDs. We then present the different works on joint text localisation and recognition, and on knowledge integration in models, which can be outside of the ID domain.

2.1 Identity Document Analysis

ID analysis involves the localisation of the fields and their recognition. Mustafina *et al.* [14] and Van *et al.* [16] perform text localisation using neural network architectures. However, due to the complexity of the background and the variability of capture conditions and of the document templates, the models tend to overdetect text fields. As discussed before, IDs of the same type share the same template. This template is sometimes used to guide the prediction of text field localisation. For instance, Attivissimo *et al.* [1] uses the template as a crop mask to directly extract the fields from the image. However, this method makes the assumption that the ID is perfectly localised and rectified. This is rarely the case due to inaccuracies from the previous steps. In addition, the authors do not evaluate the CER when the localisation is not correct. Also, ID printing process incorporates field position variations from one document to another, making the direct application of the template impossible. Bulatovich *et al.* [3,4] perform text detection via morphological analysis and then apply a template alignment algorithm to extract the fields. Nevertheless, this alignment balances between the different kinds of templates and need to be hand-tuned in an *ad hoc* manner, using labelled samples for dealing with new ID types. This is particularly difficult as, in case of new document types, only the template may be available. Text recognition in the cropped field images is usually performed using a Convolutional Recurrent Neural Network (CRNN) [1,14,16].

[1] https://gitlab.inria.fr/tneittho/midv2020-rectified-photo.

A two step approach propagates the errors from the localisation to the recognition task. To address this problematic, we seek to perform jointly the localisation and the recognition. asa well as leveraging the templates in this joint context. In the next sections, we present works outside of the ID domain for joint text localisation and recognition as well as knowledge integration.

2.2 Joint Text Localisation and Recognition

Recent works have addressed this joint tasks by performing implicit localisation using attention mechanism. Indeed, Bluche *et al.* and Coquenet *et al.* [2,8] perform paragraph level and page level recognition on ancient documents by leveraging the attention mechanism to deal with line breaks or to select the current line to read. Furthermore, methods from Yousef *et al.* [17] and Coquenet *et al.* [7] guide the network to unfold the multiple lines of the document into a single sequence. However, these methods deal with single column documents where IDs present complex layouts and backgrounds on the entire document.

Coquenet *et al.* [9] addresses this problem by learning the reading order of the text blocs in the document with the use of Transformers. However, this method heavily relies on the templates seen at training time, from real data and generated synthetic data. Generating real samples of synthetic IDs is a very hard task which is out of the scope of this work. The lack of real examples limits the use of such kind of methods, especially when dealing with new ID types.

Another way to address the problematic is the design of a multitask framework. In [6] and [15], the authors leverage a multitask network to perform ancient document recognition. A first part of the network is dedicated to text localisation as object detection. The detected areas are used to pool the feature maps of the backbone which are then given in input to the second part, dedicated to text recognition. The backbone is shared by the two parts. This kind of structure allows to focus the recognition only on the relevant parts of the document.

2.3 Knowledge Integration

Administrative documents, similarly to IDs, usually follow a structure that can be used to guide the recognition. We can distinguish two kinds of approaches for using *a priori* knowledge. The template approach is commonly used in current industrial pipelines. A mask or a graph [11] is created and adjusted to the document image. The full-text approach defines a number of logical rules based on detection and recognition results and the content is associated to the related semantic value (Couasnon *et al.* [10] and Guerry *et al.* [12]). However, these methods are not optimized based on the available knowledge in a end-to-end manner.

Neural networks can benefit from this knowledge to obtain better performances. In [13], a feature vector representing the possible classes is concatenated to the features of the network to guide disease classification. Such proposition is interesting as it allows to condition the predictions by the input knowledge.

Currently, no acceptable proposition have been made to answer text localisation and recognition in the context of new ID types. Traditional solutions in the ID domain rely on a two step approach that easily propagates errors from the localisation to the recognition. In addition, the alignment algorithms proposed to leverage the template at disposition need to be fine-tuned using labelled samples of the new ID type. Propositions from Carbonel *et al.* [6] and Kushibar *et al.* [13] are interesting in our context but need to be further explored.

3 Proposed Methods

Based on the State of the Art, we aim to integrate the knowledge inside a multitask network. We further develop those ideas and adapt them to the analysis of new ID types. In this section, we formalize the paradigm of knowledge integration in neural network by defining its core concepts. We then present our proposals to leverage them to address new ID types recognition without any additional data. Finally, we give details about our architectures.

3.1 Definition of Knowledge Integration Inside Neural Networks

We propose to divide Knowledge Integration inside neural networks into three core concepts that complements one another: the **knowledge representation**, the **knowledge connection** and the **integration operation**.

The **knowledge representation** corresponds to the form of the knowledge used as input of the network. For example the template is integrated as a graph or as a mask and what kind of graph or mask (Sect. 3.2). We call **knowledge connection** the way the knowledge representation is processed, and where it connects to the feature maps from the input (Sect. 3.3). The **integration operation** is the method used to mix the features from the template and the ones from the input (Sect. 3.4).

3.2 Knowledge Representation

We propose to give as input to the model an ID image and the corresponding template. The template contains the knowledge about the fields localisation and type (first name, emission date, ...). To remain in the same modality as the image, we consider the template as a mask of the same dimensions as the input image. We explore two representations (Fig. 3a)). A **single-channel** mask, which is a binary mask to discriminate between the text fields and background and a **multi-channel** mask where each channel corresponds to one type of field.

3.3 Knowledge Connection

We choose to integrate the template inside the backbone in order to extract features combining the template and the input image. As the backbone is shared by the localisation and recognition heads, the knowledge integration (i.e. the

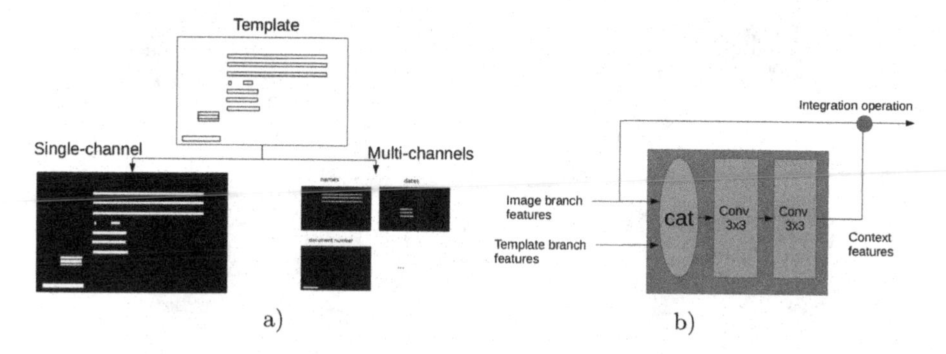

Fig. 3. a) Template representation as a single-channel or a multi-channel mask. b) Context Block (CB) for template integration. The features of both image and template branches are concatenated. Two convolutions are used to build feature maps that are then given to the integration operation (see Sect. 3.4).

template) will impact the predictions of the whole system. Thanks to that, the network is able to focus on the important features based on the template and it can be easily used with new document types, that we called **unseen** ID types.

Knowledge Connection as an Additional Channel. First, an intuitive way is to consider the template as another channel of the image. That way, the network can extract correlations between the channels and process directly the template in the input representation. Here, the backbone is the same than the one presented in Fig. 4. We refer to this architecture as **Naive**. However, doing so might lead the neural network to consider the template as part of the graphical structure of the image and thus misuse it.

Knowledge Connection as an Auxiliary Branch. A second approach is to process the template and the image in two parallel branches that are then connected one to another. To connect the two branches, we introduce a **context block** (CB) presented in Fig. 3b). This block is composed of two convolutions and takes as input the concatenation of the image branch features and the template branch features. Thus, a CB is used to combine the two sets of feature maps and to output context features that will be mixed with the image branch. In the following paragraphs, we explore three variants based on this method.

In a first architecture (Fig. 5), the template branch follows the descending stage of the image branch, so that the *a priori* knowledge guides the construction of more complex features. We refer to this architecture as **Multitask with Knowledge Integration (MKI)**. Nevertheless, the reconstruction stage might benefit from the template branch, this is why we propose two other variants.

The second architecture **MKI-S** (Fig. 6) uses skip connections from the template branch in each up-sampling block in the reconstruction stage. Thus, the representations of the template at different resolutions guide the reconstruction.

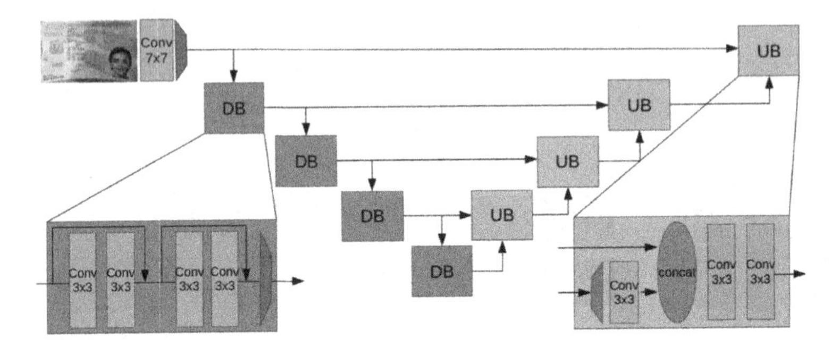

Fig. 4. Our multitask network backbone based on a UNet-like architecture. The descending stage is made of Down Blocks (DBs) and the ascending stage of Up Blocks (UBs). UBs take the feature maps of the previous stage (lower arrow) and the skip connection from the DB (upper arrow). The red trapezoids are maxpooling operations and the green ones are upsampling with cubic interpolation. (Color figure online)

The last variant **MKI-D** (Fig. 7) extends the template branch to match the image one, creating a double UNet structure. This structure makes the template branch to compute specific features for the reconstruction stage. However, it makes the branch harder to train as the template branch depth increases.

3.4 Mixing the Knowledge Features into the Image Branch

As introduced in Sect. 3.1, the knowledge integration is also determined by **integration operations**. They specify the way to mix the feature maps from the context blocks into the image branch (Fig. 3b)). In our context, the system should focus on the important elements according to the template. It is indeed important in case of IDs due to large and various background parts that don't refer to text fields. We propose a template self-attention operation on the image features, based on the features coming from the context block (see Fig. 3b)).

3.5 Implementation Details

The different architectures follow the figures presented in Sect. 3.3. The backbone is a U-Net like structure starting with a 7×7 convolution with a stride of 2 and 64 output channels. It is composed of four stages down and four stages up. After each down stage a maxpooling operation reduces by two the dimension of the feature maps and the number of output channels is multiplied by two. The up stages take as input the previous stage feature map as well as a skip connection from the stage of the same resolution. The previous stage feature maps are upsampled following a bilinear policy before going through a 3×3 convolution. The two inputs are then concatenated and processed by two 3×3 convolutions.

In this work, field localisation is performed by semantic segmentation. Our localisation head is composed of three 3×3 convolutions with a sigmoid activation to discriminate each pixel as either background or text field. Our localisation

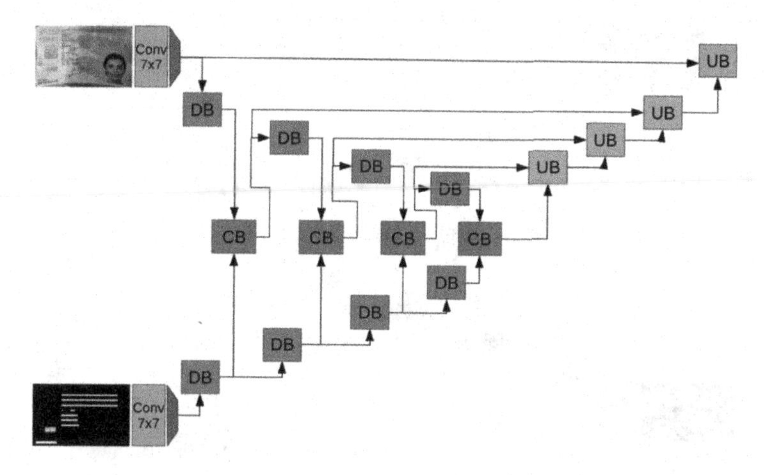

Fig. 5. Backbone of the MKI architecture. The upper branch is the document image branch and the lower branch the template branch. The Context Blocks (CB) in between allow the information of the template branch to flow to the document image branch. The DBs and UBs are the same as in Fig. 4.

head is trained using a binary mask as label and the balanced cross-entropy (BCE) loss to take into account the class imbalance. Then, we compose the bounding boxes by reducing the connected predicted pixels to rectangular boxes. We filter out boxes with an area under a threshold. The boxes are then used to pool the corresponding areas in the feature maps produced by the backbone.

The recognition head is a traditional CRNN-like module. It takes as input the list of pooled features corresponding to the detected fields. It is composed of two 3×3 convolutions, two BLSTM and finally a linear layer to output the character probabilities. This recognition head is trained using the well-known CTC loss. As we deal with both localisation and recognition tasks, we associate the target differently from a traditional text recognition task. If a predicted bounding box overlaps with a ground truth box, then the expected label is the one associated to the ground truth box. If the predicted bounding box does not overlap with any predicted ground truth box, the expected label is the empty string.

We balance the two losses by linearly interpolating the BCE and the CTC loss. The system is trained for 150 epochs on different GPUs with the Adam optimizer and a learning rate of 1×10^{-4}.

4 Databases

We now present another contribution: a new dataset and task on MIDV2020 database, that evaluates both field localisation and text recognition.

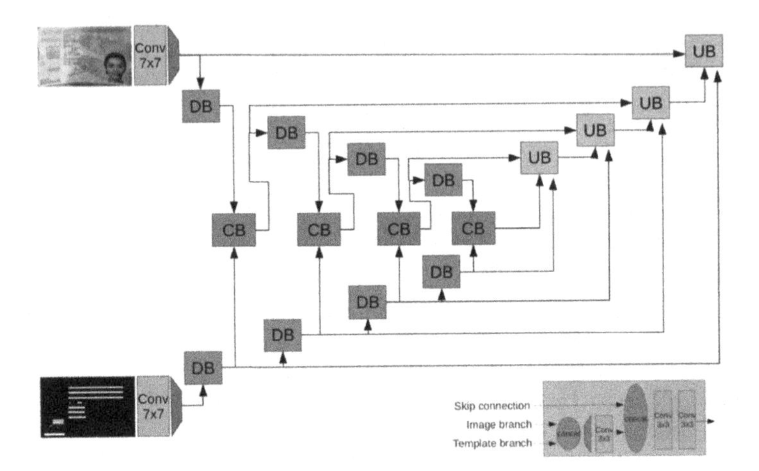

Fig. 6. Backbone of the MKI-S architecture. The skip connections propagate template information to the reconstruction stage. The UB is modified to take the incoming skip connections from the template branch in input.

4.1 New "*rectified photos*" Dataset and New Task for MIDV2020

The main public databases for IDs are the MIDV databases with the latest version being MIDV2020 [5]. The dataset "*templates*" of MIDV2020 contains the images of the different documents in full page with the annotations of the position of the fields as well as their corresponding text label and field type. However, this dataset does not reflect real capture conditions, such as lightning conditions or deformations. Furthermore, MIDV2020 contains a "*photos*" dataset, made of ID in real capture conditions, but that does not contain the annotations at field level. To that end, we propose a new dataset for MIDV2020 and a new task to evaluate Text Localisation and Recognition on Unseen ID Types (TLR-UIDT).

To create the new dataset, each document image, cropped from the "photos" dataset, is rectified to the dimensions of the same document image from the "templates" dataset. The annotations of text field positions coming from the "templates" dataset are then rectified to match the localisation of text fields from the resized document. We also design the template of each ID. We obtain a new dataset called "*rectified photos*", composed of the same 1000 document images as in the MIDV2020 dataset. It contains 10 ID types with their template and 100 document images per type. The text field positions and transcriptions for each document are known for training and test.

Using this "*rectified photo*" dataset, we propose a new task: TLR-UIDT that addresses text localisation and recognition of new ID types that we call *unseen ID types* : this refers to images from ID types (and templates) never seen in training. The rectified document images can be done in input of a network which is trained to output the text localisation, the name of the field and the textual content. Table 1 illustrates the training, validation and test set. To evaluate systems on unseen IDs types, we reserve one type of document for validation and another

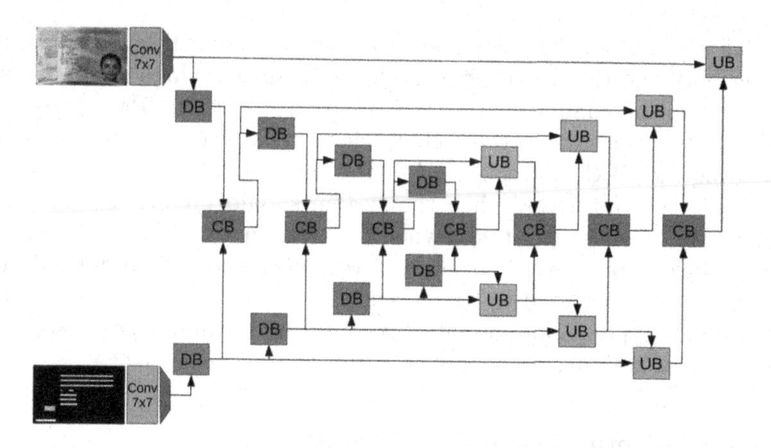

Fig. 7. Backbone of the MKI-D architecture. The template branch follows the reconstruction stage to guide the reconstruction step.

Table 1. Organisation of the *TLR-UIDT* task, for the public "*Rectified Photo*" dataset and the private dataset.

"*rectified photos*"	Types of ID	Nb of images	Image distribution
Train	8 ID types	640	80 images/ID type
Validation	8 seen ID types + 1 unseen ID type	180	10 images/seen ID type + 100 images from unseen val ID type
Test	8 seen ID types + 1 other unseen ID type	180	10 images/seen ID type +100 images from unseen test ID type
Private			
Train	7 ID types	6629	947 images/ID type
Validation	4 seen ID types + 4 unseen ID types	4912	614 images/ID type
Test	5 seen ID types + 13 unseen ID types	2147	113 images/ID type

one for test. We also add examples in validation and test of the types seen in training to evaluate the model prediction and prevent overfitting. We propose to perform a 10-fold cross-validation, following the experimental setting shown in Table 1 so that each ID type is considered unseen once. For example, fold 0 would keep the *Serbian passport* for the unseen ID type in validation and the *Slovak id card* for the unseen ID type in test, fold 1 the *Albanese id card* and the *Greek passport*, and so on. Both the new dataset "*rectified photo*" and the splits for the new task TLR-UIDT are publicly available.

4.2 Metrics

To evaluate the performance of field localisation, we use common metrics: the recall, the precision, and the Intersection over Union (IoU). A value of IoU equal to 0.8 is used in order to evaluate the quality of text field localisation (for computing both the recall and precision). Using such an high value of IoU for extracting boxes and computing the Character Error Rate (CER) may not reflect the quality of text recognition due to box rejection based on the threshold. Thus, for computing the CER, we select an IoU threshold equal to 0.1. We distinguish three cases: i) true positives, when the system should predict the label of the associated field; ii) false positives, when the system should predict an empty string as no field is present; iii) true negatives, when the system did not predict the box then it is considered to have predicted an empty string. Each metric is computed for ID types (and templates) seen during training (*seen* case) and for those not seen at training time (*unseen* case).

4.3 Private Identity Document Dataset

In the following experiments, we also apply our work on a private dataset (Table 1). It is made of real world samples. We study a subset of 24 ID types. The dataset is split into a *train_set*, a *val_set* and a *test_set*. Similarly to the new task on MIDV2020, we keep unseen ID types in validation and test. For each set, the number of images per ID type is balanced following Table 1:

- *train_set* seen ID types: Albanian Passport front, Austrian Passport front, Czech Identity Card front, Czech Identity Card back, Swiss Identity Card front, Swiss Identity Card back, Swiss Passport front;
- *val_set* unseen ID types: Belgian Residence Permit front, Belgian Residence Permit back, Belgian Passport front, Bulgarian Passport front;
- *test_set* unseen ID types: Austrian Driving License front, Belgian Identity Card front, Belgian Driving License front, Bosnian Passport front, Beninese Passport front, Croatian Identity Card front, Croatian Passport front, German Residence Permit front, Greek Passport front, Lithuanian Passport front, Polish Identity Card front, Polish Identity Card back, Ukrainian Passport front.

This dataset is harder than the MIDV2020 dataset as the samples present more capture variations and some of them have ID localisation errors. Indeed, as the ground truth of the ID localisation is not available, perfect rectification is hard and can distort the field structure. As this is an industrial dataset that contains sensitive information, we cannot provide it to the community.

5 Results

We now evaluate our proposals for knowledge integration: the different architectures for **knowledge connection** and the **knowledge representations** on performances. Finally, we compare our best method to the state of the art approaches. In our experiments, we do not use any data augmentation strategy nor synthetic data as we focus on evaluating the impact of the different approaches.

Table 2. Comparison of knowledge connection architectures on the TLR-UIDT task of MIDV2020. The mean and deviation are computed over the ten folds.

	Architecture	Naive	MKI	MKI-S	MKI-D
SeenID types	cer ↓	0.46 ± 0.14	0.17 ± 0.02	$\mathbf{0.14} \pm 0.02$	0.17 ± 0.03
	recall ↑	0.94 ± 0.01	0.94 ± 0.01	0.94 ± 0.02	$\mathbf{0.96} \pm 0.01$
	precision ↑	0.93 ± 0.02	$\mathbf{0.95} \pm 0.01$	0.94 ± 0.01	0.94 ± 0.01
	IoU ↑	0.87 ± 0.01	0.87 ± 0.01	0.87 ± 0.02	$\mathbf{0.88} \pm 0.01$
UnseenID types	cer ↓	0.81 ± 0.16	0.34 ± 0.06	$\mathbf{0.31} \pm 0.07$	0.47 ± 0.11
	recall ↑	0.80 ± 0.12	$\mathbf{0.86} \pm 0.09$	0.84 ± 0.08	$\mathbf{0.86} \pm 0.08$
	precision ↑	0.84 ± 0.08	0.84 ± 0.10	$\mathbf{0.87} \pm 0.09$	$\mathbf{0.87} \pm 0.10$
	IoU ↑	0.77 ± 0.09	0.79 ± 0.06	0.79 ± 0.05	$\mathbf{0.80} \pm 0.05$

5.1 Comparison of the Knowledge Integration Architectures

We evaluate the different architectures for **knowledge connection** (see Sect. 3.3) on the new MIDV2020 TLR-UIDT task (Table 2). We observe that the Naive variant is not efficient. For both seen and unseen ID types, it presents a really high CER (0.46/0.81). This shows that the use of an additional channel is not the best way to integrate the template.

For the seen ID types, we note that the localisation performances are similar for every architecture with a recall/precision of 0.94/0.94. MKI-S obtains the lowest CER (0.14) although the localisation performances are equally good. This may indicates that the skip connections from the template branch help produce more generic features and better propagate gradients in the template branch. For the unseen ID types, the localisation is improved when the template is connected to the reconstruction stage.

To evaluate the capacity of the recognition, the recognition head was trained on the ground-truth segmented text fields. We obtain a CER of 0.33 for the unseen ID types which is higher than the MKI-S architecture (0.31). This shows that MKI-S is able to extract relevant features in the backbone for text recognition. In the next experiments, we use the MKI-S architecture.

5.2 Selection of the Knowledge Representation

We study the impact of knowledge representations defined in Sect. 3.2 on the MKI-S architecture. Results are presented in Table 3. On the seen ID types, the multi-channel approach performs better on the localisation and present similar performances for text recognition. This representation seems to hold more information to help separate the fields. However, regarding the unseen ID types, the single channel presents a lower CER.

Taking into account the standard deviation, selecting the best representation is not obvious. The single-channel version reaches a lower CER and should be able to handle low representation of some field types. Thus, the single-channel representation is used to design our multitask model for knowledge integration.

Table 3. Comparison between the different knowledge representations on the TLR-UIDT task of MIDV2020 for MKI-S{att}.

	Seen ID types		Unseen ID types	
Template	single	multi	single	multi
cer ↓	**0.14** ± 0.02	0.15 ± 0.03	**0.31** ± 0.07	0.35 ± 0.11
recall ↑	0.94 ± 0.02	**0.96** ± 0.01	0.84 ± 0.08	**0.88** ± 0.10
precision ↑	0.94 ± 0.01	**0.96** ± 0.02	**0.87** ± 0.09	0.86 ± 0.11
IoU ↑	0.87 ± 0.01	**0.89** ± 0.02	**0.79** ± 0.05	**0.79** ± 0.05

5.3 Comparison Against Existing Methods

We now compare our method to three existing systems presented in Sect. 2 on the MIDV2020 dataset (Table 4). As the source code is not available for some of them, we have re-implemented the state-of-the art methods. The **Chained** system is inspired from the sequential approaches of Mustafina et al. [14] and Van et al. [16]. To make the comparison fair, we use our backbone for the text field localisation; in a second step, our CRNN, used in the recognition head, is applied on the detected text line images for text recognition. We call **Adapted-** [6] the adaptation of Carbonell et al. [6]: it refers to our multitask network without knowledge integration. **Similar-** [1] is our implementation of [1]. This system directly applies a large template on the image as a crop mask. We extend the margins of the template of TLR-UIDT task to match their experimental conditions. We then apply our CRNN for text recognition.

Table 4 shows that Similar- [1] suffers from poor localisation performances that heavily impact the recognition. Indeed, this system is designed for IDs without deformations. This shows that the simple application of a template is not sufficient to perform text localisation and recognition.

Regarding the seen IDs types, MKI-S significantly outperforms the other approaches in localisation and presents a better recognition. The localisation performances of the **Chained** and **Adapted-** [6] are equivalent but the last one is better than the **Chained** system for text recognition. This shows that the text recognition head benefits from shared features in the backbone. The gap between our MKI-S model and the other ones is even more significant on the unseen IDs types. The Chained and Adapted- [6] localisation scores drop by around 20 points of percentage which also impact the text recognition. By training with the templates, our MKI-S model is able to adapt to new templates.

5.4 Generalisation to the Private Dataset

Finally, we perform the same comparison as in Sect. 5.3 on the industrial private database (Table 5). This database contains real world samples and a higher number of documents and types. Thus, those results give a better overview of the performance of each system in the real world context.

Table 4. Comparison of methods on the TLR-UIDT task of MIDV2020. The line "template" indicates whether the template is used.

		Chained	Adapted- [6]	Similar- [1]	MKI-S (ours)
	template	-	-	✓	✓
SeenID types	cer ↓	0.21 ± 0.04	0.18 ± 0.02	0.98 ± 0.04	**0.14** ± 0.02
	recall ↑	0.93 ± 0.01	0.93 ± 0.01	0.40 ± 0.03	**0.94** ± 0.02
	precision ↑	**0.94** ± 0.01	0.93 ± 0.01	0.41 ± 0.03	**0.94** ± 0.01
	IoU ↑	0.86 ± 0.01	0.86 ± 0.01	0.57 ± 0.01	**0.87** ± 0.01
UnseenID types	cer ↓	0.70 ± 0.18	0.72 ± 0.14	1.01 ± 0.04	**0.31** ± 0.07
	recall ↑	0.72 ± 0.14	0.75 ± 0.06	0.41 ± 0.14	**0.84** ± 0.08
	precision ↑	0.67 ± 0.17	0.73 ± 0.11	0.42 ± 0.14	**0.87** ± 0.09
	IoU ↑	0.63 ± 0.11	0.69 ± 0.08	0.57 ± 0.07	**0.79** ± 0.05

Table 5. Evaluation of our system on the private dataset.

	Seen ID types				Unseen ID types			
	Chained	Adapted- [6]	Similar- [1]	MKI-S (ours)	Chained	Adapted- [6]	Similar- [1]	MKI-S (ours)
cer ↓	0.17	0.24	1.23	**0.03**	1.15	0.87	1.22	**0.17**
recall ↑	0.88	0.89	0.05	**0.98**	0.36	0.38	0.08	**0.79**
precision ↑	0.86	0.88	0.05	**0.97**	0.30	0.42	0.08	**0.81**
IoU ↑	0.77	0.83	0.42	**0.91**	0.39	0.41	0.40	**0.76**

On the seen ID types, the MKI-S architecture reaches better performances than on the MIDV-2020 dataset with a recall/precision of 0.98/0.97 and CER of 0.03. Despite the dataset being harder, our method takes advantage of the higher number of training samples. This shows how important is the template integration even on seen ID types. On the unseen ID types, MKI-S achieves the best results and the gap with the other methods is even greater. MKI-S heavily benefits from the integration of the template to deal with new ID types.

6 Conclusion and Perspectives

In this paper, we address the task of text localisation and recognition in the context of unseen ID types, i.e. when a type has not been used for training. The template of ID types, describing the field structure, is always known even on new types. We leverage this knowledge by integrating it inside the neural network. We define the core concepts of knowledge integration in neural networks and propose template self-attention to guide the predictions in the context of unseen ID types with no annotated data. We compare three different architectures to leverage template self-attention and integrate the template. We show that the use of an auxiliary branch is crucial. Besides, skip connections from the template branch helps to compute relevant features and to generalize better. To evaluate our methods, we design a new dataset for the MIDV2020 database as well as a new task to perform recognition on unseen IDs types and make them public.

We show that our approach benefits from IDs template integration to achieve better results than state-of-the-art methods on the new dataset as well as on a private industrial database composed of real world samples. Contrary to state-of-the-art works, our approach is able to deal with new ID types. Going ahead, we plan to improve the unseen IDs types analysis by integrating the template into Transformer-based architectures via attention and querry conditionning.

References

1. Attivissimo, F., Giaquinto, N., Scarpetta, M., Spadavecchia, M.: An automatic reader of identity documents. In: IEEE International Conference on Systems, Man and Cybernetics (SMC), pp. 3525–3530 (2019)
2. Bluche, T.: Joint line segmentation and transcription for end-to-end handwritten paragraph recognition. In: Advances in Neural Information Processing Systems, vol. 29 (2016)
3. Bulatov, K.B., Bezmaternykh, P.V., Nikolaev, D.P., Arlazarov, V.V.: Towards a unified framework for identity documents analysis and recognition. Comput. Opt. **46**(3), 436–454 (2022)
4. Bulatov, K., Arlazarov, V.V., Chernov, T., Slavin, O., Nikolaev, D.: Smart IDReader: document recognition in video stream. In: ICDAR, vol. 6, pp. 39–44. IEEE (2017)
5. Bulatov, K.B., Emelianova, E., Tropin, D.V., et al.: MIDV-2020: a comprehensive benchmark dataset for identity document analysis. CoRR, abs/2107.00396 (2021)
6. Carbonell, M., Fornés, A., Villegas, M., Lladós, J.: A neural model for text localization, transcription and named entity recognition in full pages. Pattern Recogn. Lett. **136**, 219–227 (2020)
7. Coquenet, D., Chatelain, C., Paquet, T.: SPAN: a simple predict & align network for handwritten paragraph recognition. In: Lladós, J., Lopresti, D., Uchida, S. (eds.) ICDAR 2021. LNCS, vol. 12823, pp. 70–84. Springer, Cham (2021). https://doi.org/10.1007/978-3-030-86334-0_5
8. Coquenet, D., Chatelain, C., Paquet, T.: End-to-end handwritten paragraph text recognition using a vertical attention network. IEEE Trans. Pattern Anal. Mach. Intell. **45**(1), 508–524 (2022)
9. Coquenet, D., Chatelain, C., Paquet, T.: DAN: a segmentation-free document attention network for handwritten document recognition. IEEE Trans. Pattern Anal. Mach. Intell. (2023)
10. Coüasnon, B.: DMOS, a generic document recognition method: application to table structure analysis in a general and in a specific way. IJDAR **8**, 111–122 (2006)
11. d'Andecy, V.P., Hartmann, E., Rusinol, M.: Field extraction by hybrid incremental and a-priori structural templates. In: DAS Workshop, pp. 251–256. IEEE (2018)
12. Guerry, C., Couasnon, B., Lemaitre, A.: Combination of deep learning and syntactical approaches for the interpretation of interactions between text-lines and tabular structures in handwritten documents. In: ICDAR (2019)
13. Kushibar, K., Valverde, S., Gonzalez-Villa, S., et al.: Automated sub-cortical brain structure segmentation combining spatial and deep convolutional features. Med. Image Anal. **48**, 177–186 (2018)
14. Mustafina, V., Ivanov, S.: Identity document recognition: neural network approach. In: International Russian Automation Conference, pp. 806–811 (2021)

15. Sarshogh, M.R., Hines, K.: A multi-task network for localization and recognition of text in images. In: ICDAR, pp. 494–501 (2019)
16. Van Hoai, D.P., Duong, H.T., Hoang, V.T.: Text recognition for Vietnamese identity card based on deep features network. IJDAR **24**, 123–131 (2021)
17. Yousef, M., Bishop, T.E.: OrigamiNet: weakly-supervised, segmentation free, one-step, full page text recognition by learning to unfold. In: CVPR (2020)

Author Index

A
Ahmad, Irfan II-5
Alaei, Alireza II-167, II-180
Alaei, Fahimeh II-167
Alhubaiti, Omar II-5
Amin, Javaria I-253
Awal, Ahmad Montaser II-302

B
Baek, Youngmin I-215
Bayer, Johannes I-192
Beurton-Aimar, Marie I-307
Biswas, Sanket I-83, I-199
Boente, Walter I-34
Boutalbi, Karima II-139
Burie, Jean-Christophe I-120, I-177

C
Calvo-Zaragoza, Jorge I-94, I-151
Carrière, Guillaume I-5
Casey, Christian I-267
Chang, Chun Chieh II-285
Chen, Jialuo I-83
Christlein, Vincent I-5, I-284
Christopoulou, Katerina II-213
Córdova, Gonzalo I-20, II-123
Córdova, Jorge I-20, II-123
Coüasnon, Bertrand II-302
Coustaty, Mickaël II-32, II-65, II-108

D
d'Andecy, Vincent Poulain II-65
Dardouillet, Pierre II-139
Decker, Franziska I-284
Dengel, Andreas I-192
Dhote, Anurag I-67
Doermann, David S. I-67
Doucet, Antoine II-108
Douzon, Thibault II-47
Dubey, Alpana I-163
Duffner, Stefan II-47
Duri, Khalid I-105

E
Espinas, Jérémy II-47

F
Fennir, Tahani I-49
Fierrez, Julian I-20, II-123
Fornés, Alicia I-83

G
Gallego, Antonio Javier I-151
Garcia, Christophe II-47
Gatos, Basilis II-213, II-272
Graliński, Filip II-94
Grande, Marcos II-123
Gribov, Alexander I-105

H
Haraguchi, Daichi II-242
Hodel, Tobias I-34
Huang, Lingxiao I-135

I
Islam, Mohammad Khairul II-195

J
Javed, Mohammed I-67
Jawahar, C. V. I-233
Joseph, Aurélie II-32, II-65, II-108
Jubaer, Sheikh Mohammad II-195

K
Kaddas, Panagiotis II-213
Kafle, Dipendra Sharma II-32

M. Coustaty and A. Fornés (Eds.): ICDAR 2023 Workshops, LNCS 14194, pp. 319–321, 2023.
https://doi.org/10.1007/978-3-031-41501-2

Khudanpur, Sanjeev II-285
Kim, Geewook I-215
Konopka, Karolina II-94
Konstantinidou, Maria I-296
Kordon, Florian I-5
Kougia, Vasiliki I-296
Kritsis, Konstantinos II-213
Kumari, Lalita II-226
Kuriakose, Suma Mani I-163

L
Laleye, Fréjus A. A. II-19
Lamirel, Jean-Charles I-49
Lamiroy, Bart I-49
Le Van, Olivier II-139
Lemaitre, Aurélie II-302
Li, Wilson I-135
Liwicki, Marcus II-257
Lladós, Josep I-199
Louis, John Benson I-177
Loukil, Faiza II-139
Lucas, Javier I-151
Luger, Daniel I-284
Lunia, Harsh I-233

M
Mahamoud, Ibrahim Souleiman II-32, II-65
Marthot-Santaniello, Isabelle I-307
Martinez-Sevilla, Juan Carlos I-94, I-151
Massip, Sylvain II-19
Mayr, Martin I-5
Moetesum, Momina I-253, II-80
Mokayed, Hamam II-257
Mondal, Ajoy I-233
Morales, Aythami I-20, II-123
Motamedisedeh, Omid II-180

N
Nederhof, Mark-Jan I-267
Neitthoffer, Timothée II-302
Nicolaou, Anguelos I-284
Nikolaidou, Konstantina I-5

O
Ogier, Jean-Marc II-65
Okamoto, Yamato I-215
Ortega-Garcia, Javier I-20, II-123
Osanai, Atsuki I-215

P
Pal, Umapada II-257
Palaiologos, Konstantinos II-213
Paparrigopoulou, Asimina I-296
Pavlopoulos, John I-296
Peña, Alejandro I-20, II-123
Penarrubia, Carlos I-94
Perera, Leibny Paola Garcia II-285
Poulain d'Andecy, Vincent II-32, II-108
Puente, Íñigo I-20, II-123

R
Rahman, Md Ataur II-195
Rakotoson, Loïc II-19
Rathore, Vaibhav Varish Singh II-226
Rauf, Sadaf Abdul II-80
Renet, Nicolas I-284
Retsinas, George II-272
Rios-Vila, Antonio I-94
Roy, Ayush II-257

S
Salamatian, Kave II-139
Seo, Sukmin I-215
Serbaeva, Olga I-307
Serna, Ignacio I-20
Seuret, Mathias I-5
Sfikas, Giorgos II-272
Sharma Kafle, Dipendra II-108
Sharma, Anuj II-226
Shivakumara, Palaiahnakote II-257
Siddiqi, Imran I-253, II-80
Singh, Sukhdeep II-226
Souibgui, Mohamed Ali I-83
Soullard, Yann II-302
Ströbel, Phillip Benjamin I-34
Sylejmani, Visar II-139

T
Tabassum, Nazifa II-195
Tabin, Julius I-267
Telisson, David II-139
Théodose, Ruddy I-120
Thomas, Eliott II-32, II-108
Tiwari, Adarsh I-199
Torras, Pau I-83
Trabelsi, Mohamed II-151
Turabi, Shabi Haider I-192
Turski, Michał II-94

U
Uchida, Seiichi II-242
Upadhyay, Abhinav I-163
Uzunalioglu, Huseyin II-151

V
Verjus, Hervé II-139
Vogeler, Georg I-284
Volk, Martin I-34
Vu, Manh Tu I-307

W
Wei, Chiching I-135
Wu, Jung-Hsuan I-135

Y
Yasin, Nehal II-80
Yokoo, Shuhei I-215

Z
Zagia, Faranak II-180

Printed in the United States
by Baker & Taylor Publisher Services

Printed in the United States
by Baker & Taylor Publisher Services